Inference and Intervention

Ryall and Bramson's *Inference and Intervention* is the first textbook on causal modeling with Bayesian networks for business applications. In a world of resource scarcity, a decision about which business elements to control or change – as the authors put it, a managerial intervention – must precede any decision on how to control or change them, and understanding causality is crucial to making effective interventions.

The authors cover the full spectrum of causal modeling techniques useful for the managerial role, whether for intervention, situational assessment, strategic decision-making, or forecasting. From the basic concepts and nomenclature of causal modeling to decision tree analysis, qualitative methods, and quantitative modeling tools, this book offers a toolbox for MBA students and business professionals to make successful decisions in a managerial setting.

Michael D. Ryall is an Associate Professor of Strategy at the University of Toronto. He holds a PhD in economics from the University of California, Los Angeles and an MBA from the University of Chicago. He is President of the Strategy Research Initiative, a scholarly society dedicated to the advancement of research in the field of management. His primary research interest is the game-theoretic foundations of business strategy and his work has been published in leading international journals. Ryall teaches courses on advanced strategy analysis and on causal modeling to undergraduate, MBA and EMBA students. Prior to obtaining a PhD and becoming a full-time scholar, he held positions in consulting, general management and finance.

Aaron L. Bramson received a PhD from the University of Michigan in 2012 in a joint program with the departments of political science and philosophy, as well as earning UM's graduate certificate in complexity in 2008. He holds an MS in mathematics from Northeastern University, as well as a BS in economics and a BA in philosophy from the University of Florida. Aaron's research specialty is complexity science, methodology for modeling complex systems, and measuring dynamics in large datasets. He is currently a researcher at the RIKEN Brain Science Institute in Japan. Previously, he worked as a research fellow in the Rotman School of Management at the University of Toronto, as a software engineer at Lockheed Martin Corporation, and has taught numerous workshops on complexity, networks, and agent-based modeling around the world.

Inference and Intervention

Causal Models for Business Analysis

Michael D. Ryall &
Aaron L. Bramson

Routledge
Taylor & Francis Group
NEW YORK AND LONDON

First published 2014
by Routledge
711 Third Avenue, New York, NY 10017

Simultaneously published in the UK
by Routledge
2 Park Square, Milton Park, Abingdon, Oxon OX14 4RN

Routledge is an imprint of the Taylor & Francis Group, an informa business

Library of Congress Cataloging in Publication Data
Ryall, Michael D.
Inference and intervention : causal models for business analysis /
Michael D. Ryall & Aaron L. Bramson.
pages cm
Includes bibliographical references and index.
1. Decision making–Mathematical models. 2. Decision making–Statistical methods. 3. Business planning–Statistical methods. I. Bramson, Aaron L. II. Title.
HD30.23.R92 2013
658.4'0101519542–dc23 2013005927

ISBN: 978-0-415-65759-4 (hbk)
ISBN: 978-0-415-65760-0 (pbk)
ISBN: 978-0-203-07683-5 (ebk)

Typeset in Berling
by Cenveo Publisher Services

Contents

List of Figures ix
Acknowledgments xiii

1 Introduction to Causal Analysis **1**

1.1 Situational Assessment 1
1.2 Managerial Intervention 4

2 Qualitative Causal Models **14**

2.1 Setting the Stage 15
2.2 Building a Qualitative Causal Model 17
2.2.1 Nodes 17
2.2.2 Links 19
2.2.3 Some Examples of Qualitative Causal Models 25
2.3 Causal Independence 31
2.3.1 Serial Triplets 32
2.3.2 Diverging Triplets 33
2.3.3 Converging Triplets 34
2.3.4 Causal Independence in General 36

3 Application: Interview Case Study **44**

3.1 Getting Started 47
3.2 Focus on the Significant Drivers 52
3.3 Seek Sources of Common Problems 56

3.4 Ask Specific Questions 59
3.4.1 Administrative Staff 61
3.4.2 Service Staff 62
3.4.3 Doctors: Generalists and Specialists 62
3.5 Bring it All Together 65
3.6 Provide Specific Recommendations 71
3.6.1 Upgrade Account Management System 71
3.6.2 Adjust to an Aging Population 73
3.6.3 Appeal to a Younger Crowd 75
3.6.4 Final Note 75

4 Quantitative Causal Models 78

4.1 Probability Basics 78
4.1.1 Variable States 78
4.1.2 Events 79
4.1.3 Probabilities 81
4.1.4 Conditional Probabilities 83
4.1.5 Joint Probabilities 84
4.1.6 System States 85
4.1.7 Bayes' Rule 86
4.2 Quantifying a Qualitative Model 86
4.2.1 More Refined Approximations 89
4.3 Working with Quantitative Models 90
4.3.1 Probabilities from Count Data 91
4.3.2 Joint Probability Tables 93
4.3.3 The Complete Advertising Model 94
4.3.4 System-Level Joint Distribution & Factorization 96
4.3.5 Marginalization 97
4.4 The Move to Causal Models 100

5 Situational Analysis 109

5.1 Marginal from Conditional Probabilities 111
5.1.1 Serial Connections 111
5.1.2 Diverging Connections 112
5.1.3 Converging Connections 113
5.2 Evidence & Inference in Causal Models 114
5.2.1 Serial Connection 115

5.2.2 Divergent Connection 121
5.2.3 Convergent Connection 123

6 Application: Modeling Business Financials 132

6.1 The Spreadsheet Approach 134
6.2 Building a Causal Model 135
6.3 Marketing Uses the "Prosecutor's Fallacy" 143
6.4 Green Ink Creates Simpson's Paradox 146

7 Single-Agent Interventions 151

7.1 One Decision, No Information 152
7.2 Multiple Decisions with Information 153
7.2.1 The Extended Model 154
7.2.2 General Solution Procedure 156

8 Application: Disrupting the Taxi Business 176

8.1 An Allocation of Resources Decision 177
8.2 Price Uncertainty and Market Research 181
8.3 Competitor Legal Response 185

9 Multi-Agent Interventions 189

9.1 Elements of a Game 190
9.2 Nash Solutions 192
9.3 Causal Form Games 195
9.4 Solving Games with Causal Models 197
9.4.1 Vicious Incumbent Entry Game 198
9.5 Games with Non-Trivial Strategies 204
9.5.1 Setup for the model 205
9.5.2 Solving the Reformulated Model 209
9.5.3 Insights from Technology Development Problem 215
9.6 Software Solutions 216

10 Data-Driven Causal Modeling 224

10.1 Causality versus Probability 225
10.1.1 Probability View 226
10.1.2 Causality View 227
10.2 Observational Indistinguishability 230
10.2.1 The Observational Indistinguishability Theorem 231
10.2.2 Causal Identification 235
10.3 Building Predictive Regression Models 241
10.3.1 From Structural Equation to Causal Models 241
10.3.2 Linear Regressions and Causal Models 242
10.3.3 Good Causal Models Imply Good Predictions 245
10.3.4 A Brief Note on Box Office Gross 248

Bibliography 255
Index 257

Figures

1.1	Robert Maxwell – causes of death	3
1.2	Feather Touch – true causal relationships	9
2.1	A simple causal model	23
2.2	Conditional certainty table for a monotonic relationship	24
2.3	Qualitative representation of non-monotonic effects	25
2.4	Causal model for market volume	26
2.5	How much to spend on advertising?	28
2.6	Including effects of other store's incentives	29
2.7	Competitors can observe your amount of advertising	30
2.8	Two paths to higher market share	31
2.9	An example of a serial triplet	32
2.10	Diverging triplet	34
2.11	Converging triplet	35
2.12	*Market Volume* is a deterministic function of its causes	36
2.13	Causal model for market share	38
3.1	Situational assessment of variables related to renewals and new accounts	49
3.2	Initial model including all initially provided information for the Hubris Health case	51
3.3	Adding relationships uncovered by digging deeper into customer satisfaction	55
3.4	The refined model after examining their business model in more detail	60
3.5	The complete model, including both modules analyzed in-depth and refined through digging deeper	66

3.6 Adding the effects of product features and industry-specific competition to the model 72
3.7 Highlighting the recommendations to Hubris within the model 74

4.1 A simple causal model 87
4.2 Consumer choice revisited and simplified 90
4.3 Consumer choice, quantified 93
4.4 Advertising model with conditional probability tables 94
4.5 Advertising model with joint and marginal probabilities 95
4.6 System-level joint probability distribution 98
4.7 Causal model of the semiconductor quality example 102

5.1 Marginal probabilities for serial semiconductor example 112
5.2 Marginal probabilities for convergent semiconductor example 114
5.3 Semiconductor example of a serial connection 115
5.4 Initial setup, serial connection 117
5.5 Updating procedure when the true state of *Test* is discovered to be *pass* 120
5.6 Semiconductor example in which *Batch* is a common cause 121
5.7 Initial setup for an example divergent connection 122
5.8 Updating a divergent connection upon discovery that *Test* = *pass* 123
5.9 Semiconductor example of a convergent connection 124
5.10 Initial setup of the semiconductor quality example with multiple, converging causes 124
5.11 Updating beliefs when *Batch* = *bad* 126
5.12 Updating beliefs when *Batch* = *bad* **and** *Error* = *no* 126

6.1 Slimtree Publishing Inc. P&L spreadsheet 134
6.2 Initial causal diagram for the determinants of *Profit* 136
6.3 Variable ranges for *RU* (retail units sold) 137
6.4 Equal probabilities on the two *RU* states 138
6.5 *RM* states correspond to variable ranges 139
6.6 Parent states associated with *RM* states via a function 140
6.7 CPT automatically generated for *RM* 140
6.8 Summary information for the initial causal model 141
6.9 Drawing inferences from operating expenses of $3.7m 142
6.10 Influence of *Marketing Effectiveness* on *Retail Units* 143
6.11 Model extended to include effectiveness of *Retail Marketing* 144
6.12 Marginal probabilities of *ME* and *RU* in extended model 145
6.13 Updated priors on *ME* given *RU* = *high* 145
6.14 Model expanded to account for production runs 147
6.15 CPT for *Run Cost* based upon manufacturing data 148

6.16 The effect of *Ink Type* on *Run Cost* when *Title* is unknown 149
6.17 The effect of *Ink Type* on *Run Cost* when *Title* is known 149

7.1 Should Apple launch a new tablet in Canada? 153
7.2 Opportunity for pre-launch market research 154
7.3 Expected *NPV* when *Perform* = *no* (and hence *Test* = *none*) and *Launch* = *no* 158
7.4 Updating the probability of *Market* given the initial configuration 159
7.5 Expected *NPV* when *Perform* = *no* and *Launch* = *yes* 160
7.6 Expected *NPV* when *Perform* = *yes*, *Test* = *bad*, *Launch* = *yes* 161
7.7 Expected *NPV* when *Perform* = *yes*, *Test* = *good*, and *Launch* = *yes* 162
7.8 The model after changing the *Launch* decision node into a new objective node and the removal of the *NPV* objective node. Also shows expected *NPV* when *Perform* = *yes* 163
7.9 Updating the probabilities of the *Test* results given *Perform* = *yes* 164
7.10 Transform *Perform* decision node into an objective node – a trivial final step 164

8.1 Model of HaHCo driver enrollment decision 179
8.2 Optimal decision: enroll 110 drivers in SF, 90 in NY 180
8.3 Spreadsheet analysis of driver allocation 182
8.4 Model with market uncertainty and research options added 184
8.5 Conditional probability table for *Result-SF* 184
8.6 Optimal allocations based upon test results 185
8.7 Model extended to include potential lawsuits 186
8.8 Conditional probability table for *Outcome* 187

9.1 Extensive form version of an Entry Game 191
9.2 R&D Game 193
9.3 Entry Game in causal form 196
9.4 *Incumbent*'s decision table completed 198
9.5 Simplified model with *Incumbent* converted into an objective node 199
9.6 Modeling a chance that the *Incumbent* is *vicious* 200
9.7 Payoffs in the Vicious Incumbent Entry Game 200
9.8 *Incumbent*'s decision table complete 201
9.9 Expected value when *Entrant* = *stay out* 202
9.10 Complete decision table for *Entrant* 203
9.11 Causal model of technology development and acquisition problem 206
9.12 Payoff table for technology development and acquisition problem 206
9.13 Transformed model – now a pure intervention problem 208
9.14 Strategy tables for both players 208
9.15 Model with decision tables and conditional probability tables added 210

9.16 Model can be solved with a single, joint decision table 211
9.17 Calculation for first cell of decision table – *strat1* and *strat1* 213
9.18 Causal model showing the results of playing *strat1* and *strat1* 214
9.19 Complete joint decision table – best replies in bold, equilibria are shaded 214
9.20 Model to compute best responses for Player 1 217
9.21 Model to compute best responses for Player 2 217
9.22 *S2-Picker* selects strategies to be played at *Player 2* 218
9.23 Best replies for Player 1 219
9.24 Best replies for Player 2 219

10.1 Historic cost and performance data on past projects 226
10.2 Conditional probabilities for $D=o$, $E=r$, $M=b$ based on likelihood 227
10.3 Parameterization of the model $D \rightarrow E \rightarrow M$ 228
10.4 Effect of $\mathbf{Do}(M=b)$ 228
10.5 Effect of $\mathbf{Do}(D=o)$ 229
10.6 Effect of $\mathbf{Do}(E=r)$ 230
10.7 Alternative models that are consistent with the data of Figure 10.1 231
10.8 All 25 possible causal models with three nodes 233
10.9 The Gatekeeper model 234
10.10 Exercise causes longevity 236
10.11 Longevity causes exercise 236
10.12 Exercise intensity and longevity caused by unobserved factors 237
10.13 Two hypotheses with an instrumental variable 238
10.14 If climate directly influences longevity, it is useless as an instrument 240
10.15 CEO *Experience* increases the rate of entrepreneurial failure 240
10.16 Parent relationships for x_3, x_4, x_5, and x_6 according to 10.1–10.4 242
10.17 Linear coefficients shown on network links 243
10.18 Here, x_3 is an *effect* of Y 246
10.19 Does foreign investment cause political oppression? 247
10.20 The causal structure inferred from the data 248
10.21 Explaining the likelihood of getting an MPAA rating of R 250

Acknowledgments

The roots of this book can be traced back to the mid-1990s, during which time, one of us (Ryall) had the great good fortune to stumble into Judea Pearl's graduate class on causal modeling, during his PhD studies at UCLA. That experience led to a dissertation exploring the implications of formal notions of causality, as conceived by Pearl and others in the artificial intelligence community, for game theory. Judea Pearl was nothing less than inspirational, as were Ryall's supervisors on the economics side – Brian Ellickson, David Levine and Bill Zame – all of whom were incredibly supportive of what was, then, a fairly unconventional line of research in game theory.

Since the 1990s, it is safe to say that causal modeling has gone mainstream. Not only are there now a multitude of books on the topic, ranging from scholarly monographs to practice-oriented texts, but so is there a multitude of software tools, ranging from very sophisticated, commercial-grade programs to simpler, open access software. Though the importance of causal modeling in business settings is clear from its widespread adoption in large corporations, we were surprised to find that there were no books aimed specifically at the business audience.

The opportunity to correct that omission arose when we were asked to teach the Integrative Thinking Practicum in the Rotman School of Management at the University of Toronto. Intended to be a capstone, first-year MBA course involving modeling of some kind, we decided that *causal* modeling was the way to go. The causal modeling approach provides an excellent framework with which students can integrate ideas learned in their disciplinary courses to solve multifaceted, real-world problems. As a result, we developed the material that follows and taught it in the core curriculum in the years 2010 and 2011, delivering it to over 500 students. We then refined and expanded our coverage which, eventually, became this book. Since then, we have also successfully taught the material as an upper-division undergraduate course as well as a second-year MBA elective.

We give special thanks to Mihnea Moldoveanu and the Desautels Centre for Integrative Thinking for their incredible support, without which this book would not have been possible. We thank our many colleagues at Rotman for their encouragement during the development of our Integrative Thinking Practicum, especially Joel Baum, Peter Pauly, Will Strange and Glenn Whyte. We benefited from discussions with many colleagues, in particular Max Chickering, Joshua Gans, Avi Goldfarb, Sarah Kaplan, and Mara Lederman. Rekha Morbia and Sally Smith provided essential administrative assistance throughout the course and book development. We also thank John Szilagyi, Manjula Raman, and the reviewers at Routledge for their assistance in bringing this book to press.

Chapter 1

Introduction to Causal Analysis

1.1 SITUATIONAL ASSESSMENT

In 2009, one of us (Ryall) had the pleasure of participating as a judge in the McKinsey Case Competition held at the Rotman School of Management, University of Toronto. Case competitions are an established feature of the student experience in modern MBA programs. The details vary from competition to competition, but the essential idea is to provide students with a brief (~15 page) description of a problem facing a particular firm, give them a week to mull it over, then have them present their recommendations and supporting analyses before a panel of experienced judges.[1]

In order to help the MBA students prepare, two Rotman alumni who had become McKinsey engagement managers (and competition judges) – Paulo Salomao and Erez Eizenman – generously gave a presentation on McKinsey's approach to business problem solving. As they explained, McKinsey uses a technique called *issue trees* to help structure some of their analyses. An issue tree takes the problem and breaks it down into issues, subissues, subsubissues, and so on. An important requirement is that the direct descendants of any issue constitute a mutually exclusive, collectively exhaustive (MECE) set of subissues. This requirement forces the analyst to think broadly about the potential issues at every level.

To illustrate the method, Paulo and Erez posed the question, "How did Robert Maxwell die?" Maxwell was a media tycoon who, in 1991 was presumed to have fallen overboard from his luxury yacht. His death was officially judged to be accidental drowning. However, questions were raised suggesting that the death may have been caused by murder or suicide. Accordingly, the root node of the issue tree they presented was labeled, "How did Robert Maxwell die?" Its direct descendants were: murder, suicide, natural causes, accident and not-really-dead

1 In this instance, an apparel manufacturer had to decide whether or not to green-light a new line of active wear.

(mutually exclusive collectively exhaustive, MECE!). Then the direct descendants of each of these nodes accounted for (all possible) potential causes of death associated with each of these categories.

Trying to think exhaustively about the issues related to a problem is clearly an important part of making any substantive business decision – and the Maxwell example provided a nice, simple illustration of how to do so. Still, there was something about this example that felt lacking. That "something" was not being told why we care about the answer to, "How did Robert Maxwell die?" Were we considering the issue from the perspective of a policeman pondering whether or not to begin a criminal investigation (is murder sufficiently likely)? From that of Maxwell's life insurance company trying to decide the payout on a claim (is it reasonable to rule out suicide)? From that of a prosecutor assembling a case (what is the theory, what evidence supports it)?

Suppose that instead of asking, "How did Robert Maxwell die?" we ask, "What was Robert Maxwell's cause of death?" Does the different phrasing change your perspective? Most people, saturated with *CSI-this* or *Law & Order-that* television programming, read "cause of death" and think in terms of a coroner's report: what was the immediate, physical event that resulted in Maxwell's expiration? Candidates might include: drowning, gunshot, heart attack, poisoning, etc.

Having introduced the notion of causality, though, one's mind naturally begins to frame the question in terms of a longer causal chain (i.e., thinking about causes of causes). If the immediate cause of death was due to a weapon, was the injury self-inflicted (suicide) or other-inflicted (murder). Of course, we do not stop there. What might cause Maxwell to commit suicide? If murder, who did it? If someone did it, what was his or her motive?

Causal thinking leads us not only up the causal chain, but down it as well. Guns cause different trauma to the body than knives. They may also cause other types of evidence to appear – gunpowder residue, bullet casings, projectile markings, etc. Framing the analysis in causal terms points us toward potentially useful sources of information. Moreover, when evidence does present itself, causal analysis allows us to incorporate it in useful ways.

Figure 1.1 gives a preview of the methodology that will be used in this book. The graph is a causal system in which the nodes represent variables that can take on various values or states and the arrows (directed links) indicate direct causal relationships. Thus, a physical cause of *gunshot* causes a certain type of trauma (*gunshot wound*) to appear on the body. A gunshot wound could be inflicted by one's self (suicide) or some other person (murder). A murderer typically has some motive that causes him or her to commit that deed. We would say that a motive is an indirect cause of death by murder.

Suppose you are a police inspector interested in determining whether someone other than Maxwell was involved in his death (murder). In our context, then, you are interested in assessing the likelihood that the true value of *People Involved*

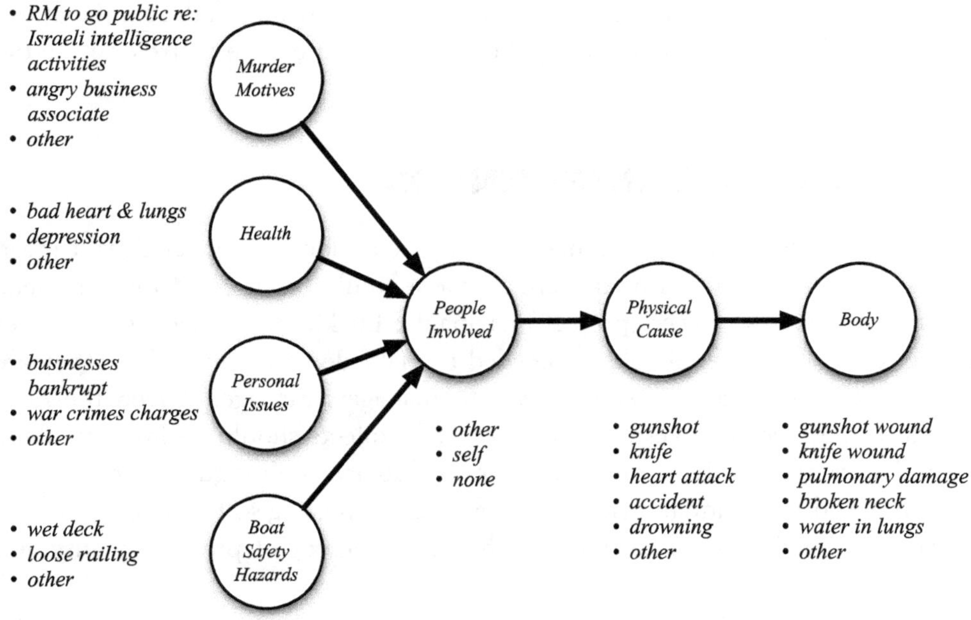

FIGURE 1.1 Robert Maxwell – causes of death

is *other*. Notice that discovering the values of other variables in this system has implications for the value of *People Involved*. For example, prior to the coroner's report, learning that the value of *Health* is *bad heart and lungs* should cause you to decrease the likelihood that the value of *People Involved* is *other*. On the other hand, discovering that Israeli intelligence may have had a motive for killing him would tend to increase that likelihood. In causal systems, evidence has upstream implications as well. For example, other things held constant, discovering that the cause of death was a heart attack should decrease the likelihood that the value of *People Involved* is *other*.

A valuable feature of causal analysis is that it permits us to factor in multiple sources of evidence. In the story of Maxwell's death, for example, all of the preceding items were, in fact, discovered (Verkaik, 2006). The muscles on the left side of his body were torn, consistent with falling over the ship's rail and dangling before falling into the ocean. Six months before he died, he was being investigated for war crime World War II. In addition, his corporate empire was on the brink of collapse; he had even illegally raided his employees' pension funds to finance his corporate debt.

As we will see, causal analysis permits all of these facts to be incorporated and used to help assess, e.g., the likelihood of murder. Thus, an important use of causal models is to assess the implications of known information. At the same time, it is worth pointing out that none of the information encoded in the McKinsey issue tree is lost. Indeed, the set of mutually exclusive collectively exhaustive states in which this system may find itself can be enumerated by taking all the combinations of the variable values and eliminating those that are impossible or irrelevant to the task at hand.[2]

1.2 MANAGERIAL INTERVENTION

Managers manage. This is, of course, true by definition. Yet, what does it mean to manage? In a world without scarcity, there would be no need for managers: firms could undertake *every* project imaginable. Luckily for managers, resources are limited and, hence, people are needed to make decisions about how they are to be used. This is a key sense in which "managers manage." Given the rates of return required by investors, which capital projects should be implemented? Should marketing funds be used in a Super Bowl ad or an online guerrilla campaign? Should the cost accountant install better financial reporting software or work with people on the shop floor to improve the measurement of product quality and resolve problems?

To manage is to face a never-ending stream of such decisions, regardless of one's area or level within the firm. Every manager is delegated some measure of control over some collection of resources. He or she must then exercise that control in a fashion designed to achieve the relevant organizational goals. The fact that certain resources are assigned to a manager in the first place immediately implies that they are both constrained and constraining – else, what is the need to "manage" them? Constraints imply trade-offs.

Sometimes, the trade-off appears explicitly in the form of a resource-allocation decision: should funds (or employee time, equipment, land, ...) be devoted to project A or project B? Other times, activities may be undertaken to loosen a constraint: employees might be better motivated, distribution inefficiencies might be eliminated, the shop floor might be reconfigured, a secondary supplier might be identified. Note well, however, that loosening constraints typically requires the application of other scarce resources – if none other than the manager's own limited time – and, hence, an allocation decision once again.

The point is that, in a world of resource scarcity, a decision on *which* objects to control or change must typically precede any decision on *how* to control or change them. In this text, we refer to a decision of the first type as a managerial *intervention*. Understanding causality is crucial to making effective interventions.

One of the purposes of this book is to help readers appreciate that understanding causality is a subtle undertaking – one prone to natural thinking traps. For example, the growing movement toward "evidence-based management" encourages managers to, among other things, base their decisions upon objective data.

2 The constraints of satisfying the MECE requirement for causal models are addressed in more detail in Section 2.2.

We wholeheartedly agree. Yet, the availability of data can lead managers into the common thinking trap of *inferring causation from statistical correlation*. To illustrate, suppose someone undertook a large-scale empirical study of the market for cold remedies and demonstrated that there was a statistically significant, positive correlation between the use of cold remedies and the number of people with colds. Now, were I to use this finding to claim that the use of cold remedies *causes* colds and that, therefore, such remedies should be banned (an intervention decision), most people would (rightly) dismiss me as a crank.[3]

Unfortunately, even slight departures from the glaringly obvious can quickly befuddle otherwise careful thinkers. For example, early observational studies of post-menopausal women found a negative correlation between taking hormone replacement therapy (HRT) and cardiovascular disease (CVD). As a result, some doctors concluded, "Hormone therapy should probably be recommended for women who have had a hysterectomy and for those with coronary heart disease or at high risk for coronary heart disease" (Grady et al., 1992). However, further investigation revealed that women on HRT tended to be from higher socioeconomic groups and, therefore, also tended to have healthier lifestyles. In other words, higher socioeconomic status was a *common cause* of higher use of HRT and lower incidence of CVD. Indeed, controlled experiments revealed that the direct effect of HRT on CVD, independent of other factors, was actually *positive* (e.g., Women's Health Initiative, 2002). Examples like this from medicine are numerous ...and most of us view doctors as a fairly smart bunch.

To see the issue in a more relevant business setting, imagine that you are the CEO of a major consumer goods manufacturing and marketing firm.[4] One of the firm's more established brands is Feather Touch (FT), a mainstay in the toilet tissue market. In an effort to improve operations, you commission TruSmartz Consulting to examine the effects of marketing decisions on sales volume for this brand.

TruSmartz conducts a major study looking at over three years of data for over 10 competitors in the market. Collecting information from TV meter records, in-store scanner data, and firm records, they conduct an empirical analysis to assess the effects of price, advertising campaigns, brand loyalty of consumers, product features, and in-store merchandising on the choice and quantity of product purchased.

In the initial draft of their report, TruSmartz finds that the significant factors affecting the volume of sales are: brand loyalty, brand choice, price, coupons, and advertising. To simplify the analysis, TruSmartz categorizes all the measures with binary values; that is, each variable takes on either a *high* or *low* value, or a *yes* or

3 At a minimum, I would be guilty of the logical fallacy of *cum hoc ergo propter hoc*. The Latin translates to: "with this, therefore because of this" and is typically used to refer to the mistake of believing that a correlation proves the existence of a causal relationship.

4 This example is inspired by Tellis (1988). My thanks to Avi Goldfarb for bringing this article to my attention.

Table 1.1
Correlation coefficients of marketing variables†

	Loyalty	Coupons	Advertising	Price	Brand	Volume
Loyalty	1.0000	0.0263	−0.0400	−0.0149	0.7729*	0.5376*
Coupons		1.0000	−0.3531*	−0.0235	0.1400*	−0.0104*
Advertising			1.0000	−0.0529*	−0.0671*	−0.0288*
Price				1.0000	0.0350	−0.1558*
Brand					1.0000	0.6902*
Volume						1.0000

† Asterisks indicate values significantly different from zero.

no value. Specifically, consumers can be *Loyal* to FT (= 1) or not (= 0); FT may offer *Coupons* (= 1) or not (= 0); FT may have an *Advertising* campaign (= 1) or not (= 0); *Price* can be *low* (= \$50) or *high* (= \$60); and unit *Volume* can be *low* (= 100) or *high* (= 150).

Suppose TruSmartz's initial report presents the correlation coefficients shown in Table 1.1. In reviewing this information, keep in mind that, for Feather Touch, *Price*, *Coupons*, and *Advertising* are all managerial decision variables (i.e., they are chosen by divisional managers). Several of the variables have the expected relationships; e.g., *Loyalty* is positively correlated with *Brand* choice and *Volume*, and *Price* is negatively correlated with *Volume*.

Other findings may strike you as odd. For example, *Advertising* and *Loyalty* appear to have no appreciable relationship. Compounding this is the finding that *Advertising* is *negatively* correlated with *Brand* choice and *Volume*. You may also note that *Advertising* and *Coupons*, both choices of your divisional managers, are also negatively correlated.

What ought we conclude from all this? Is advertising causing low unit sales? It *is* possible – perhaps Feather Touch's "Enjoy the release!" campaign is actually offensive to consumers and causing lower sales in the periods when the ads are run. As an evidence-based decision-maker, you know better than to jump to conclusions. Therefore, you ask the consultants at TruSmartz to drill down into the data by running a multivariate linear regression of volume on the other variables.[5]

TruSmartz returns with information presented in Table 1.2. The top row details the regression parameters. In other words, the estimated model (rounding the

5 Note that we are not claiming that linear regression is the best way to analyze this data. There are other statistical analysis techniques that could be used here, and some would be improvements to linear regression – actually we'll see one such improvement later in this chapter. We are using this example to make a point about causal thinking, so we invite you to not worry about the technical details for the moment.

Table 1.2
Estimated regression coefficients for unit sales

	Intercept	*loyalty*	Coupons	*Advertising*	*Price*	*brand*
Coefficient	150.9089	−1.5194	−15.1094	−1.8454	−0.8227	34.8035
Average	–	0.3333	0.0333	0.7833	56.6000	0.2667

coefficients) is:

$$q = 150.9 - 1.5l - 15.1C - 1.8A - 0.8P + 34.8b$$

where q is unit sales, l is whether or not the customer is loyal to FT, C indicates whether coupons are running at the time of the purchase, A is whether or not an advertising campaign is running at the time of the purchase, P is the price and b is whether the customer chooses to buy the FT brand. Managerial decision variables are identified with capital letters. These are the variables that can be set to specific values by order of the CEO – resulting in a management intervention on his part.

Notice the odd fact that the signs on the coefficients for l, C and A seem to be going in the opposite direction. For example, increasing advertising should increase the quantity sold, not decrease it.

The second row provides the average values of these variables as found in the actual data. This says that roughly one third of consumers are loyal to FT but FT is the brand of choice only 26% of the time. Coupons run about 3% of the time, while advertising campaigns are run 78% of the time. Prices average $56.6. In addition, you are told the average unit sales over the three year period is 111.2.

If we plug the average values into the regression equation, we see the predicted unit volume is, indeed, very close to 111.2. Ultimately, you are interested in profitability. Profits (π) are known to be computed as

$$\pi = Pq - A \tag{1.1}$$

where P is the price, q is the unit volume and A is the cost of advertising ($0 if there is no campaign and $200 if there is a campaign). To keep things simple, we assume unit costs and the cost of couponing are zero. Using the above regression model we have an expected profit of $6,135.63.[6] The actual average profit over this period

6 Estimate expected q as

$$\bar{q} = 150.9 - 1.5\bar{l} - 15.1\bar{C} - 1.8\bar{A} - .8\bar{P} + 34.8\bar{b}$$

where the expected value is just the average observed value: $\bar{l} = 0.3333$, $\bar{C} = 0.0333$, etc. Then, $\bar{q} = 111.2$ and expected profit is computed as $\bar{\pi} = \bar{P}\bar{q} - 200\bar{A}$, which comes to $6,135.63 (allowing for minor rounding error).

was $6,120. This seems to imply that the predictive power of the regression model is quite good, despite the counter-intuitive coefficient values.

Based upon these regression results, what is the appropriate intervention? First, let us assume that your time is valuable. Your firm is big and has many issues demanding your attention. Moreover, Feather Touch has been doing things the same way for a long time, and managers there are fairly entrenched in their views on how best to run their division. As a result, your judgment is that you can convince FT managers to modify their approach with respect to only *one* of the three key marketing components (coupons, advertising, or price). Which one should it be?

Suppose you decide that trying to manage prices from the top down is not a wise idea. That leaves imposing either a new coupon or advertising policy. The negative coefficients on the regression clearly indicate you should: (i) drop coupons altogether; or (ii) stop advertising. Numerically, these translate into setting, respectively, $C = 0$, or $A = 0$. When one plugs each of these into the regression model (while holding the other values fixed), it produces the following results:

- $C = 0$ results in profits increasing to $6,208
- $A = 0$ results in profits increasing to $6,417

Based on these results, the optimal intervention appears to be telling FT managers to halt their ineffective advertising.

Overcoming initial resistance, you eventually succeed in imposing the no-advertising policy. According to your regression model, profits under this new policy were expected to increase by roughly 5%. However, over the next year, profits average only $6,158. This is an increase of less than 1%. Moreover, you note that there were no significant changes in competitor behavior, costs, or market conditions. FT managers develop an I-told-you-so attitude and you realize that you will have to pull teeth to get them to adopt any future policy changes.

What went wrong? Of course, the clue lies in the counter-intuitive coefficient values. The ultimate answer lies in the distinction between causality and correlation. It turns out that the data used to compute the averages and regression estimates used in this example were based upon an actual causal model. That model is shown in Figure 1.2.[7]

What does this model tell us? As we know from our earlier profit equation, profit is a direct function of price, unit sales, and advertising. Here, an arrow from one variable to another indicates a direct effect. A positive (negative) sign on a link means that an increase in a variable results in a direct increase (decrease) in the variable to which it points. For example, running an advertising campaign (turning advertising from 0 to 1) directly results in a $200 decrease in profit (due to the cost

7 The relationships shown here are roughly consistent with the real-world findings of Tellis (1988).

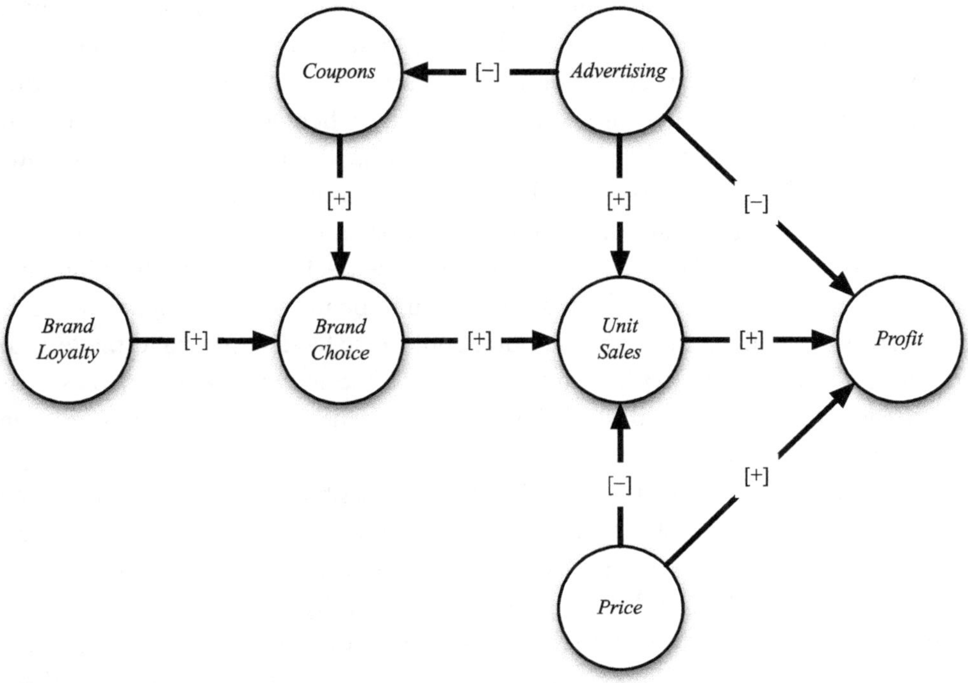

FIGURE 1.2 Feather Touch – true causal relationships

of advertising). Notice that prices affect profits in two ways. Holding everything else constant, raising prices results in a direct increase in profit. However, higher prices also have a direct negative effect on unit sales (because demand curves are downward sloping). Thus, price has a positive direct effect on profit but, at the same time, a negative indirect effect.

Notice that the decision to advertise has a direct negative effect on the decision to run coupons. How can this be? As it turns out, FT managers are quite enamored with running big advertising campaigns. Coupons are not viewed as particularly helpful. Thus, the way decisions actually get made is: (1) managers decide whether or not to run advertising in a particular period; and (2) if not, and only then, are coupons issued. Thus, coupons only run when there is no advertising (and then, as it turns out, only some of the time).

Also interesting is that one feature of this industry is that advertising only affects volume purchased *once a customer has already decided to go with the FT brand*. This is in direct conflict with the belief held by FT managers that advertising builds brand loyalty. As it turns out, brand loyalty is difficult to influence. Instead, it is possible to get non-loyal customers to try the FT brand ... but, the best way to do that is with coupons!

Thus, the relationships in the data reflect the fact that the decision process at FT creates a negative correlation between advertising and coupons. This has an indirect negative effect on brand switching, lowering the effect of advertising on volume purchased and, hence, sales overall. When advertising runs, coupons do not – and the negative effect of advertising on coupons more than offsets its positive effect on quantity. Moreover, because coupons run only when advertising does not, the data also reflect a negative correlation between coupons and sales volume (because the positive effect of advertising is absent when coupons run).

The central point is that a linear regression does not demonstrate causality – it only quantifies *correlations* present in the data. The decision about which variables to make "dependent" and which to make "independent" is a choice made by the regression analyst. For example, one could just have easily have made price the dependent variable, with quantity as one of the independent variables – the regression procedure would work fine by identifying the coefficients consistent with the correlations in the data.

If one knows the true causal model, then it is possible to use regression analysis to make more accurate estimates of the effects of interventions (knowing the true causal model is a very tall order – we have more to say about this in later chapters). One technique is called multi-stage regression. Basically, using the same data used to estimate the coefficients in Table 1.2, we can break the single-stage regression into parts. That is, we regress: *Coupons* on *Advertising*; *Brand Choice* on *Brand Loyalty* and the **estimated** value for *Coupons*; *Unit Sales* on estimated *Brand Choice, Price* and *Advertising*; and *Profit* on *Advertising, Price* and estimated *Unit Sales*.[8]

Suppose we do that. The estimates are:

$$C = 0.15 - 0.15A \tag{1.2}$$

$$c = 0.02 + 0.30C + 0.72l \tag{1.3}$$

$$q = 146.8 + 32.8c + 0.4A - 0.7901P \tag{1.4}$$

The estimated unit volume *without any intervention* is the same as under the naive model (i.e., 111.2). Knowing the true causal model, we know that the divisional practice of making coupons dependent upon advertising decisions is a big mistake. Suppose you adopt the policy of letting FT managers run advertising as they see fit (on average, 78% of the time, as we know from Table 1.2) but insist they run coupons all the time, regardless of whether or not an advertising campaign is running. Working through the calculations (using the expected values from

8 The purpose of this example is to illustrate the distinction between causality and correlation, not to divert readers into a tutorial on multi-stage regressions. In our experience, those with a basic understanding of regular, single-stage regressions find the intuition behind this technique intuitive – which is all we are shooting for here.

Table 1.3
Regression estimates vs. true values

	Profit		
	Single-stage regression	**Multi-stage regression**	**True**
Do nothing	6,135.4	6,135.4	6,120.0
Set C = 1	5,352.4	6,679.4	6,951.8
Set $A = 1$	6,112.5	6,079.0	6,157.9

Table 1.2 of 0.33 and 56.6 for l and P, respectively),

$$\mathrm{C} = 1$$

$$c = 0.02 + 0.30 \times 1 + 0.72 \times 0.33 = 0.56$$

$$q = 146.8 + 32.8 \times 0.56 + 0.4 \times 0.78 - 0.79 \times 56.6 = 120.7$$

$$\pi = 120.7 \times 56.6 - 200 \times 0.78 = 6{,}679.4$$

The multi-stage regression model is not perfect (for reasons we will learn about later). However, it does a much better job than the naive model. What the multi-stage model gets right is the negative effect of advertising on coupons. Table 1.3 compares the estimated effects of each approach versus the truth.

The important thing to note is that the single-stage regression incorrectly suggests that increasing coupons is the *worst* option. Under the multi-stage regression (designed to capture the true causal relationships), the assessed value of increased coupons is significantly greater – indeed, it is correctly assessed as the optimal choice. The message is this: causality matters. It matters especially when the decision is about choosing an intervention. Running regressions without understanding causal structure can cause serious miscalculations. The preceding example is a preview of things to come. First, however, we will begin with some simple, qualitative modeling techniques in the next chapter.

KEY CONCEPTS: CHAPTER 1

- Causal modeling is useful for both situational assessment and managerial intervention. The former is achieved through evidence-based reasoning and the latter through intervention decision analysis.
- Inferences made through causal thinking lead us both ways along the causal chain; toward potentially useful sources of information, and to the evaluation of evidence for variable likelihoods.

- To make good intervention decisions requires a deep understanding that statistical correlation does not imply causation.
- Regression analyses reveal correlation relationships among variables in data generated from some process. Because intervention changes the process, one needs a causal model to properly interpret the effects of those changes.

CHAPTER 1 REVIEW QUESTIONS

1.1 What are the two major uses of causal modeling?

1.2 A large retail corporation is (as usual) trying to increase its profit, but it is not sure what is the best approach. It has collected data about its operation and determined that the following elements are key: *Advertising* (A), *Cost Per Unit* (C), *Selling Price* (P_i), *Competitor Price* (P_j), *Unit Sales* (Q), *Number of Customers* (N), and *Profit* (π). They hire a very expensive analyst to produce a multiple regression model to help them produce better predictions of profits in the future. Now management wants to uncover the implicit causal structure of the multiple regression model … and they want *you* to do it. Assuming the following (multi-stage) regression model accurately reflects the causal structure in this situation, reconstruct the underlying qualitative causal model that embodies the same relationships (e.g., indicate nodes, links and [+] and [−] indicators for all the variable links as was done for Figure 1.2).

$$P_i = 0.33 + 1.23C$$

$$N = 1.73 - 0.32P_i + 0.28P_j + 4.91A$$

$$Q = 44.02 + 0.74N + 2.70A$$

$$\pi = 3.81 + 12.24Q + 2.83P_i - 3.11C - 22.59A$$

1.3 In order to find ways to increase its revenue, a major automobile maker does a regression analysis of annual car sales on both the fuel efficiency (mileage) of the cars and the number of sales associates. The company discovers that the coefficient on the variable for mileage is large and negative, while the coefficient on the number of sales people is positive. They are both statistically significant. Based on this information, which of the following must be true?

(a) The company should decrease the fuel efficiency of its cars to ramp up sales.
(b) Dealerships that sell more cars can afford to hire more employees.
(c) The commission system encourages the car sales people to push the high-profit SUVs and luxury sedans that also have terrible fuel efficiency.
(d) None of the above.

1.4 In the causal model for a car manufacturing process, the node representing *Color* has four states: *red, blue, silver,* and *black*. Assuming that the causal diagram is properly and completely specified, what must we assume about yellow cars from this manufacturing process?

1.5 A national credit card corporation wants to test the relationship between a card's *Credit Limit* (e.g. *bronze, gold,* or *platinum* level) and the *Level of Debt* carried on the users' card. To do this, the company calculates a regression equation relating the *Limit* to the actual credit used and finds a large and statistically significant positive coefficient. What inferences that relate a person's credit card limit to the amount of debt carried on their card are *consistent* with this finding?

1.6 An established market research firm produces a linear regression model incorporating all of the market data they can find. They use this model to make predictions about sales in other markets and in submarkets and it performs fairly well. They keep hearing about the importance of some "causal modeling" technique, so they hire an expert to build a multiple regression model informed by the causal relations they believe exist among the variables generating the data. They then plug their market data and some test data into the causally informed multiple regression model and it delivers the same predictions. Is there something wrong? Shouldn't the "improved" model deliver better predictions?

Chapter 2

Qualitative Causal Models

We now introduce a very intuitive modeling framework known as *causal modeling*. This approach is useful for tackling a wide range of problems in many settings. It can be "scaled" depending upon the requirements of the issue at hand. It can be used as: (1) a qualitative guide to help think through the consequences of various decision options; (2) as an "informal quantitative" tool to broadly assess how environmental variables respond to managerial interventions; or (3) as a very sophisticated technique by which to provide a precise quantitative analysis of a strategic decision problem.

Causal modeling is sufficiently general to be useful in the four primary categories of decision analysis: logic audit, explanation, exploration, and prediction. These techniques come to us from a very rich stream of formal research in the field of artificial intelligence called *Bayesian networks*. The decision analysis community has augmented Bayesian networks to include interventions and objectives under the name of *influence diagrams*. As a result, not only are there many good technical references on the analysis approach, there are also software resources (some of which we will see later). There are even techniques available to discover underlying causal relations from data (i.e., for situations in which the causal relations are uncertain – we discuss this in more detail in Chapter 10). Many settings can be represented via causal models, including interactive decision analysis (aka game theory – see Chapter 9).

The methods we include in this text are best used in settings in which the causal relationships are safely assumed to be stable over the decision horizon. Also, while it is possible to construct causal models of systems involving feedback loops ($A \rightarrow B$ *and* $B \rightarrow A$), doing so is cumbersome. For most of the chapters we will only consider cases in which the causal relations are stable and contain no loops.

Note on Causality

Causality is not a well understood, or even well defined, phenomenon. Its meaning and significance have been hotly debated for thousands of years (at least going back to Aristotle). This debate is beyond the scope of our discussion. What we do know is that the human mind has a natural tendency to view the world through a causal lens. Indeed, causal relations come much more easily to us than, say, estimating joint probability distributions. Thus, as supreme court justice[1] Potter Stewart said about hard-core pornography (in Jacobellis v. Ohio, 1964)[2] so will we say about causality: it may be hard to define, but we know it when we see it. That said, not just any perceived causal relationship will hold water; as we will see throughout this book, there are specific technical requirements that a relationship must meet in order to be properly considered causal.

2.1 SETTING THE STAGE

Before getting into the details of a model, it is helpful to give some thought up-front to what the model is intended to achieve and, with that in mind, to its scope (i.e., which elements will be included and which left out). In order to facilitate construction of the model, give some thought to the following items:

- *Objectives*: There are two levels of objective that should be considered. The first level is the objective of constructing the model. In the real world, the cost of building a model must be offset by some tangible benefits. Typically, the cost will be calculated in terms of your own limited time and the benefit in terms of better decision making. Therefore, before diving in to build a model, give careful thought to what, specifically, you are trying to achieve. What is the decision at hand? What are your objectives? Can these objectives be quantified or are they qualitative in nature? Is your objective the sort of thing that can be optimized (like maximizing profits or minimizing costs)? How do the other elements in your proposed model relate to your objective in building it?

  The second level is the objective, or objectives, of the decision-makers who will be represented in the model as strategic agents. If the model is to be used for situational assessment, then there are no objectives that will be represented at this level. If the model will take the form of a single-agent influence diagram, then take some time to think about the consequences associated with the various choices available to the decision-maker and how those consequences would be ranked in order of preference by the decision-maker. Often, the decision-maker represented in a single-agent influence diagram is you and the second-level objective is the same as the first-level objective (i.e., the model is

1 http://en.wikipedia.org/wiki/Potter_Stewart
2 http://en.wikipedia.org/wiki/Jacobellis_v._Ohio

designed to help you achieve your objective). If you are considering a multi-agent influence diagram, then you will want to give some thought to how those being represented as strategic agents in the model assess the consequences of the interactions represented in the model.

- *Strategic Agents*: If the purpose of the model is to analyze a decision, ask yourself what individuals or organizations can take actions that may affect the consequences of your choices. A "strategic" agent is one whose actions may affect your ability to achieve your own objectives (which typically includes yourself), and so these agents should be included in your analysis. Strategic agents may be direct competitors, alliance partners, large buyers, employees, critical suppliers, and so on. In addition to identifying strategic agents, you must give some thought to *their* objectives in the situation you intend to model.
- *Strategic Options*: For each of the strategic agents included in the model, enumerate the significant actions they might take in order to achieve their own objectives. For example, a competitor may be able to retaliate against you by cutting their prices. Or you may be designing an incentive program to get employees to increase effort. Or a potential alliance partner may attempt to free-ride on your efforts. For simple models, focus on "big-ticket" actions (i.e., the actions they might take that could have the greatest effects on your objectives).
- *Environmental Factors*: Typically, your ability to attain an objective will be affected by a number of things that are not the result of actions taken by strategic agents. For example, the success of your project may depend critically upon international exchange rates. These are determined by massive global economic forces, not the actions of a few strategic agents.
- *Causal Relations*: Once you elaborate your objectives, strategic agents, strategic options, and environmental factors, then you must consider how these elements interact with one another. Early on, think about how changes in environmental factors and strategic options affect one another. Does a change in one element result in a change, or the possibility of a change, to another element? What does each agent know at the time they choose their strategic option? Do some elements influence others indirectly (through intermediary elements)? Going through this process, you may uncover missing causal links that should be added to the model.

The distinction between Environmental and Strategic components of a model is usually straightforward. For example, if your firm manufactures jet airliners (i.e., a b2b application), your customers are few and significant. They should probably be represented as strategic agents in any model concerning sales. If you are managing a product launch in a mass retail market, then a single consumer is not a strategic agent (because such individuals have little ability to affect your overall revenue). In this case, the entire set of consumers would be modeled as part of the environment (e.g., via the use of a demand function).

As already mentioned, causal models can be constructed with or without strategic agents. Models without strategic agents are called Bayesian networks and are useful for situational awareness; i.e., *evidential reasoning* and *causal inference* over the values of variables involved in some process. Chapters 4 and 5 deal with these inferential models. If the model will be used to help a single agent achieve an objective, then this is a *decision problem*. In Chapter 7, we look at situations in which you are the sole strategic agent optimizing your objective. Then, in Chapter 9 we will learn how to add strategic agents to solve interactive decision problems; also known as *games*. Causal models with at least one strategic option and objective are known as *influence diagrams*.

Definition 2.1 *Bayesian network: a causal model that only includes environmental and system factors connected by causal relations.*

Definition 2.2 *Influence diagram: a causal model that includes strategic agents, strategic actions, and an objective.*

Though you are strongly encouraged to put much thought into these elements before you begin building your model, you do not need to get everything in one shot. Building models is almost always an iterative process. Our advice is to start with the most slimmed-down model possible, something with the fewest number of moving parts, and then add more components only as you actually need them. Using this incremental process you will understand your final model better, and if (when) strange things start happening in your model along the way, you'll know at which point they were introduced.

2.2 BUILDING A QUALITATIVE CAUSAL MODEL

The essential object of a causal model is a *directed graph*. A directed graph is a picture, or "network diagram", consisting of a set of nodes linked together by arrows going from one node to another node. The nodes represent variables in the system and are represented by different shapes depending upon the type of variable it captures. The arrows linking nodes represent the causal relationships and are unidirectional (i.e., they point in one direction only). We now go into more detail on the node and link types.

2.2.1 Nodes

The nodes of the graph represent key variables in the model: objectives, the options available to strategic agents, and environmental factors. Different types of variables

Table 2.1
Node types in a causal model

Variable type	Node shape	Description
Probabilistic	Oval/Circle	Chance variables, uncertain quantities, environmental factors, and other elements outside the *direct* control of strategic agents in the model.
Objective	Hexagon	Payoff, profit, value, desirability, or utility of model outcomes. Decisions are made to optimize the objective.
Strategic option	Rectangle	Decision point, choice variable, value *directly* controlled by a strategic agent.
Function	Chevron	Value is a deterministic function of values of other variables (like an equation).

are represented by different shaped nodes. There are four basic types of variable, each of which is described in Table 2.1.

Bayesian networks typically contain only probabilistic nodes, though they may also have function nodes. Influence diagrams *must contain both an objective node and at least one strategic option node* – you cannot have one without the other. All influence diagrams used in this text will contain exactly one objective node. Strategic option nodes will also be referred to as "decision nodes", and probabilistic nodes will sometimes be called "variables" or "variable nodes" reflecting the role they play in the causal system. Influence diagrams are a causal version of decision trees that are also more compact because they do not need to maintain a tree structure. We describe how to solve decision problems with influence diagrams in Chapter 7.

The variables we consider will always have a finite number of *states*, meaning the number of values that can be taken by the variable.[3] The set of states should be mutually exclusive and collectively exhaustive (MECE). It is not uncommon to include an "other" category as a catch-all state to ensure that the list is collectively exhaustive. For numerical data, it is often possible to be truly exhaustive; e.g., a variable *Price* can be broken down to <$5 and ≥$5. We may then interpret these categories as *low* and *high* without having a catch-all value.

Furthermore, in many cases there are impossible or irrelevant values once a problem domain is specified (such as negative prices of goods). These can be left out once the domain assumptions have been explicitly defined. To keep things simple and clear, we typically use variables with only two or three values, and they usually take the form of categorized numerical data (e.g., *low*/*medium*/*high*) or

3 It is worth pointing out that, although we do not do so in this text to keep the exposition clear and the calculations simple, causal models can contain continuous variables.

Boolean variables (e.g., *true*/*false* or *yes*/*no*). This allows us the further convenience of referring to values using binary values (e.g., *no* = 0, and *yes* = 1) or ternary values (e.g., *low* = 0, *medium* = 1, and *high* = 2). The running assumption throughout is that values which are not listed are impossible.

2.2.2 Links

Causal relations between nodes are represented by directed links between them. By "directed link" we mean an arrow *from* one node *to* another indicating the direction of influence. An influence link from a node representing some variable A to another node representing variable B means the following:

1 There is some configuration of states of the nodes other than A and B such that, held constant, changes in the state of A result in a change in the likelihoods with which the states of B occur; and
2 Changes in the state of B do not affect the likelihoods with which the states of A occur regardless of the states of other variables in the system.

Taken together, these two conditions capture what we mean by, "A is a cause of B." The first item allows for situations in which A only influences B under special circumstances. For example, suppose we have a causal system composed of three variables: (1) *Coin Flip*, which has states *heads* and *tails*; (2) *Die Roll*, with states *one* through *six*; and (3) *Score*, with states *zero* through *six*. Assume the state of *Score* is: *zero* if the *Coin Flip* is *tails*; and equal to the state of *Die Roll* if the *Coin Flip* is *heads*. In other words, if the *Coin Flip* is *heads* and the *Die Roll* is *four*, then the *Score* is *four*. However, if the *Coin Flip* is *tails*, the *Score* is *zero*, regardless of the *Die Roll*. The point being that there is a state of *Coin Flip* under which the state of *Die Roll* affects the likelihood of *Score* states. Therefore, an influence link must be placed from *Die Roll* to *Score*.

When building a model, the criterion for whether to add a link is straightforward. Ask: (1) whether there are circumstances under which an *isolated* change in A – meaning a change in A holding the states of all variables in the model besides B fixed – can change the probabilities with which the states of B occur; and (2) whether changes in B have no effect on A under any circumstances. If the answer to both is yes, then there should be an influence link from A to B.[4]

Some additional terminology will be helpful in referring to relationships among nodes. When there is an arrow from node A to node B, we say A is the **parent** of B, and B is the **child** of A. A graph contains a **path** between A and B if there is a sequence of links connecting A to B, *regardless of the direction of arrows* on the path. A path is a **directed path** from A to B if there is a sequence of links connecting A

4 We use the terms "causal link" and "influence link" interchangeably.

to *B* such that $A \rightarrow \ldots \rightarrow B$ (i.e., all arrows point toward *B*). We may also refer to a directed path as a **causal path**.

Node *B* is a **descendant** of node *A* if there is a directed path from *A* to *B*. All children are descendants of their parents because the path is the direct connection from parent to child. But not all descendants of a node are its children because they could be the children of children, etc. Clearly all directed paths are paths, but not all paths are directed. For example, $A \rightarrow B \leftarrow C$ is a path from A to C, but not a directed path. A **root node** is a node without any parents. A **leaf node** is a node without any children.

Definition 2.3 *Parent: if there is an arrow from node A to node B, then A is a parent of B.*

Definition 2.4 *Child: if there is an arrow from node A to node B, then B is a child of A.*

Definition 2.5 *Path: a sequence of links connecting two nodes regardless of direction.*

Definition 2.6 *Directed (or causal) path: a sequence of links connecting two nodes in the direction of the arrows.*

Definition 2.7 *Descendant: node B is a descendant of node A if there exists a directed path from A to B.*

Definition 2.8 *Ancestor: node A is an ancestor of node B if there exists a directed path from A to B.*

Definition 2.9 *Root node: a node without any parents.*

Definition 2.10 *Leaf node: a node without any children.*

Directed links indicate one of two types of relationship: **informational** and **non-informational**. If node *A* is the parent of node *B*, and node *B* is a strategic option, then (and only then) the link between *A* and *B* is informational. Because strategic options are choice variables for strategic agents, the parents of a strategic option node indicate that the agent making the decision represented by that node does so knowing the states of its parent nodes. Of course, changing an agent's information may very well affect her choices.[5]

5 If knowing the states of the parent variables does not change the agent's choice under any circumstances, then they should not be parents of the decision variable! They represent useless information with respect to that decision.

> **Definition 2.11** *Informational link: an arrow pointing into a strategic option node.*

Another important convention is that strategic options are always assumed to be made in a particular sequence (*decision 1*, *decision 2*, ...). The sequence is indicated by the arrows from one option to the next. If the order of choices does not matter (e.g., when they are made simultaneously), then any directed path among those decision nodes is fine.

It is also assumed that strategic agents do not forget things. Specifically, all information links pointing to a strategic option of an agent also point to all of that option node's descendant options. In other words, if an agent makes decision A and, at some later point, decision B, then there should be a link $A \rightarrow B$. In addition, all nodes pointing to A are also assumed to point to B (at the time of making decision B, the agent remembers what he knew at A as well as the actual decision made at A). To economize on links in our diagrams and keep them as uncluttered as possible, redundant information links are not shown. In our example, only new information at B would be indicated with links and it would be implicitly understood that everything pointing to A also points to B.

All arrows besides the ones pointing into decision nodes indicate causal influence relationships. Specifically, if node A is the parent of node B, and B is not a strategic option, it means that a change in the value of A can result in a change in the value of B. We will require of our causal diagrams that they contain no cycles (the graphs must be **acyclic**). This means there are no directed paths such that $A \rightarrow \ldots \rightarrow A$. Note that feedback loops can still be included in our models via time-stamping of the variables (e.g., $A_t \rightarrow \ldots \rightarrow A_{t+1}$) so that they act like ordinary distinct variables.

Causal effects of influence links

Qualitative causal effects between variables can be indicated on influence by putting "[+]" or "[−]" on the link. A "[+]" on a link means that increasing the state of the parent variable increases the likelihood that the state of the child indicated by that link is more likely to be high; i.e., increases in the value of the parent tend to result in increases in the value of the child, and decreases in the parent tend to result in decreases in the child. Conversely, a "[−]" on a link means that likelihoods of states have an inverse relationship; i.e., increases in the value of the parent tend to result in decreases in the value of the child, and decreases in the parent tend to result in increases in the child variable. Since agents are assumed to have free will, *informational links (the ones pointing into decision nodes) are never signed.*

Note well that in order to speak of "increases" or "decreases" in the states of a node, one must first be able to order the states in a way that this idea makes sense. If the states are numeric, such as the values of a die roll or various ranges of profit, then the order is obvious. If the states are *heads* and *tails*, then there is no order. Sometimes, we can assign an order by labeling the states appropriately. For example, *heads* = 1 and *tails* = 0. Then, in the coin-die-score example above, we would put [+] on the link between *Coin Flip* and *Score*: the *Score* is *zero* if *Coin Flip* = 0 and *one* or more if *Coin Flip* = 1 (i.e., the likelihood of a higher value of *Score* is increasing in the value of *Coin Flip*).

One of the critical lessons of this text is that correlation **does not** imply causation. However, a causal relationship between two values can make them correlated. It is important to keep in mind throughout that the meaning of [+] and [−] is the nature of the causal link between those variables, not merely a description of how their values change with respect to one another. The importance of this distinction will become increasingly clear as we progress through more complicated (non-linear and categorical) relationships between variables.

As described above, in most of our examples the variables are assumed to be in states 0 or 1 (with appropriate meanings attached such as *low* and *high* values). When this is the case, the meaning of [+] is unambiguous: discovering that the value of a parent is 0 increases your certainty that the value of the child is also 0, and discovering that the parent is 1 increases your certainty that the child is also 1. Conversely, if [−] is indicated, then being told the value of a parent is 0 increases your certainty that the value of the child is 1, and finding out that the parent is 1 increases your certainty that the value of the child is 0.

We will now introduce some mathematical notation to allow us to write things down more compactly. Let "Δ" mean "change" in the certainty that a variable's value is 1, and read "⇒" as "implies". Then, a causal relationship from x to y shown with a "[+]" arrow means that

$$+\Delta x \Rightarrow +\Delta y, \text{ and}$$

$$-\Delta x \Rightarrow -\Delta y.$$

Conversely, a relationship from x to y shown with a "[−]" arrow means that

$$+\Delta x \Rightarrow -\Delta y, \text{ and}$$

$$-\Delta x \Rightarrow +\Delta y.$$

A simple example will help clarify the causal modeling way of representing these relationships. Let's say you are researching the purchase of new office equipment for your company. You can buy equipment with a *Quality* that is either *low* or *high*. High quality equipment can perform more tasks, and produces better results with less work, but the parts for this equipment are more expensive. The quality of equipment therefore affects its *Maintenance Costs*: the greater the quality the higher the maintenance costs.[6] The most natural encoding of these states is for

6 Quality in this case refers to the output of the equipment (e.g., professional vs home office quality) rather than the care taken in the design and construction of the equipment – though these may be related as we'll see later.

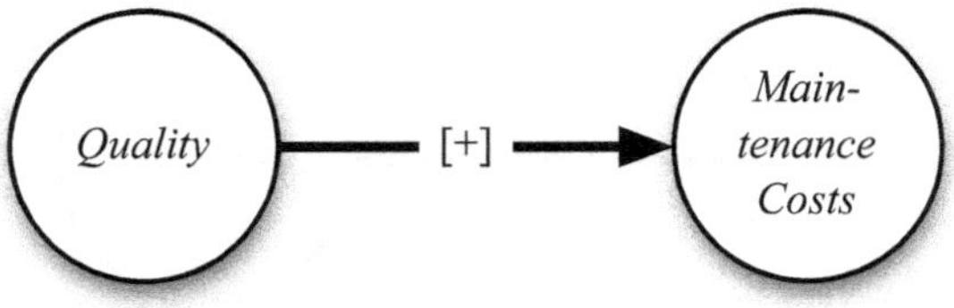

FIGURE 2.1 A simple causal model.

both *Quality* and for *Maintenance Costs*: *low* = 0 and *high* = 1. This relationship is therefore a positive one [+], and the causal model for this relationship is shown in Figure 2.1. Note that if we instead assign *low* = 1 and *high* = 0 for *Maintenance Costs* or *Quality* (but not both), then the relationship changes to a [−], but the same relationship is captured. If both variables use *low* = 1 and *high* = 0, then the relationship is again [+].

A **monotonic relationship** is one in which moving a variable in one direction causes the other variable to also move in one direction (either up or down). So a relationship in which the child variable tends to increase when the parent tends to increase is monotonic. More accurately, what this means is that as higher values of the parent become more likely, higher values of the child also become more likely. The converse type of relationship, in which the child variable tends to decrease when the parent tends to increase, is also monotonic (just in the other direction). Thus, our qualitative causal models, which use [+] and [−], allow us to capture monotonic relationships. Many, but not all, relationships fit this description.

More complicated relationships can be represented using **conditional certainty tables**. A conditional certainty table is associated with a child node. Its rows correspond to the various combinations of parent node states and its columns correspond to the states of the child with which it is associated. The entries in the table are qualitative: "+", "−", and "0". The entries indicate the direction in which one should adjust one's assessment of the likelihoods of the child states given information about the states of the parent variables. The adjustment is with respect to one's belief prior to having any information about parent states.

To make this procedure explicit, extend the previous example by introducing intermediary quality and cost values so that for both *Quality* and *Maintenance Costs*: *low* = 0, *med* = 1, and *high* = 2. It may still be the case that these variables have a monotonic relationship that we can capture with a [+]. Such a table is presented in Figure 2.2. The way to interpret the "+" and "−" signs in this table is that being told that the value of *Quality* is *low* increases your certainty that *Maintenance Costs* is also *low*. Simultaneously, your certainty that *Maintenance Costs* is *med* or *high* decreases. A similar interpretation would follow for the other rows. The relationship is still monotonic because quality increases correspond to increases in maintenance costs.

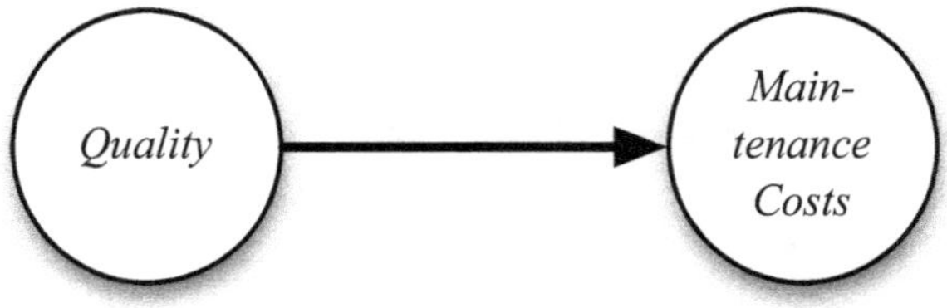

	Maintenance Costs		
Quality	*low*	*med*	*high*
low	+	–	–
medium	–	+	–
high	–	–	+

FIGURE 2.2 Conditional certainty table for a monotonic relationship.

In order to interpret the table in Figure 2.2 properly, one must ask, "being told that the value of *Quality* is *low* increases your certainty ...*relative to what*?" The answer is: relative to what it would have been without knowledge of the state of *Quality*. For example, it may be that, without any knowledge of the state of *Quality*, our belief is that all three levels of *Maintenance Costs* are equally likely (i.e., occur with probabilities 1/3, 1/3, 1/3). Then, the conditional certainty table in Figure 2.2 says that, if you learn the state of *Quality* is *low*, then you think the likelihood of *Maintenance Costs* being *low* is greater than 1/3, and the likelihoods of *med* and *high* are each less than 1/3.

The table becomes more important when the example becomes more complicated. Let's change our story a little bit. Low quality equipment still has low maintenance costs because you can pick up the parts, or complete replacements, at the local office supply store. And let's say that the high-quality professional equipment more likely results in either high costs (if it does break) or medium costs (because it breaks less often). But let's say that medium quality equipment is a bit of a gamble; if you're lucky the thing that breaks is a cheap modular part, but certain aspects are custom made and if/when they break you have to order them express from the manufacturer (which is expensive). This situation is represented by the table in Figure 2.3.

There is no simple [+] or [–] relationship from one variable to the other here.[7] Instead, we attach a conditional "it-depends" table to the child node delineating the effects on its certainty of changes in the parent node's value. If there are multiple parent nodes, then the table is set up so that each row corresponds to a unique

7 If the parent variable's values are categorical, then you can sometimes create a monotonic relationship by rearranging the order of the rows. If the values have a natural ordering (i.e., some are lower than others like with money, weight, size, and quality) then such a reordering could make the table show monotonically related relationship, but if the ordering actually means something then you obviously shouldn't reorder them.

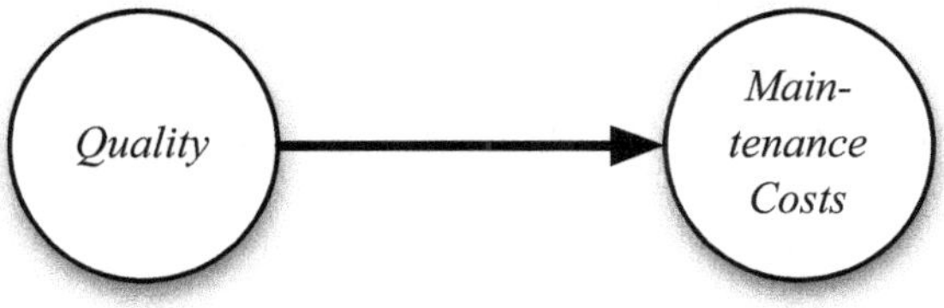

	Maintenance Costs		
Quality	*low*	*med*	*high*
low	+	–	–
medium	+	–	+
high	–	+	+

FIGURE 2.3 Qualitative representation of non-monotonic effects.

combination of parent states, one state from each parent variable. The rows should exhaust the possible combinations of parent states, but you really only need to include the *possible* combinations of parent variables. We'll see some examples like this later on.

> **Definition 2.12** *Monotonic relationship: as higher values of the parent become more likely, higher values of the child either consistently become more likely or consistently become less likely.*
>
> **Definition 2.13** *Non-monotonic relationship: as higher values of the parent become more likely, the probabilities of values of the child change in inconsistent ways.*

2.2.3 Some Examples of Qualitative Causal Models

Recall from the previous chapter that we said causal models are particularly useful for two types of analysis: *situational assessment* and *managerial intervention*. Let's illustrate these ideas with some simplified business examples, starting with the former type and segueing into the latter.

Example 2.1: Basic Causal Reasoning

Suppose you are asked to make an assessment of the size of the market for some good, such as an appliance or consumer electronic product. For concreteness, think of it as a new kind of super-high-def-3D television. For now assume that you are not personally involved in the industry, and so you are not trying to make some decision and optimize your outcome. You just want to know what the total market volume will be for some market sector, and how it changes in response to a given set of variables.

As a first pass, you decide the system can be represented by five environmental variables:

- *Price*: average price per unit (*low*=0, *high*=1);
- *Advertising*: the amount of money spent on advertising products (*low*=0, *high*=1);
- *Number of Customers*: a count of how many individuals enter related retail stores or online sales platforms (*low*=0, *high*=1);
- *Media Hype*: whether independent media sources report on or display related products (*no*=0, *yes*=1);
- *Market Volume*: the total amount of goods sold for your product category (*low*=0, *high*=1).

Figure 2.4 presents a causal model that illustrates your hypotheses regarding the influence relationships in the model. The meaning of this model should be quite intuitive. The model assumes that the price has a negative effect on the number of customers who will shop for the TV. This assumption is indicated by the [−] shown on the link between *Price* and *Number of Customers*. If the *Price* is *high* (= 1), then your certainty that the *Number of Customers* is *high* (= 1) *decreases*. Conversely, if the *Price* is *low* (= 0), then your certainty that the *Number of Customers* is *high increases* (while holding everything else constant).

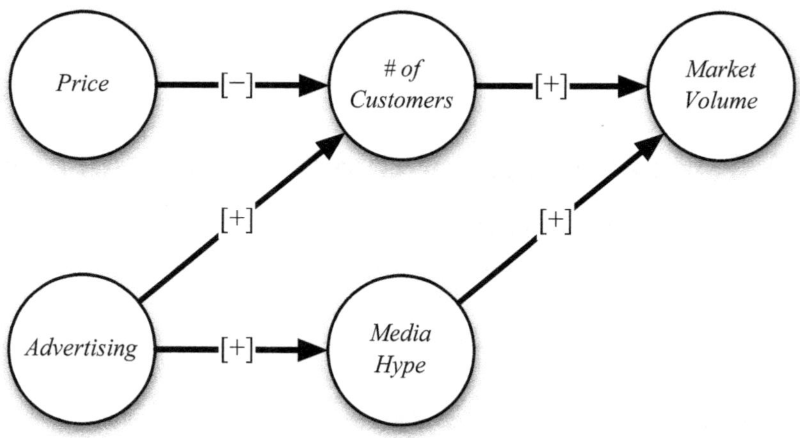

FIGURE 2.4 Causal model for market volume.

Even though this model is entirely qualitative and quite simple, we can already use it to perform some analysis. For example, as your certainty that the price is high increases, your certainty that the total market volume for the TVs will be large decreases. To see this, follow the causal chain: increasing your certainty that *Price* = 1 decreases your certainty that *Number of Customers* = 1; and lower certainty that *Number of Customers* = 1 implies lower certainty that *Market Volume* = 1 (because [+] implies movement in the same direction).

We can also use the model to reason inductively (backwards up the causal chain). For example, if your friend tells you that a large number of TVs were sold (*Market Volume* = 1), then you should conclude that the causes of this outcome are more certain (a large *Number of Customers* and some *Media Hype*). *The signs on the links are valid in both directions.* Thus you can furthermore reason that the amount of *Advertising* was *high* and that the *Price* was more likely *low*.

Example 2.2: Influence Diagram

Suppose that, instead of trying to do a situational assessment of the market for super-high-def-3D televisions, you are actually in charge of a retail chain that will carry these devices and you are trying to decide for yourself how much to spend on advertising for these products. Your objective is a high market share (the proportion of sales through your retailers compared to the total number sold). Before you decide to advertise or not you will know the manufacturer's suggested retail price.

You develop the simple model in Figure 2.5. In this model, *Advertise* is a decision variable set by you. The intervention decision here is simplified so that you can only choose either: (1) to run an advertising campaign featuring the super-high-def-3D TVs; or (2) you cannot do that. The link from *Price* to *Advertise* is an information link. It indicates that you will know the *Price* at the time you set *Advertise*. More advertising leads to a greater certainty of a larger number of unit sales. However, for a given level of advertising, higher prices tends to result in lower sales. Of course, *high Sales* lead to greater certainty of a *high Market Share*. In this example your *Market Share* is the objective and, hence, it is shown as a hexagon.

Other things being equal, higher sales will certainly lead to a greater market share (by definition), but the model only shows that your *certainty* that your market share will be *high* increases as your sales numbers increase. It is important to remember that the relationships in these models are about how changes in the value of a parent affect the relative certainties of the values of its children – i.e., rather than simple deterministic changes from node to node along the chain. It is especially easy to lose sight of this fact when working with qualitative causal models, but you will be rewarded in later chapters if you can keep the distinction straight.

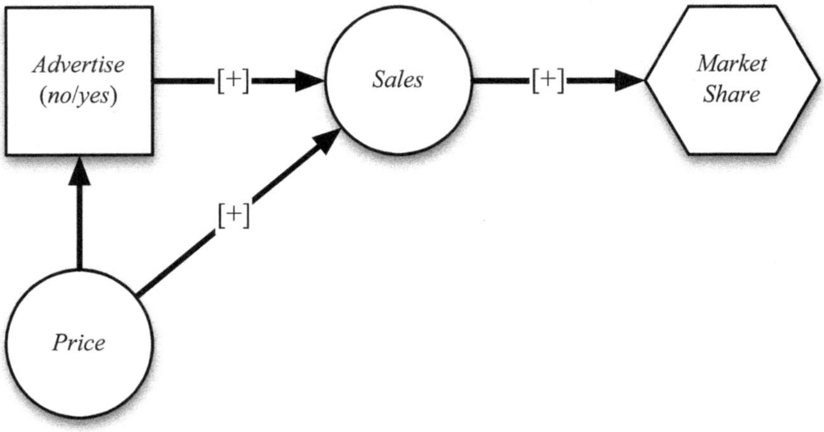

FIGURE 2.5 How much to spend on advertising?

Example 2.3: Influence Diagram Extended

The qualitative message from the previous example is simple: increase advertising spending to raise your market share. However, as we all know, things are not always so simple. You realize that your market share depends on the sales of everybody in the market. Even if your sales increase, if your competitors provide greater incentives and attract more customers, then the greater their sales will be as well, and as a result your market *share* may not increase – it may even decrease!

Moreover, since everybody observes the manufacturer's suggested retail price (MSRP), higher prices will cause all competitors to offer more incentives to buyers. You realize you need an augmented model and come up with the one in Figure 2.6. Notice that you have chosen to model the aggregate coupons and sales of the other stores as an environmental variable to keep things simple (i.e., you are not including each other retailer in the market as a strategic agent, which would require you to insert a separate decision variable for each of them).

The model in Figure 2.6 is useful. It tells you that the MSRP has both direct and indirect effects on your market share. The direct effect of an increase in *Price* is to make your advertising less effective. That is, holding *Advertising* constant and increasing *Price* lowers your sales numbers. The model also warns that thinking about the price only in terms of its effect on your sales will lead to unexpectedly bad performance. *The simpler, incomplete model of Figure 2.5 may well cause you to advertise too little, thereby failing to achieve your desired market share.*

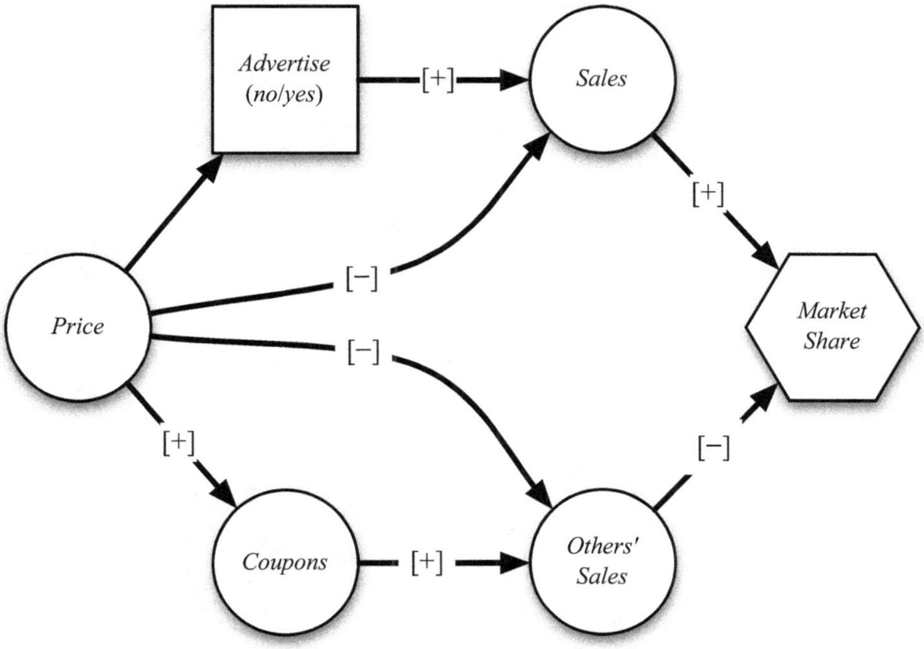

FIGURE 2.6 Including effects of other store's incentives.

Example 2.4: One More Twist

Now we add another element to enhance the analysis. Suppose you are the only store considering a large advertising campaign while everyone else plans to only use coupons as an incentive to buyers. Coupons can be offered in a much shorter time-frame than organizing and putting out a series of print and billboard ads. As a result, if you advertise then the other stores will notice, and then they are more certain to offer coupons. This extension to the model is illustrated in Figure 2.7.

Now, we see something interesting. Your advertising expenditure has two effects on your market share: a direct, positive effect and an indirect, negative effect. The indirect results form the following causal chain:

$$+\Delta Advert \Rightarrow +\Delta Coupons \Rightarrow +\Delta Others' Sales \Rightarrow -\Delta MarketShare.$$

This gives you something more to think about. Some of your advertising spending "bleeds off" by provoking your competitors to offer coupons and, as a result, increase the total market volume for the product (thereby decreasing your share of the total). If the indirect effect is small, then you may not care much about this link. If it is huge, then it may actually be optimal to not advertise at all!

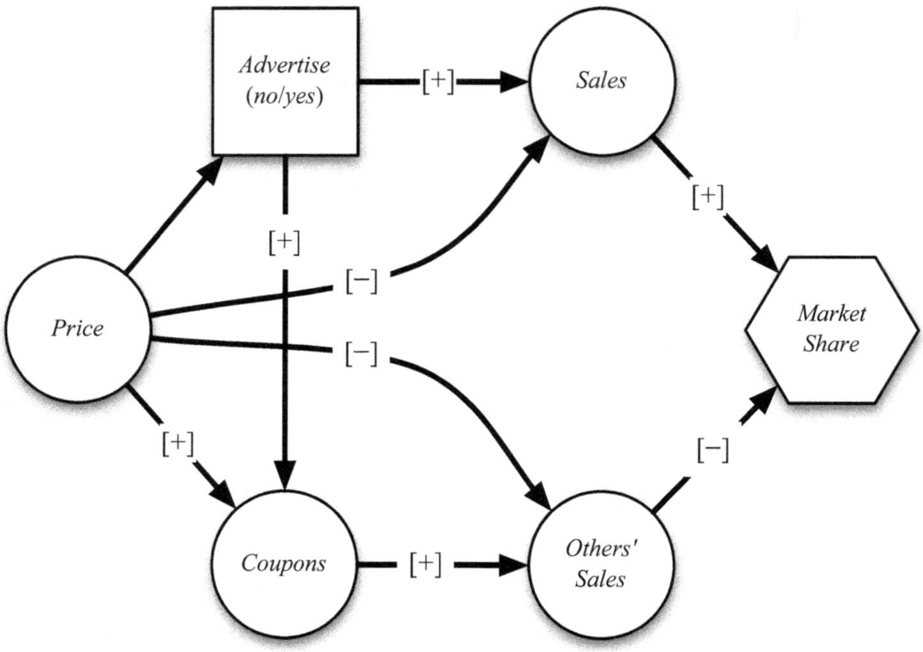

FIGURE 2.7 Competitors can observe your amount of advertising.

A causal dilemma

In the model of Figure 2.7, you face a dilemma. You would like to advertise to increase sales and assure yourself of a higher market share, but any such effort may be self-defeating in that it will trigger a coupon response by your competitors that both raises the total market volume and possibly decreases your own sales. Both of which could wipe out the positive effect of advertising on your market share.

This is frustrating. Yet, this is where thinking about situations in a structured way is helpful. The model's structure and implications beg certain questions. Are you missing anything? Are there other options that could be used to effect desired outcomes? Can the tension identified in the model lead to a creative reframing of the situation?

As you sit and think about it, you realize that there are other actions you could take. For one thing, you might advertise for several products together, thereby keeping competitors from knowing that you are targeting the new TV technology (though, this has implications for the other products and the efficacy of your advertising). Or, you could advertise using mailers and online banner ads that are difficult for your competitors to monitor, so they wouldn't know the full scale of your campaign.

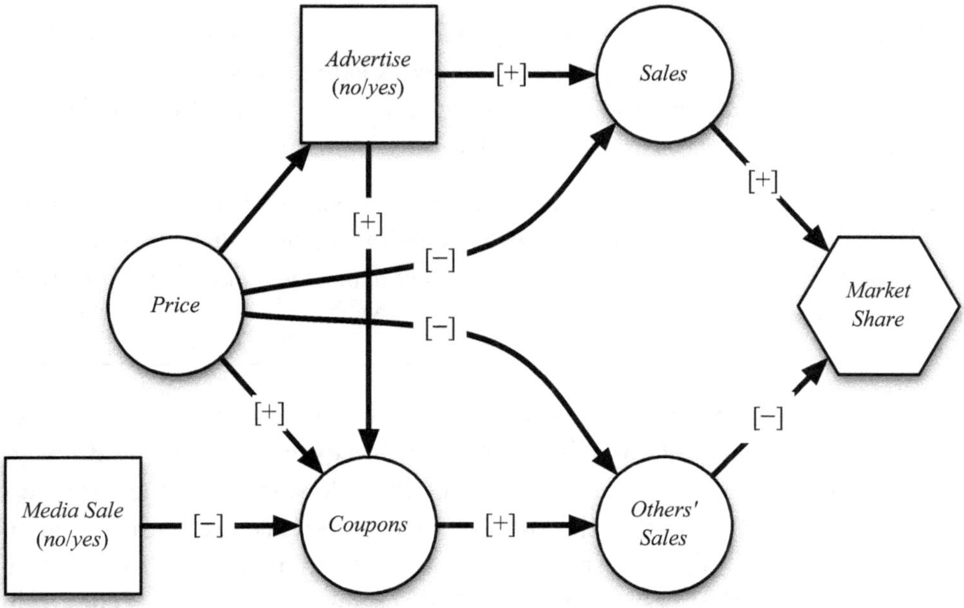

FIGURE 2.8 Two paths to higher market share.

Alternatively, right around the time of the product release, you might throw a huge sale on movies and music. Faced with this immediate substitute marketing tactic, the relative value placed by your competitors on offering a discount on the new TVs may drop. The model including this option is shown in Figure 2.8. Note that all the other variables (*Price*, *Sales*, etc.) still refer to just the TV product, and the *Media Sale* may generate sales and market share outside your current TV considerations, and that these factors are left out of the current model. Suppose you cannot afford to both have the huge sale and spend a ton on advertising. Which intervention would be most effective?

2.3 CAUSAL INDEPENDENCE

The preceding series of examples were simple, and so was the causal reasoning involved in their analysis. The analysis was simple because it was rather easy to follow the chain of influence effects forward and backward once we were given information about a variable. Causal models are not typically so simple, and so we now wish to present some rules for evaluating the flow of causal effects in these models. To assist in that task, it is helpful to introduce the notion of causal independence.

Definition 2.14 *Causal Independence: Given knowledge about the values of some variables in a causal system, two variables with unknown values are said to be "causally independent" if discovering information about the value of one implies nothing about the value of the other.*

Definition 2.15 *Causal Dependence: Given knowledge about the values of some variables in a causal system, two variables with unknown values are said to be "causally dependent" if discovering information about the value of one implies a change in our assessment of the other.*

Causal independence is useful when the task at hand is gathering data in order to make inferences about variables of interest. For example, one would not pay to obtain data about a variable that is causally independent of the variable of true interest. As it turns out, the rules of causal inference are both simple and surprisingly general. They are built upon the three foundational structures one finds in causal models: (1) **serial triplets**; (2) **diverging triplets**; and (3) **converging triplets**.

2.3.1 Serial Triplets

The first foundational causal structure we will describe is the **serial triplets**. Recall our Bayesian network example 2.4 from the last section ...before we made it into a decision problem. In that example, *Price* has an effect on the *Number of Customers* which, in turn, has an effect on *Market Volume*. Using our notation we have the following relationships:

$$+\Delta Price \Rightarrow -\Delta Number\ of\ Customers \Rightarrow -\Delta Market\ Volume$$

and

$$-\Delta Price \Rightarrow +\Delta Number\ of\ Customers \Rightarrow +\Delta Market\ Volume$$

Those positive and negative relationships can be captured with a serial triplet in a Bayesian diagram as shown in Figure 2.9.

FIGURE 2.9 An example of a serial triplet.

If we learn that *Price* = *low*, then according to this model we are more certain that *Number of Customers* = *high* and, hence, more certain that *Market Volume* = *high*. That chain of reasoning follows easily through the diagram. Suppose, however, that the first thing we learn is that the *Number of Customers* = *high*. Then, knowing more about the true value of *Price* does not change our assessment of the *Market Volume*. This is an implication of the structure of the causal model because the effect of *Price* on *Market Volume* is only indirect through the *Number of Customers*. When the value of *Number of Customers* is known, *Market Volume* and *Price* become *causally independent*, meaning that changes in the certainty of one no longer affect one's certainty in the other.

In other words, if you are told the *Number of Customers*, then your certainty in both *Price* and *Market Volume* must change because (as you probably recall) when making causal inferences, the effects roll in both directions. However, if you already know the true value of *Number of Customers*, then learning the true value of *Price* will not affect your certainty with respect to *Market Volume* AND learning the true value of *Market Volume* will not affect your certainty of *Price*.

One way to think about this is to imagine evidence about the *Number of Customers* as "blocking communication" between *Price* and *Market Volume*. That is, knowing the value of *Price* does not influence one's assessment of *Market Volume* when one already knows the value of *Number of Customers*. *More generally, two variables in a serial chain are causally dependent unless the state of any variable in-between those variables is known (in which case, they are causally independent)*. In our example, we would say that *Price* and *Market Volume* are **causally independent** given *Number of Customers*. Naturally this generalizes to any serial connection and is not specific to the variables in our example – it is a feature of the serial structure of the causal influences among any variables.

2.3.2 Diverging Triplets

The second basic structure is the **diverging triplet**. An example is shown in Figure 2.10. A divergent triplet is characterized by a parent node and two children of that parent node. Intuitively, knowing the value of the parent should change our certainty with respect to the children. Less obvious is whether knowing the value of a child affects one's assessment of the other child.

Let us continue by taking a portion of the Bayesian network from our example 2.4. Recall that in this example, the amount of advertising affects the number of customers and the likelihood of independent media hype, but that media hype has no direct effect on the number of customers (media hype instead changes the purchasing behavior of the customers further along in the full model to determine total market volume).

Suppose we do not know the amount spent on *Advertising*. Then, being told that the true value of *Media Hype* is *high* increases our certainty that *Advertising*

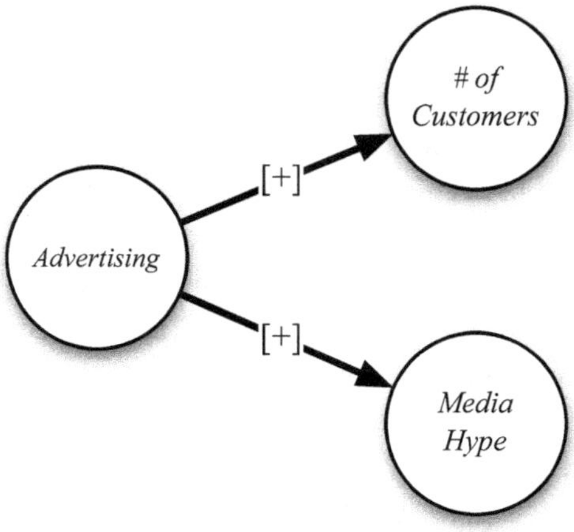

FIGURE 2.10 Diverging triplet.

was *high* and, therefore, that the *Number of Customers* was *high*. Again, *high Media Hype* does not cause *high Number of Customers*, but the causal structure supports this inference through the divergent connection.

However, if we already know that *Advertising* was *high*, then discovering whether there was *high* or *low Media Hype* or not does not change our assessment of the *Number of Customers. In general, the children of a common parent are causally dependent* ***unless*** *the state of the parent is known (in which case, they are causally independent)*. In our example, *Media Hype* and *Number of Customers* are causally independent given *Advertising*. This again has nothing to do with our variables and everything to do with the inferential structure of a divergent causal triplet.

2.3.3 Converging Triplets

The third, and final, basic structure of causal models is the **converging triplet**. An example is shown in Figure 2.11. Here, we have two parents influencing a single child. These structures are inherently different (and a bit more complicated) than the previous two. The reason is that changes in the parent variables independently influence the value of the child. Without knowing the value of the child, it is impossible to use the value of one parent to draw inferences about the other.

Isolating our attention to this triplet, being told that *Media Hype* is *high* clearly increases our certainty that *Market Volume* is *high* as well. Similarly, if we discover that the value of *Market Volume* is *high*, then our belief that *Media Hype* and *Number of Customers* are both *high* also increases. However, learning that *Media Hype* is *high* – with no further information – tells us nothing about the *Number of Customers*. According to Figure 2.11, these variables do not influence each other. Logically, however, if we additionally find out that the true value of *Media Hype*

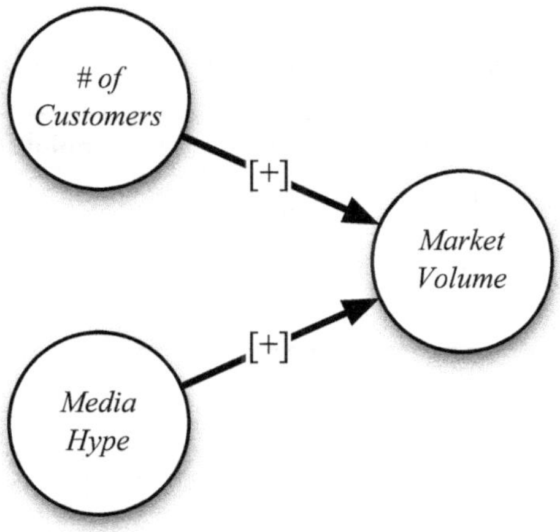

FIGURE 2.11 Converging triplet.

is actually *low*, then our certainty that *Number of Customers* is *high* must increase. Think about it. If you know the causal structure is as shown, being told that *Market Volume* is *high* implies increased certainty that the value of both parents is also high. If you then learn that *Number of Customers* is *low*, your certainty that *Media Hype* is *high* should increase even more.

We flesh that out a bit more with an extended example below (and A LOT more in section 5.2.3), but here is the general rule: *two parents of a common child are causally independent,* ***unless*** *the value of the child is known (in which case, they are causally dependent).*[8] In our example, *Media Hype* and *Number of Customers* are **causally dependent** given *Market Volume*, while *Media Hype* and *Number of Customers* are **causally independent** without additional information about the true value of *Market Volume*.

Now extend the example a bit more. Assume the *Market Volume* can be *low*, *medium*, or *high*. For ease of reference, let's attach numerical values to these states of 0, 1, and 2, respectively. For *Media Hype* and *Number of Customers* we'll use *low* = 0 and *high* = 1 as before. Suppose we know that the value of *Market Volume* is determined as shown in Table 2.2.

In other words, increasing the *Number of Customers* and/or increasing the *Media Hype* increases the *Market Volume*. In fact, what Table 2.2 says is that the relationship between the *Market Volume* and its causes is actually a simple deterministic function, namely:

$$\textit{Market Volume} = \textit{Number of Customers} + \textit{Media Hype}$$

We indicate that the relationship is deterministic by changing the shape of the *Market Volume* node to a chevron, as illustrated in Figure 2.12.

8 Recall that a descendant variable is a child, or a child of a child, or a child of a child of a child, etc.

Table 2.2
Functional relationship among variables

Media Hype	and	*Number of Customers*	influence	*Market Volume*
0	&	0	⇒	0
0	&	1	⇒	1
1	&	0	⇒	1
1	&	1	⇒	2

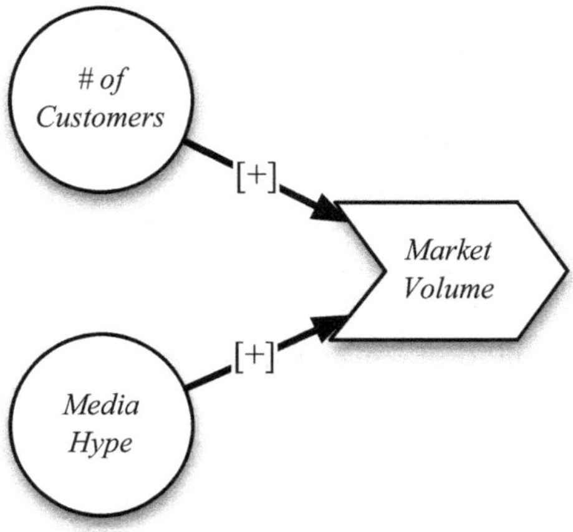

FIGURE 2.12 *Market Volume* is a deterministic function of its causes.

Suppose you discover that there is media hype on the product (*Media Hype* = 1). This tells you that the *Market Volume* must be either *medium* or *high* (1 or 2). However, at this stage we cannot say anything about the number of customers. To convince yourself of this, note that evidence that *Media Hype* = 1 puts the model into one of the two bottom rows of Table 2.2. This means that the *Number of Customers* can still be low or high. Thus, information about one cause in a converging connection tells you nothing about the certainty of another cause.

Now, suppose you are told that *Market Volume* = 1. Then, you **know** that *Number of Customers* = 0. Similarly, if you are told that *Market Volume* = 2, you **know** that *Number of Customers* = 1. With a convergent connection, information about the child renders its causes causally dependent. Put differently, evidence about a child opens communication between its parents.

2.3.4 Causal Independence in General

The three cases just discussed account for all the possibilities with respect to assessing the effects of how changes in information can affect one's assessment of a variable in a causal model. We can now synthesize these ideas into a single definition of causal independence, applicable in all causal models.

> **Definition 2.16** *Causal-independence: Two distinct variables x and z in a causal model are causally independent (or c-independent) if, for all paths between x and z, there is an intermediate variable y (distinct from x and z) such that either:*
>
> 1 *The connection is serial or diverging and the value of y is known; or*
> 2 *The connection is converging and there is no change in certainty with respect to y or any of y's descendants.*

Two variables that are not c-independent are said to be *c-dependent*.[9] We will give some examples of this definition in action in a moment. First, however, let us emphasize why this definition is important. Let us say you have a causal model relating many variables, including one variable in which you are particularly interested. *If one variable is causally independent with respect to the variable of interest, then getting more information about the state of the former will tell you nothing about the state of the latter.*

This is a striking result because, as we will see later, it holds in the quantitative versions of these models. We have a qualitative condition, one that can be checked simply by examining a diagram, that has important quantitative implications (including the expected value of new information about variables, robustness of values, and so on). These implications will be explored fully as the book progresses.

Example 2.5: Causal Independence

Let us illustrate definition 2.16 by returning to our Bayesian network from the preceding examples, as shown in Figure 2.4 and repeated here as Figure 2.13 .

Suppose, *beginning with no knowledge about the state of the variables in the system*, you discover that the price was high. How does this change your certainty with respect to the amount spent on advertising? Remember that a "path" is a sequence of links from one node to the next without regard to direction of the arrow. There are two paths from *Price* to *Advertising*. The first path is:

Price → *# of Customers* ← *Advertising*.

The second path is:

Price → *# of Customers* → *Market Volume* ← *M Hype* ← *Advert*.

9 The Bayesian network literature refers to definition 2.16 as "*d-separation*." The original definition is due to Pearl (1988), where the *d* denotes *directional*. We take liberty and alter this label for mnemonic purposes.

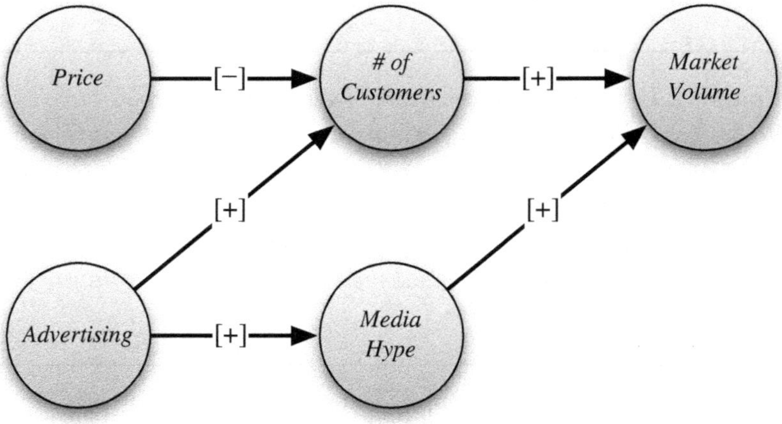

FIGURE 2.13 Causal model for market share.

Along both paths we have converging connections on nodes with no change in certainty – i.e., at this point, we have no new evidence or information regarding the state of *# of Customers* (on the first path) or that of *Market Volume* (on the second path). Therefore, by item 2 of definition 2.16 and the information at hand, *Price* and *Advertising* are causally independent. Thus new knowledge about the state of one (knowing *Price* = *high*) tells us nothing new about the state of the other.

Keep in mind that you *only need to have communication along one path to break the independence*. In other words, if we were to learn the value of **either** *# of Customers* or *Market Volume*, then the value of *Price* would cause us to update our certainty with respect to *Advertising*.

To see how item 1 of definition 2.16 works, suppose you know the true values of *# of Customers* and *Advertising*. Then, learning the true state of *Price* has no effect on your assessment of *Market Volume*. Once again, there are two paths to consider. The first path is:

Price → *# of Customers* → *Market Volume*.

The second path is:

Price → *# of Customers* ← *Advert* → *M Hype* → *Market Volume*.

On the first path, new knowledge about *# of Customers* blocks the serial connection. On the second path, this same knowledge about the *# of Customers* opens the path from *Price* to *Advertising*, permitting communication between them. However, the divergent path with *Advertising* as the source is blocked by the information about *Advertising*, so information about *Price* that flows through

# of Customers to *Advertising* is stopped there. This analysis reveals that both paths from *Price* to *Market Volume* are blocked given information about *# of Customers* and *Advertising.*

KEY CONCEPTS: CHAPTER 2

- Causal models may include the following elements: (1) Objectives, (2) Strategic Agents, (3) Strategic Options, (4) Environmental Factors, and (5) the Influence and Informational Relations among those elements.
- It is assumed that all the causal relationships are stable over the decision horizon and contain no loops.
- "A is a cause of B" means: (1) changes in A may result in changes in B; and (2) changes in B do not affect A. In this case we write "A → B", and we say A is a *parent* of B and B is a *child* of A (and also a *descendant*) of A.
- The parents of a strategic option node (rectangles) indicate that the chooser knows those values at the time of the decision. The inflowing informational relations do not have [+], [−], or other such qualitative effects.
- Evidence in a *serial chain* may be transmitted along the chain **unless** the state of a variable in-between those variables is known.
- Evidence about one child of a parent in a *diverging connection* may be transmitted to the other children of the parent **unless** the state of the parent is known.
- Knowing the true state of one cause of a variable in a *converging connection* does not alone change our certainty with respect to other causes. However, communication between upstream causes may occur if there are changes in certainty of the child (or any other descendants).
- Two distinct variables x and z in a causal model are *causally-independent* if, for all paths between x and z, there is an intermediate variable y (distinct from x and z) such that either: (1) the connection is serial or diverging and the value of y is known; or (2) the connection is converging and there is no change in certainty with respect to y or any of y's descendants.

CHAPTER 2 REVIEW QUESTIONS

2.1 A nationwide credit card company is evaluating the level of credit card usage by the card holders' credit limit. They analyzed the amount of *Debt Carried* from month to month as a percentage of the credit limit (*zero, low, medium,* and *high*) for each of the four levels of *Credit Limits* (increasing in the order *bronze, silver, gold,* and *platinum*). For personal users they discovered the following:

1 A direct negative relationship between percentage *Debt Carried* and *Credit Limit*; i.e., *bronze* limit cards increase the likelihood of state *high*

and decrease the likelihood of other states, *silver* limit cards increase the likelihood of state *medium* and decrease the likelihood of other states, and so on.

2 Many *gold* cards are held by businesses for travel expenses, and due to administrative inefficiencies they tend not to be paid on time, so the likelihood of a *high* amount of carried debt is also likely for *gold* cards.

3 The other exception is that holders of *bronze* cards are also likely to have a *zero* debt balance percent because they never use the card.

For a causal model with an arrow from *Credit Limit* to *Debt Carried*, generate a table for that arrow capturing the non-monotonic causal relationships described above.

2.2 Examine the influence diagram in the figure below and report the following sets of nodes:

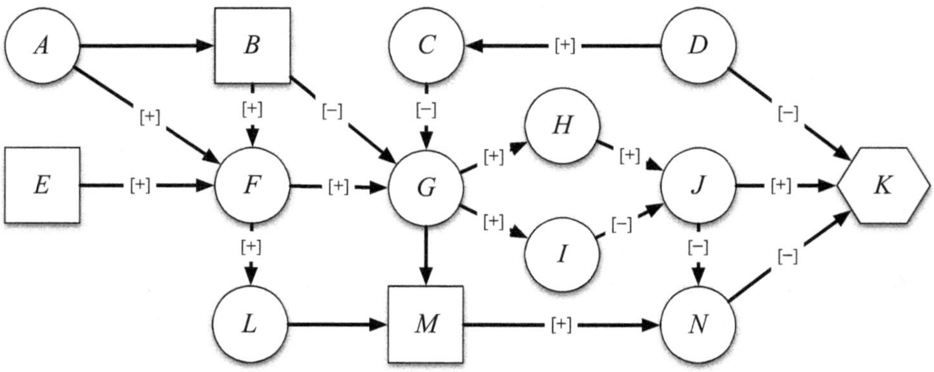

(a) The *parents* of *B*.
(b) The *parents* of *H*.
(c) The *children* of *J*.
(d) The *children* of *F*.
(e) The *descendants* of *D*.
(f) The *descendants* of *H*.
(g) The nodes along one *directed path* from *A* to *N*.
(h) All the *root* and *leaf* nodes of the graph.

2.3 Use the causal model diagram from Question 2.2 to answer the following questions:

(a) If you want to make *K* as high as possible, should you decide to make *M high* or *low*?
(b) If we increase our certainty that the value of *I* is *high*, will that have a decreasing, increasing, or some other effect on our evaluation of *K*?
(c) If you wish to be as certain as possible that your evaluation of *K* will increase, and you can only set one of the three decision nodes, which one should you set?

2.4 Use the model presented below and assume that the values of *E* and *J* are both known.

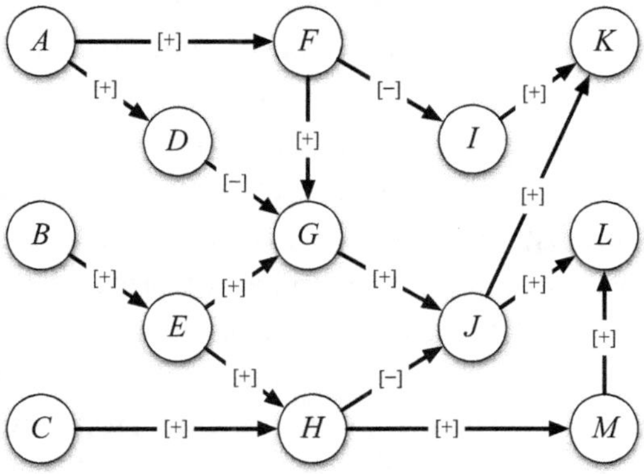

For each of the following pairs of nodes, determine whether the first is causally independent of the second. If they are causally *dependent* then report an open path between them; if they are causally *independent*, then demonstrate that all paths are blocked.

(a) From *B* to G.
(b) From C to *K*.
(c) From G to *L*.

2.5 Again using the same causal model, but now assuming that the values of *H* and G are both known, answer the following questions:

(a) Which nodes are causally dependent with *A*?
(b) In order for evidence about C to influence your assessment of *M*, which other nodes (in addition to *H* and G) would you also need to receive evidence about?

2.6 In the causal diagram presented below the value of nodes *D* and *J* are known. In this situation, which nodes, if any, are causally dependent with node G?

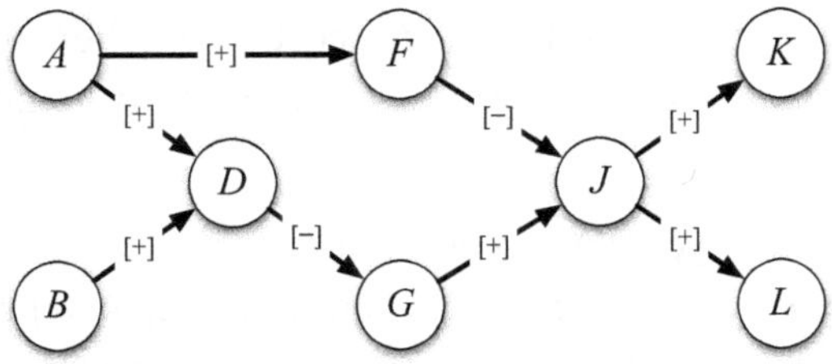

2.7 Consider the following situation:

- A manufacturing company produces either *few*, *some*, or *many Units* based on whether *Expected Sales* are *high* or *low*; the higher the *Expected Sales*, the greater the number of *Units* that are produced. If a *low* value of

Expected Sales is expected then the factory manager produces either *few* or *some Units*, and if *high Expected Sales* are thought to be coming, then the manager produces either *some* or *many Units*.

- The *Expected Sales* report is also sent to another division to secure *Warehousing* facilities for the merchandise. When *Expected Sales* are *low*, a *small Warehouse* is rented, otherwise a *large Warehouse* is rented.
- Because the divisions don't like to share detailed accounting information, as upper management you only have access to the *Combined Unit Costs* (fixed + variable) for producing and storing the goods (*low*, *medium*, or *high Combined Unit Cost*). Each pair of values for *Warehouse* size and number of *Units* can only result in one possible *Combined Unit Cost* level. With a *large Warehouse*, the more units stored there, the lower the per unit cost because the higher fixed costs are distributed over many units. *Small Warehouses* yield *low Costs* when *some Units* are made, but *medium Costs* when *few* are made. When *many Units* are made, a *small Warehouse* cannot hold them all. As a result one must either rent extra space, throw some units out, or sell them at below cost ... all of which lead to higher *Combined Unit Cost* (assume that all you care about is the effect on costs and not which of those options you would choose).
- Finally, producing more *Units* increases *Profit*, while having higher *Costs* decreases *Profit*.

Use this description to complete the tasks and answer the questions below:

(a) Draw a qualitative causal diagram representing the variables and their relationships in this situation. (Hint: two of the arrows need tables to capture the relationship they represent.)

(b) Given the policies described for the number of units and the warehouse size, would *high* combined unit costs ever be observed?

(c) If you find out that *some* units were produced, does that information allow you to alter your evaluation of the warehouse size?

(d) If you know the combined unit cost is *low*, then does getting information that *some* units were produced allow you to infer the expected sales level?

2.8 Your boss hands you an envelope and tells you to ship it in the cheapest possible way (i.e. to minimize the shipping expense) and break ties with the fastest method. The price of sending an envelope depends on three factors: *Size*, *Weight*, and *Speed*. The *Size* can be either *normal* = 0 or *oversized* = 1. The weight can be either *normal* = 0 or *overweight* = 1. Finally, the envelope can be sent via *regular* = 0, *priority* = 1, or *express* = 2 mail. The base price for sending a *normal* sized, *normal* weight, envelope via *regular* mail is \$1. Being *oversized* or *overweight* costs an additional dollar each. Sending it *priority* mail doubles the *Price*, and sending it *express* mail triples the *Price*.

(a) Draw a qualitative causal diagram for this situation.
(b) Write out the equation that determines the value of the *Price* variable.
(c) Given this pricing scheme, under what conditions would you send it *express* mail?
(d) If the post office offers new services such that the overweight fee is waived for *priority* mailing, and the oversize fee is waived for *express* mail, would that make it better to send an envelope faster than regular mail? If so, when? (Hint: use a table to determine the price of each option combination.)

2.9 You are consulting for a car manufacturer who has completed development of both a new SUV and a hybrid sedan; however, due to unexpected constraints, they can only actually produce one or the other vehicle. They have hired you to help them decide which vehicle to go ahead with. You are supposed to use the following factors to help them make this decision. Naturally, you wish to build an influence diagram of the situation from the information they provide.

- The figures they have for car sales are by family (rather than individual), and the number of cars sold increases when the mean family income does.
- When the price of fuel is high, then (*ceteris paribus*) hybrid cars sell more units than gas guzzling SUVs do because the expected fuel expenditures are less.
- Clearly a higher selling price for a car reduces the number of cars sold, and the selling price increases with increasing manufacturing costs. The price also increases with increasing competitor prices (which can be either low or high). If the manufacturing prices are high, then the company is forced to sell the vehicle at a high price, but otherwise, they match the competitors' price for similar products.
- The manufacturing costs (including materials and construction) for an SUV are always at a medium level, but the manufacturing costs of a hybrid car depend on the availability of lithium. When lithium is abundantly available the cost to make a hybrid is low, however when lithium is rare the manufacturing costs of a hybrid are high.
- The manufacturing costs clearly have a negative effect on profits, while the number of vehicles sold and their price both have a positive effect.

Create a qualitative causal diagram for this situation with [+], [−], or a table on the appropriate causal links. Assume a variable takes the values *low* and *high* unless otherwise stated.

Chapter 3

Application

Interview Case Study

Qualitative causal modeling is especially useful in situations where the time available to conduct the analysis is limited. A great example of such a situation is the case-based analysis used by the major management consulting companies in their interviews. It is essentially a structured brainstorming session, and the skills involved are the same for starting any new analysis and getting a handle on a problem. The case analysis portion of an interview occurs after passing an initial screening process. Everybody who gets this far has a chance at getting their intended job. Whether they proceed beyond this point depends upon the quality of their analysis of the case with which they are presented.

In an interview case analysis, the job candidate is presented with some summary information about a substantive business problem that the consulting firm actually encountered as part of an earlier client engagement. The candidate is given some time to read through written material about the situation and think about it. Then the candidate is asked a series of questions by the interviewer designed to assess various dimensions of the candidate's problem-solving skills, including analytical, quantitative, and conceptual abilities. Throughout the discussion the candidate is encouraged to ask the interviewer questions. This not only helps advance the analysis, but also allows the interviewer to assess the quality of the candidate's ability to identify useful information … even when it isn't provided.

A successful case interview requires a precise and thorough analysis of the questions posed by the interviewer. Since the business settings and issues involved vary widely, candidates must approach this challenge with a flexible, systematic framework. Ideally, the framework should permit: (1) explication of key relationships among many factors; (2) flexibility in adjusting the analysis as new information becomes available; (3) reasoning forward to reach conclusions and backward to identify what additional information might be useful; and (4) formulation of recommendations with clear quantitative implications. Qualitative causal models deliver on all of these requirements.

In addition, the causal modeling approach fits well with the evaluation criteria of consulting firms. The more sophisticated and logically coherent a candidate's analysis, the more positive the candidate's assessment will be. They are not looking for specific answers in case study interviews, they are evaluating your ability to explore a situation, identify salient information, structure that information, and use your analytical abilities to address a problem (often vaguely stated). The key point is how you analyze the problem; whatever conclusions and recommendations you end up with need to be well thought through and supported by your analysis of the information you determined was relevant.[1]

There is no particular technique or analytical tool that you have to use, and sometimes multiple approaches will be effective in different parts of the problem. However, they are looking specifically to assess your analytical, quantitative, and conceptual abilities. Your demonstration of these abilities can range from simple and direct to complex and containing interdependencies. Thus your assessment will improve if you go beyond obvious relationships and can present the interviewer with a chain of relationships linking a cause to an non-obvious effect. It is even better if you can demonstrate a complex network of cause and effect relationships – exactly the kind that the causal models presented in this book reveal. Of course you will also have to derive implications of the model, synthesize information from multiple areas into a coherent narrative, and use the model to structure necessary quantitative components to your analysis.

Interview cases typically start with a clear business problem: eroding profitability, decreasing sales in some regions, increasing per-unit or per-customer costs, and so on. Occasionally, a candidate will be asked to perform a situational assessment such as determining whether a new product would be successful, or whether the client should invest in a new technology, or whether to buy out a competitor, or something along those lines. The objective (i.e. the desired end result) will be clear, but the "problem" you are given is really just a downstream symptom of the "real problem" you are trying to find and offer solutions for. In addition to a stated objective, the initial material will include sufficient facts to start structuring the analysis in a coherent way.

At the start of the case interview, you will be presented with some material (typically three or four paragraphs) providing a rough outline of the immediate problem. Once you have read the information, you should proceed by laying out the elements of a causal model that captures key features of the situation in broad strokes. Begin by asking what the problem is and use the answer to this question to

1 You do not have to take our word for it. Marc Cosentino has an excellent book on how to succeed in these interviews: *Case in Point: Complete Case Interview Preparation* (Cosentino, 2007). Another source for detailed case interview advice is Victor Cheng's book *Case Interview Secrets: A Former McKinsey Interviewer Reveals How to Get Multiple Job Offers in Consulting* (Cheng, 2012). Furthermore, each of the major consulting firms has their assessment criteria on their website, and though they vary slightly in specifics, they all emphasize the importance of communicating your logical and fact-supported analysis.

define the objective node. Typically, the immediate problem at issue will be stated directly (management feels market share is too low, costs are too high, advertising does not seem to be working, and so on). One way to proceed is to start at the end – that is, turn the stated problem into an objective node – and, from there, "walk back" the causal chain to include several layers of causal influence. Walking back the model induces you to think about the main causes of important variables in a very natural way. A good rule-of-thumb is to develop an initial model with two or three causal layers (i.e., paths in the model will have three or four nodes, including the objective).

Because the case interview is structured so that new information is released as the interview proceeds, we often find it useful to begin with *Firm Profit* as the initial objective node, even though that is not the problem initially presented. For example, the stated concern might be low unit sales – ultimately, a possible cause of the more significant problem: low profitability. In this instance, unit sales should certainly be a node in the model. However, because attempts to influence unit sales are not cost-free (e.g., increasing advertising spending), the best analytical strategy is to have unit sales as a variable pointing to profit, the real objective. Then, as you think about the options available to increase unit sales, the model will naturally help you keep in mind the effects of those options on the firm's profitability.

Given the time constraints imposed by the interview format, if the case begins with a very involved discussion on a narrow area of concern (e.g., a deep description of market share issues, including details on clients, prices, sales force logistics and advertising), it may be necessary to postpone the introduction of the profit objective node in order to move the analysis along. Instead, start with the narrower objective (e.g., market share). Still, you should keep overall performance or profit in mind as the ultimate objective. In a later stage of the analysis, it is likely you will wish to introduce it explicitly.

Once the objective node is placed, begin to flesh-out the model by adding the direct causes of that variable. For example, profit is a function of unit sales, prices, variable costs, fixed operating costs, and so on. Then go to the next level in the chain: what are the causes of unit sales, prices, etc.? As mentioned, backing the analysis up two or three levels in the causal chain is usually sufficient for the first stage of analysis.

As you lay out the model, it will become apparent that additional information is required to fully flesh-out the problem. This is important – remember, evaluating a candidate's requests for additional information is an integral part of the assessment process. By using common business sense, and using the responses to your questions by the interviewer, you will build a sufficiently comprehensive model to get at the heart of the problem in hand.

Let us walk through an example to illustrate. Although what follows is purely fictional, its form and structure are representative of actual cases used by large consulting companies as part of their interview process. We will proceed in the same, step-by-step fashion you would encounter in a real-world interview.

3.1 GETTING STARTED

At the start of an interview case, you will be presented with a number of potential issues, a substantive number of facts, as well as historic and contemporaneous data of various types. Your job at this point is not to offer any advice, but rather to understand the factors at play, structure your thinking, and identify the questions that need to be asked. Not all of the initial information provided to you will be relevant. However, until you have a concrete grasp of the business situation, you will not know what is useful and what is superfluous. Read through the case material carefully (a few times if possible) and begin to sketch out your model. As you do, you will invariably get to places where further information is required in order to proceed. Make note of these – asking for the missing pieces will be crucial to providing good answers and recommendations. Assume you are presented with the following information.

Initial Case Information

This case analyzes the situation of the firm Hubris Health Incorporated, the largest – and consistently most profitable – healthcare company in its region. Hubris Health makes arrangements with employers (mostly large companies and universities) to be the exclusive healthcare provider for their employees. Customers are insured through Hubris Health, and Hubris Health also provides direct healthcare services through its physicians and specialists. Client companies pay Hubris Health a fixed amount for each employee and, in return, Hubris Health handles all the employee's health needs.

Hubris Health currently has a combined 1.5 million people enrolled in its plan from 600 different institutions in the region. The number of doctors directly employed varies from year to year, but currently Hubris Health employs 2,000 medical service providers (including general practitioners, specialists, surgeons, dentists, etc.) and another 1,000 staff members (including management, records, administrative tasks, etc.).

Over the past five years, Hubris Health has been experiencing reductions in renewals by companies, and for the past two years, a decrease in new corporate accounts. This has led to an overall drop in Hubris' number of client companies. Although they have adjusted their medical and administrative staff proportionately, the average cost per enrollee has been increasing. You are being tasked with understanding why this may be happening and what can be done to reverse these trends.

In this case, the stated objective is to help Hubris Health understand and improve their worsening situation with respect to: (1) a reduced number of renewals, (2) fewer new corporate accounts, and (3) increasing average cost per enrollee.

We are not told (at this stage) that the firm's goal is to improve its current profit performance, but it is a good assumption that this is their eventual objective. After all, if they were somehow making more money with fewer accounts then why would they want to reverse this trend? For now, however, let us focus upon the issues and data points presented.

As discussed above, the *Number of Renewals* and *New Accounts* should be distinct variables because they have different timing, and therefore they either have different causes and/or different conditional responses to shared causes. Then, begin the walk-back process. Using your knowledge of business, complemented with a healthy dose of common sense, ask yourself what variables are likely to be the most important causes of these variables. A few items come immediately to mind. First, *Competition*: Hubris Health is the biggest health provider around, but it is surely not the only one. Stronger competition from other companies would reduce the numbers of both renewals and new clients. Second, the *Local Economic Condition*: it is possible that Hubris' decline mainly reflects a worsening economic environment. If local companies are going broke, leaving the area, and not being replaced, then Hubris would see these declines even if it were maintaining its market share.

Moving on, *Customer Satisfaction* is clearly a determining factor of *Renewals*, but not of *New Accounts* (potential customers do not have any basis upon which to be satisfied or dissatisfied). However, it is possible that the satisfaction of current customers affects the company's *Brand Image*. This, in turn, may affect whether a potential client will sign on. It is important to keep in mind that the individual employees are not the ones who sign up for the service, rather the employers sign a contract with Hubris and all their employees are enrolled in the plan. Because of this, the customer base whose satisfaction matters for renewals and brand image is the large corporations that signs on to the service. Thus, the relationship between patient/employee satisfaction and their employer's decision to sign on and/or renew is ambiguous (note for follow-up question).

The model thus far is summarized in Figure 3.1. Because the model neither contains an objective node nor nodes for managerial decision variables, the best we can do at this point is provide a situational assessment of the causal system in Figure 3.1. If the initial case reading had ended with, "You are being tasked with understanding why this may be happening," then we would be ready to ask our questions and provide some answers. However, that last sentence went on with, "… and what can be done to reverse this trend." In order to give advice with respect to what Hubris Health should *do*, it is necessary to identify what variables management can influence, what their direct effects are, and how they relate to Hubris Health's ultimate objective.

We are not told Hubris Health's ultimate objective, only that they are concerned with drops in renewals and new accounts as well as increasing costs. Of course, Hubris Health is a business, and the default goal of every company is to make

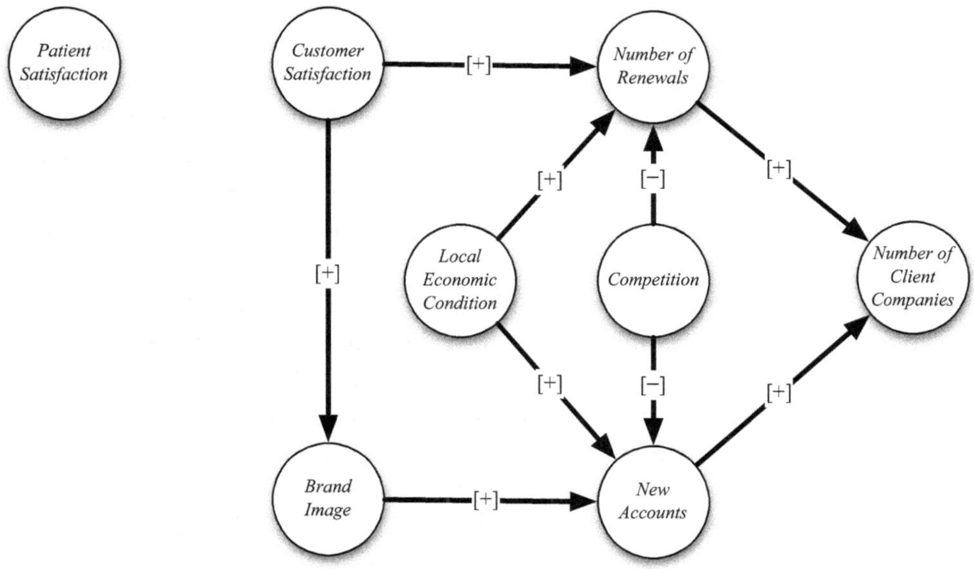

FIGURE 3.1 Situational assessment of variables related to renewals and new accounts.

money – that is, to maximize profit. Therefore, now is a good time to add *Profit* as the objective node (because we are not told that the objective is something else). This node must be properly situated within the model. Noting that profits are defined as total revenues minus total costs, insert a *Revenue* and a *Cost* node as direct parents of the *Profit* node.

This setup invites you to consider the influences on revenues and of costs. The Hubris Health plan described in the information box is simple: "the client company pays Hubris Health a fixed amount for each employee, and in return, Hubris Health handles all their health needs." That is, *Revenues* are generated by a *Fixed Fee* paid by employers for each *Enrolled Employee*. Add these nodes and relationships.

Now walk it back another layer. What influences the number of *Enrolled Employees* and of the *Fixed Fee*? The number of employees is certainly affected by the number of client corporations, though this relationship is presently ambiguous. For example, it is possible to have lots of small companies with total enrollment lower than one large corporation. We are told that Hubris services "mostly large corporations and universities." Thus, more such clients probably correlates positively with more enrolled employees.

Although we are not told anything about how Hubris sets their fee structure, a basic understanding of how businesses work should tell you that they know they have to cover their costs and incorporate cost information into their fees. Maybe they use estimated cost per enrolled employee, and there needs to be an information link from that node as well. It is certainly possible to break out costs into fixed and variable costs, although you are currently not given information with that breakdown; adding these nodes would prompt you to ask for this in a structured, logical, and supported way. Also, because the clients are large firms themselves, actual fees may be the result of some bargaining process. All of that said, we must be mindful that the setting here is a job interview – time is limited and our goal is to capture the situation in broad strokes, capturing information that is actually provided (holding off on the myriad possible ways it might be). Therefore, we choose to treat setting the value of the fixed fee as one of Hubris Health's decision variables.[2]

Setting the fee is probably not the only decision Hubris can make to effect their situation. When a consideration such as brand image or public opinion appears in a model, as it does here, it usually makes sense to add at least one decision node to represent the firm's marketing activities designed to influence it. Therefore, let us add an *Advertise* decision node as a parent of *Brand Image*.

In the initial information, we are given the *Number of Medical Personnel* and the *Number of Staff Members*, so it is likely (but not necessary) that these values will be important later on. It is certain that these people all get paid, so these variables need to feed into the *Cost* node. It is also clear that Hubris has control over how many and which types of people it employs. Therefore, decision nodes should be added for each type of employee.

Might these variables also influence *Customer Satisfaction* and/or *Brand Image*? For example, perhaps long delays on the phone, or long lines at clinics, are frustrating the employees and as a result upsetting clients, thus giving Hubris Health a bad reputation. The fees might also be a source of client dissatisfaction. At this point, these effects are just possible, so including them now would be leaping ahead of the facts and analysis. Still, these are excellent points upon which to begin a line of inquiry with the interviewer.

Figure 3.2 presents the model thus far. At this stage, it is still incomplete. On some dimensions this is due to missing information. For example, one salient, open question is how patient satisfaction fits into the picture. Relatedly, you also know virtually nothing about what actions, if any, Hubris Health can take to affect *Customer Satisfaction*. On other dimensions, the incompleteness is by choice. For example, there is no link connecting *Advertising* to *Costs* – despite the fact that everyone knows it is not free to advertise. This might be realistically justified if the advertising costs are negligible compared with those of staff and doctor salaries. For our purpose, it is justified for practical reasons. If the diagram gets too crowded with incidental connections and variables (for the sake of completeness), then it becomes

---

2 Breaking down the situation's cost and revenue structure differently would not be wrong as long as it were done in a logical way that was communicated effectively to the interviewer. Going down that path would produce a different causal model and a different downstream analysis through a different set of salient facts and questions. However some paths are more justified than others given the information provided, so choose wisely and do not overstretch what *might be* just because you can build a plausible causal model of that situation.

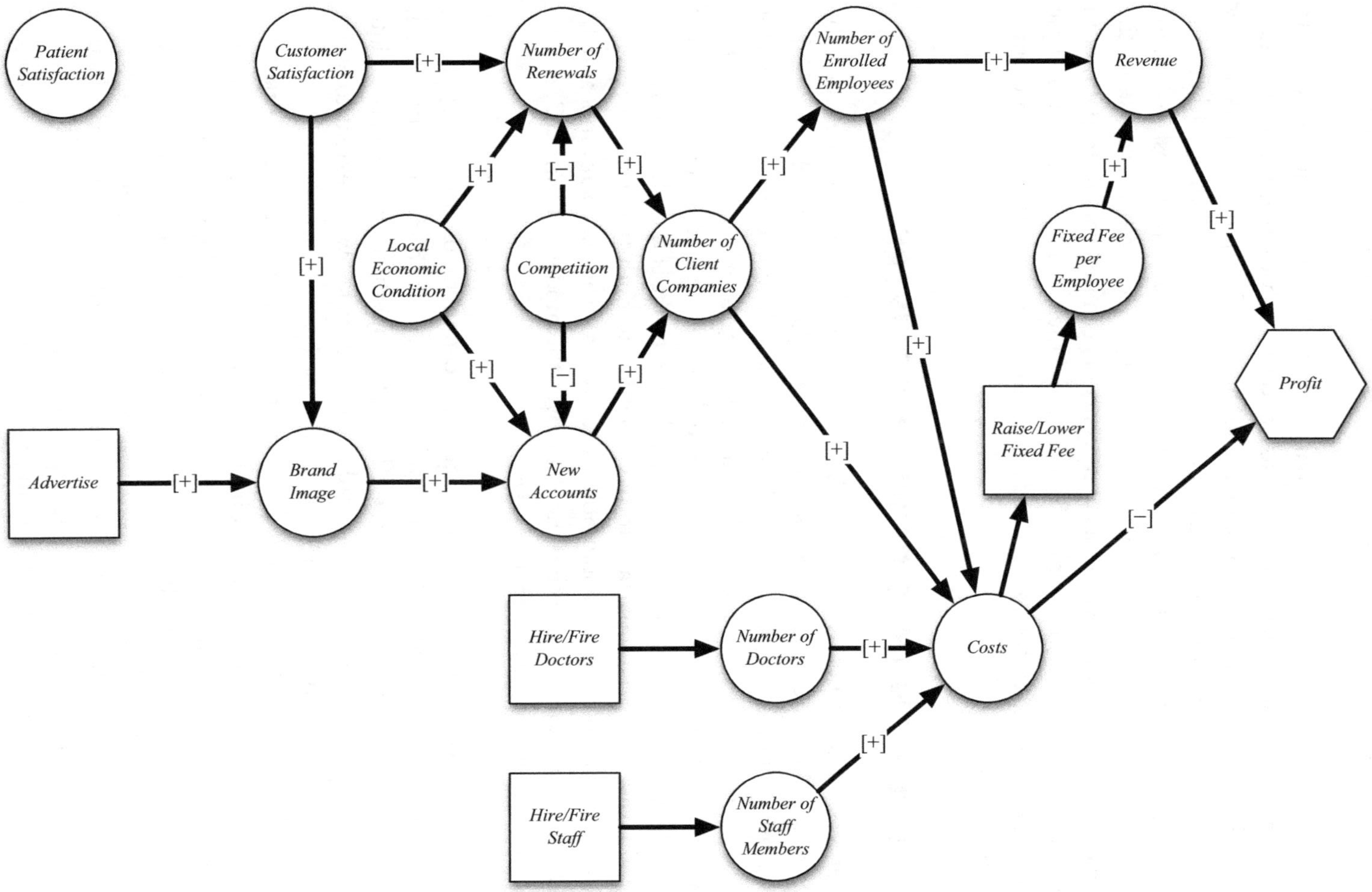

FIGURE 3.2 Initial model including all initially provided information for the Hubris Health case.

too unwieldy to use in an interview. Your goal is to construct a qualitative model that hits the main points and leads the discussion around to the most important aspects of the situation.

Even though the model in Figure 3.2 only includes the information provided to you in a short initial information box plus some additional variables and connections suggested by basic economic reasoning, a sophisticated causal system is already evident. Seeing the relationships this way should make it clear that, without the aid of a graphical representation of the key variables and relationships among them, it would be easy to lose focus and consistency during the course of the interview. Other frameworks and analysis concepts can be brought to the task to lead your thinking in developing the situation, and the causal modeling technique will reinforce the logical consistency and rigorous thinking sought by your potential employer (and beyond).

3.2 FOCUS ON THE SIGNIFICANT DRIVERS

The process of constructing this model has raised several important questions, the answers to which are required to provide meaningful advice to Hubris Health. Asking questions is an integral part of the interview – the quality of your questions will be viewed as an indication of your suitability to a career in consulting. Therefore, in deciding what to ask, you should focus on a small number of the most salient issues suggested by the model. When profit is the ultimate issue, as is usually the case, it is important to pay attention to both the cost and revenue sides of the equation; i.e., ask about both sides even if the initial issues appear to be weighted toward one or the other. Looking over the model in Figure 3.2, several important questions present themselves:

1 To what extent are local and/or national economic conditions responsible for reductions in renewals and new accounts? According to your model, economic conditions confound the effects of competition. Before delving into questions about competition, it would be helpful to know the extent to which performance is driven by general economic cycles.
2 Does Hubris Health know what drives customer satisfaction? Using situational assessment, upon being told that both renewals and new accounts are falling, your model implies that customer satisfaction must also be dropping.
3 What is the relationship between the satisfaction of companies contracting with Hubris Health and the satisfaction of the client companies' employees with the actual healthcare they receive? Having identified customer satisfaction as a problem, its causes should be fleshed out.
4 Is there anything in common among the companies that are not renewing? The purpose of this question is to flesh-out additional factors that might be influencing renewal rates.

Suppose you ask the first question and the interviewer tells you that general economic effects have been stable. Take this as a signal that she or he is not interested in further analysis down that path. A word of advice: be careful not to get too sidetracked from the big picture. If the interviewer suggests a particular area of exploration is not relevant, drop it and move on. In our example, you could remove the *Local Economy* node from your model to keep it from distracting you from more promising lines of reasoning.

Upon hearing your other questions, the interviewer smiles and hands you the second part of the case. Bingo! You are on the right track.

Digging Deeper into Customer Satisfaction

Hubris Health performs quarterly surveys to assess the quality of their services. An agent of Hubris Health calls the appropriate manager at each client institution and asks a series of questions aimed to uncover any aspect of their service requiring improvement. The responses are provided on a spectrum from 1 (strongly disagree) to 5 (strongly agree). The questions include:

1 The fixed fee per employee is reasonable.
2 It is easy to manage our employees' health benefits accounts.
3 Our employees are satisfied with their healthcare services.
4 I would recommend Hubris Health to other companies.

The average score for the fixed fee question is 4.6. The responses to this question do not correlate with renewal rate. Client companies score Hubris Health low on the account management question – averaging 1.8. However, these scores are also uncorrelated with individual company renewal rates. The average score on account management has declined in step with the declining renewal rate. This is despite the fact that Hubris Health has not changed its account management system for 20 years.

The variance in employee satisfaction scores is large. In particular, companies that report an employee satisfaction score of 3 or lower are less likely to renew. The score on the recommendation question also varies widely. These scores are correlated with: (1) the account management question, (2) the employee satisfaction question, and (3) the renewal rate. The overall (average) satisfaction score has been slowly dropping over the past five years, despite the fact that the clients scoring them the lowest are leaving.

What does this new information tell you about your model? Most clients are fully satisfied with the fees they are paying for healthcare services and, furthermore, the ones that are slightly less satisfied with the fees are not consistently the ones that leave the service. Thus, you can safely ignore any connection between *Fixed Fee* and *Customer Satisfaction*. Clearly, companies do consider the cost when contemplating renewal, but here it does not appear to be a significant driver of the dropping renewal rate (or for similar reasons would probably not be affecting the new accounts either).

The reported relationship between low employee satisfaction and a low renewal rate appears significant. On the surface, one might be tempted to assume a direct relationship. However, your model makes it clear that this need not be (and probably is not) the way it is. When people do not get the care they want, they complain to the employers. These employers are the ones who make the decision to renew or not. Thus, your model helps you see that employers *are* responsive to the desires of their employees. You should, therefore, include an arrow positively connecting *Patient Satisfaction* to *Customer Satisfaction*. If there are enough complaints, the companies respond by looking elsewhere for health insurance and healthcare. You make *Patient Satisfaction* a parent of *Customer Satisfaction* with a positive relationship. Similarly, *Account Management* should also be a parent of *Customer Satisfaction* with a positive relationship.

Since renewals are declining with the decline in overall satisfaction, you cannot tell what is happening with *Competition*. Competition may be intensifying and making the decline in renewals worse than it otherwise would be. Or, it could be staying constant, with all the decline due to falling customer satisfaction. Or, it could be that Hubris Health's standing relative to its competitors is improving but not enough to offset the effect of deteriorating customer satisfaction. One thing that is fairly safe to assume is that competitive intensity with respect to price has not increased (because clients say they are satisfied with the fees they are paying). The status of the competition, particularly with respect to the area of customer satisfaction, is worth exploring with the interviewer – so make a note for later.

The client recommendation question opens the door to think about the *Brand Image* portion of the model. You can use the score on this question as a proxy for the general perception among companies of the value and quality of Hubris' services. Suppose the interviewer tells you that: (1) the region has a limited number of large companies (Hubris' customer base), (2) these companies all need health services, and (3) the client managers who set up healthcare deals for their firms are public knowledge. She goes on to ask whether you think the firm should launch an advertising campaign to improve their brand image in order to improve new accounts and the number of renewals.

Having just been told that all sales are made person-to-person at the management level of client companies, you can see no reason to advertise. Moreover, although the model indicates that *Brand Image* is correlated with both *Number of Renewals* and *New Accounts*, your model shows that the common cause of all these effects is *Customer Satisfaction*. By your reckoning, changes in *Brand Image* should have no influence on *Number of Renewals*. Your interviewer indicates that this is correct.

Given the expanded model and what you now know about the values of some of the variables, you are now in a position to tackle part of the original question,

"Why are renewals and new accounts dropping and what can be done to reverse the trend?" The first part of the answer is that *Customer Satisfaction* is dropping due to declining satisfaction with *Account Management Services* as well as to employee/patient satisfaction with their healthcare. *Customer Satisfaction* has a direct effect upon *Renewals* and an indirect effect on *New Accounts* through *Brand Image*.

The second part of the answer follows immediately: if Hubris Health wants to reverse the decline, it must address the customer satisfaction problems. You know that Hubris hasn't changed their *Account Management System* for 20 years – modernizing it will certainly affect ease-of-use. At the same time, a system update will likely have a positive effect on patient satisfaction by ensuring more timely and accurate reimbursements or by requiring less paperwork on the part of patients. Performing an overhaul of the account management system is certainly within the control of Hubris' management, so you make these additions to the diagram. The expanded part of the model covering customer satisfaction can be seen in Figure 3.3.

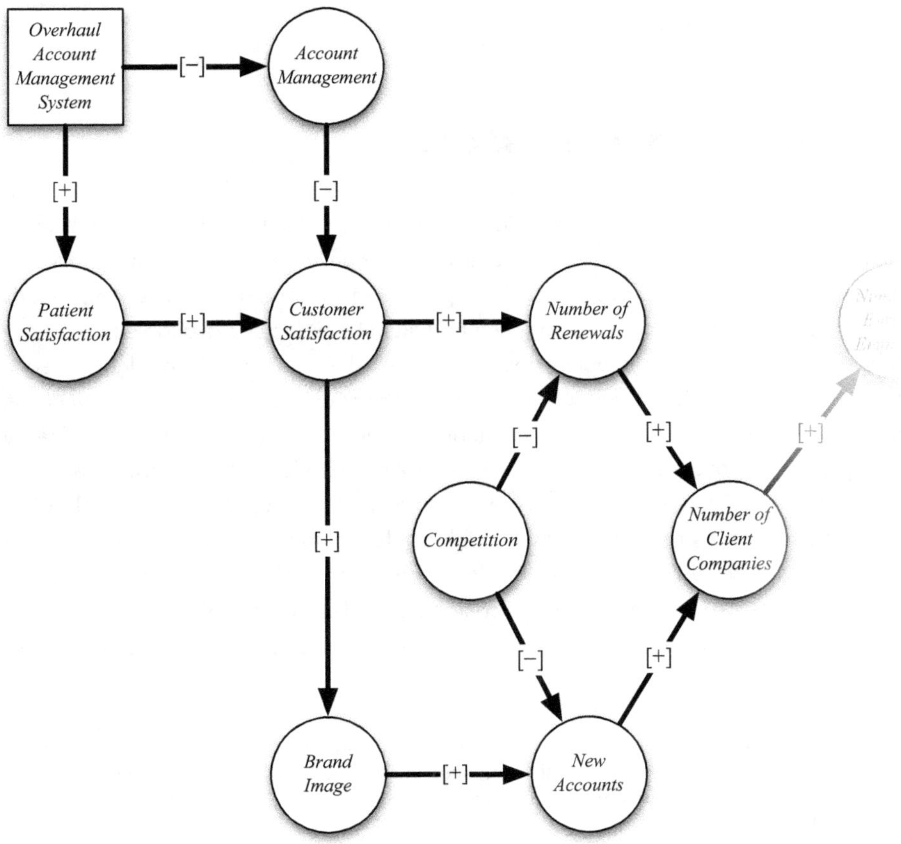

FIGURE 3.3 Adding relationships uncovered by digging deeper into customer satisfaction.

3.3 SEEK SOURCES OF COMMON PROBLEMS

The previous section addressed the causes of the declining renewal and new account problems, so now we focus our attention on the problem of increasing average cost per enrollee reported in the initial information box. In many of the cases encountered in interview situations, the operating and financial problems of the focal firm are due to a lack of perspective. What we mean by this is that the managers in the problematic firm are operating under a business model that contains a basic flaw. Perhaps, the managers don't see this flaw because they are too close to the current system, or are blinded by their corporate culture.[3] Either way, identifying the basic flaw typically leads to straightforward policy suggestions. Thus, the trick is identifying the flaw (which, itself, is rarely obvious).[4]

Your advice regarding the dated account management system is solid. Yet, one has the sense this is not the whole story. Is there a basic flaw elsewhere in the business model? Might satisfaction be linked to the number of doctors and/or staff, and hence to costs? Is the rise in costs derived from one area more than others? Is this cost increase reasonable in light of other changes or anomalous? You ask for more information about the financial aspects of the business and you are provided with the following information.

Scrutinizing their Business Model

Currently Hubris Health has 1.5 million employees enrolled in its plan who are serviced collectively by 2,000 medical service providers in the Hubris Health System. The healthcare providers are either generalists (e.g., internists and family practitioners) or specialists (cardiologists, radiologists, dentists, ophthalmologists, etc.). Doctors are recruited into Hubris with annual contracts covering a portion (often full-time) of the doctor's available hours.

When patients wish to see a doctor for any reason, they call their local Hubris Health customer service representative and talk to an account specialist to make an appointment. A patient's needs are first broken down into requiring general care or specialist care. When a patient requires care not covered by the Hubris' team, they are referred to an outside specialist (which is still handled through the Hubris Health administration system). Hubris refers the patient to an associated provider and pays a referral fee. Hubris coordinates all aspects of the patient's service, including monetary transfers, maintenance of medical records, etc.

Hubris Health deals exclusively with large companies and universities (those with more than 300 employees) using annual contracts. Currently they have 600 clients, some very large.

3 In Victor Cheng's book (Cheng, 2012) he recounts a common practice of consulting companies merely by interviewing a client's base-level employees and reporting the results to management. The problems identified and solutions provided by these interviews were already within the company, but were not known to the people in charge.

4 In the full complexity of the real world, solutions are rarely as straightforward as they are in case interviews. Be that as it may, the purpose of these exercises is to assess the quality of your analytical thinking – and therefore the cases are explicitly set up to facilitate that assessment.

Each company is assigned one Hubris agent that deals (almost) exclusively with that client. An additional 50 administrative staff members also handle tasks such as searching for new clients and hiring more administrators. Corporate management and functional areas, such as accounting and marketing, are also considered administrative staff for present purposes.

Based upon this new information, you realize you need to refine the financial model portion of the causal model in Figure 3.2. The revised model should lead your thinking to new questions – always keeping in mind the stated problem of rising average cost per employee and the goal of maximizing profit. The information provided is not stated directly in terms of causal relationships, so it will take some work to tease out the relevant variables and their relationships.

First, look at the revenue side. Were you given any new information relevant to incoming funds? The only source of revenue is still the fixed fee received per enrolled employee, so those linkages remain unchanged. It is worth noting in the model that *Revenue* is a simple function of the fee times the number enrolled, and therefore it is best represented by a functional node rather than a variable.

Now look at the cost side. Revenues are determined by a fee per enrolled employee. This fee must be greater than the average cost per enrolled employee basis in order for total revenues to be greater than total costs. Thus, the average cost per enrollee is the information that Hubris needs in order to assess the adequacy of its fee structure. Moreover, it is the increase in average costs per enrollee that Hubris identified as a problem. Therefore, it makes sense to relabel the *Cost* node to *Cost per Enrolled Employee* to maintain an appropriate focus on this central issue.

You are told that both *Renewals* and *New Accounts* are decreasing, which the model tells you implies a reduction in the *Number of Enrolled Employees*. For the average cost per enrollee to be increasing, it must be that total costs are (1) increasing, or (2) decreasing more slowly than the *Number of Enrolled Employees*. Assuming costs are increasing is the obvious response and will not impress anybody. Understanding that Hubris may be cutting costs in the wrong places, or in the wrong ways, or not enough, etc. provides a deeper answer and opens up more avenues of inquiry.

You have already explored what is generating the reduction in enrollment (with respect to customer satisfaction), so the remaining step is to look at the cost side. In the earlier model, the numbers of patients, clients, doctors, and staff all directly pointed to *Costs*, but now you know that setup is too simplistic. How are costs actually determined? What are the variables with a direct cost component?

The population of 2,000 doctors employed by Hubris Health is broken down into generalists and specialists. Although specific numbers are not provided, this may be important information for thinking about their cost problem, so the *Number of Doctors* node from the model in Figure 3.2 should be replaced with two nodes: *Number of Generalists* and *Number of Specialists* nodes. The number of each type is determined by management. For increased clarity without loss of information, it is sufficient to make one decision node for *Hire/Fire Doctors* that points to the nodes for both types of doctors. The doctor distribution could also be captured by a *Number of Doctors by Type* variable that does double duty in capturing the values across categories. Sometimes that is simpler, sometimes more complicated. If it later becomes important that some variables be considered separately or together instead, then it can be changed at that time.

Before moving on to other cost drivers, consider what information may be used by management when deciding the number and mix of doctors. If the medical staff has not been properly adjusted for diminishing enrollment, then that could (at least partially) explain the increasing cost problem, so this is important. We are told in the first information box "the number of doctors directly employed varies from year to year" and that they have been adjusted proportionately. The newest information reveals how this is done: both the doctors and the clients are on annual contracts. This allows Hubris to use the *Number of Enrolled Employees* in making the doctor hiring decisions (which means that the node is best labeled *Doctor Contracts* rather than *Hire/Fire Doctors*). This information link is added to the model.

The next paragraph of the information box is concerned with the service staff and patient care. Service staff are needed to accepts calls and make appointments for enrolled employees needing care. The number of these *Service Staff Members* is also controlled by management. This decision is also likely made with the number of enrollees in mind, although you are not told whether they can be hired and fired with equal flexibility. You can add these nodes to the model, and link the *Service Staff Members* to the *Total Costs* variable as well. Might inflexibility in hiring and firing service staff be part of the cost problem? Make a note of this question for now and complete the assimilation of the new information before treading down that path.

According to the case, patients request either general or specialist care (the same breakdown as the doctors). Therefore, the number of actual treatments provided can be separated into two nodes: *Number of Generalist Treatments* and *Number of Specialist Treatments*. This separation is important because only specialist treatments are ever referred to a doctor outside of Hubris, which incurs an additional referral fee. The *Number of Referrals* clearly depends on the types and the *Number of Specialists* employed by Hubris. This realization should prompt some additional questions: Has the number of patients requiring specialist treatment increased? Has the number of referrals increased? Has the referral fee increased? This information has not been provided, so make a note of these questions and move on to the final paragraph of new information.

The last paragraph deals with the administrative staff, and provides some specific numbers: 600 clients and at least one staff member is assigned to each one. This tells you that there are separate variables for the *Service Staff* and *Administrative Staff*. You also find out that the number of *Administrative Staff* members that Hubris decides to employ is informed by the number of clients. You can now add these nodes regarding the administration to the model and indicate that more staff means greater costs.

Take a closer look at the numbers provided, they often provide deeper insight. We are told in the initial description that Hubris Health employs 1,000 staff members total, and now we know that 650 of them work on the account administration side. So there are 350 staff members on the customer service side, and therefore each one must handle (on average) over 4,000 enrolled employee's needs.

Is that a large number? What metrics could be used to make that determination? The only time an employee calls Hubris Health is when he or she is ill, or wants a check-up, or needs to pay a bill, and so on. The number of staff needed is likely correlated with number of enrolled employees, but it is really the number of treatments (of both kinds) that determine the number of calls each administrative staff actually handles. If the number of service staff members is insufficient for the load, then you would expect to see long phone wait times for patients requesting service. Is that happening? Thinking back to the last section, this would likely result in observed higher levels of patient and client dissatisfaction.

How can you relate the staff numbers to increasing costs? You are told that each client is handled by a specific agent within Hubris. The number of client companies is shrinking, and you are told that Hubris adjusts staff appropriately – according to their 20-year old system. If Hubris is maintaining the 1:1 ratio of representatives to clients, might this be contributing to increased average costs on a per-enrollee basis? What information would you need in order to determine that? This section has built up a sophisticated causal model from the information provided as well as provided fertile ground from which to expand your analysis in many directions. The cost-related portion of the model can be seen in Figure 3.4.

3.4 ASK SPECIFIC QUESTIONS

In the process of analyzing the information and building the model in the previous section, several questions came up that bear directly on the problem facing Hubris Health. In order to provide specific advice to Hubris Health, you need some specific answers to some specific questions. Below is the list of questions brought out through the above analysis.

1 What are the percentages of generalist and specialist doctors?
2 How flexibly can Hubris hire and fire service staff?
3 Has the number of patients requiring specialist treatment increased?
4 Has the number of referrals increased?
5 Has the referral fee increased?
6 How has the ratio of enrolled employees to administrative staff changed?

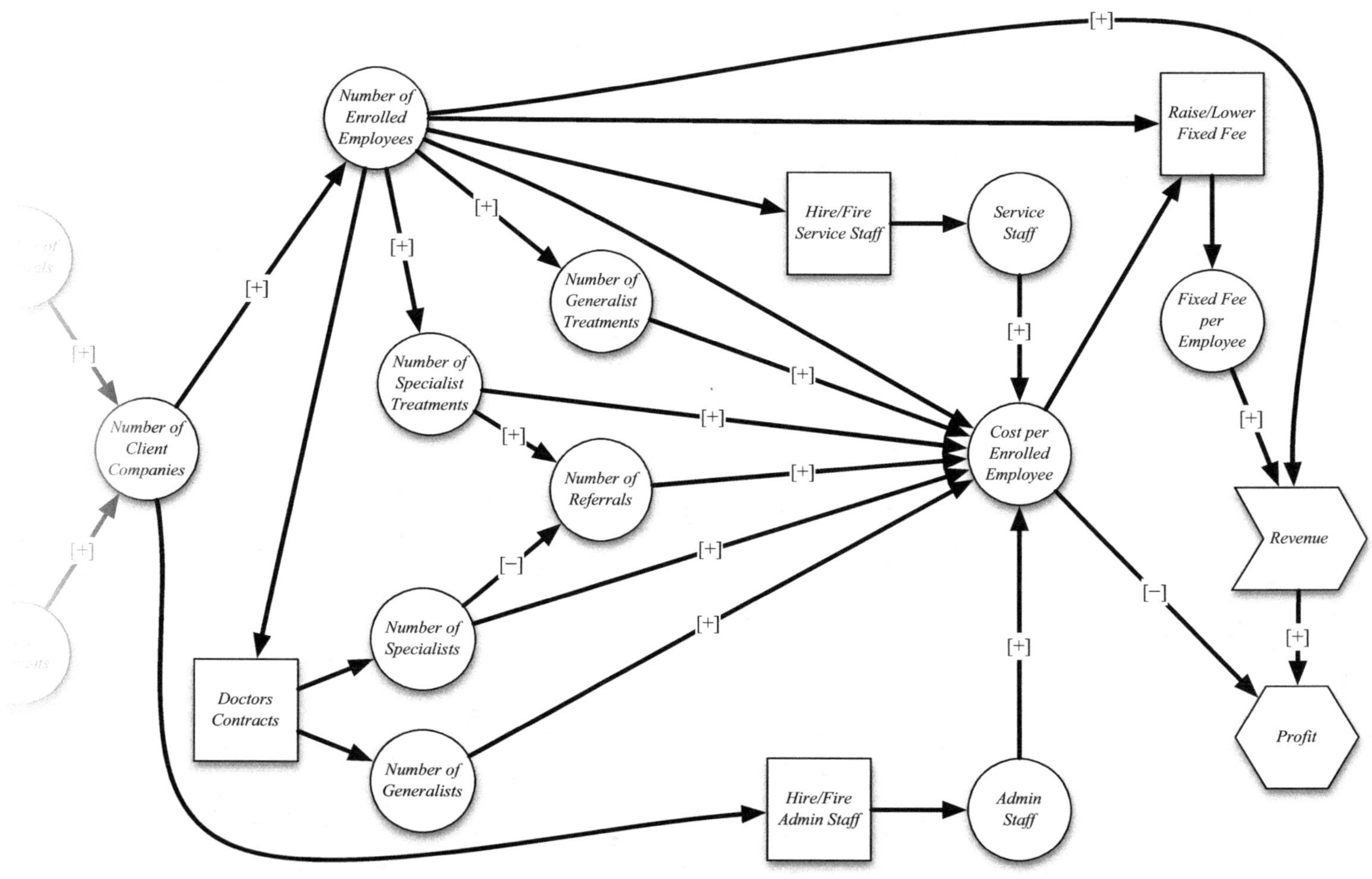

FIGURE 3.4 The refined model after examining their business model in more detail.

You can ask all these questions, perhaps a few more, but you are not likely to get all the answers you wish for. In real-life business problems, the pieces of data you need are not always available, or it is not in the most convenient or obvious format. Irrelevant data is also sometimes included with the useful. And be careful what you ask for, the answer to the question that obviously sprang from the earlier analysis may not be the information that you actually need. Use the logic and the causal model to continue to structure your analysis and turn data into relevant information.

After asking the above list of questions of the interviewer, you are provided with the following information/data.

Crunching the Numbers

Both the administrative and service staff are kept at appropriate numbers and mixes as determined by Hubris Health's 70 years of experience in providing quality health services.

- There is a core staff of long-term, full-time staff members assisted by flex-time, part-time, and temporary workers. Hiring more workers is fast and easy, cutting back on service staff hours to accommodate recent losses in enrolled employees has not been a problem.
- The exact ratio of administrative staff per enrolled employee has fluctuated as companies leave and enter the Hubris Health system. The average ratio over the past five years has decreased −0.2%.
- Of the 2,000 doctors working for Hubris Health, 500 are generalists and 1,500 are specialists. This 1:3 ratio is maintained as they scale their number of doctors to match the number of enrolled employees.
- The number of patients requiring specialist care has increased 8% each year for the past 5 years.
- The number of referrals has increased 5% each year for the past 5 years.
- The referral fee has remained constant for many years.

There are three major, and mostly disjointed, sources of increasing costs per employee being explored here: (1) administrative staff; (2) health services staff; and (3) doctors. In practice you will only have time to seriously analyze one of these options, but further investigation is necessary to determine where is the best place to look.

3.4.1 Administrative Staff

In the previous section you uncovered the possibility that the "one customer representative per client company" policy could be increasing average cost per enrollee. This would happen if the average number of enrolled employees per company remaining in the Hubris Health system were decreasing, and thus increasing the ratio of administrative administrators (the cost) to the number of enrolled employees (the revenue). As it turns out, this ratio is actually improving, meaning that the average company size within Hubris is (probably) getting larger.[5] In terms of uncovering another driver of increasing costs, this line of thinking is a dead end.

However, it did uncover another hint that may be useful for further analysis: what may be causing and what are the likely effects of increasing enrollments per company. Time in an interview (and space in this book) are limited, so it suffices to point out that Hubris seems to be managing their administrative staff members well. Insofar as you are being tasked with finding and fixing problems (rather than just providing a detailed picture of the situation), it would help you to remove the administrative staff segment of the model. This will clean up the diagram and keep you focused on the variables and decisions that matter.

Furthermore, because the cost effects of administrative staff are being ignored, the only role that the *Number of Client Companies* variable is playing in Figure 3.2 is to influence the *Number of Enrolled Employees*. The connections leading from *Number of Renewals* and *New Accounts*, through *Number of Client Companies*, to *Number of Enrolled Employees* are all positive; therefore you can eliminate the *Number of Client Companies* node to simplify the diagram and keep you focused on relevant elements of the situation.

3.4.2 Service Staff

The role of service in increasing costs can also probably be dismissed given this new information. You are told directly in the first bullet that appropriately accounting for drops in enrollment has not been a problem. It could still be, as was speculated earlier, that some other change has made it so that what used to be appropriate no longer is. And if people were calling more and making more appointments, then this could create long waiting times and this could contribute to dissatisfaction. That is a great story, but that is not the story unfolding. You have to let the information lead the analysis, not the other way around. Let it go. Pursuing service staff costs is not what the interviewer wants, so you can remove this portion from the model too.

3.4.3 Doctors: Generalists and Specialists

In order to determine whether Hubris Health's mix and number of doctors is appropriate for their customer base, you need to have some benchmark against

5 Larger enrollments per company is the most likely direct explanation given the information you have and the structure of relationships you have uncovered, although it could be that Hubris has laid off some accountants or other administrative staff – you do not know precisely what the case is here. As you will see in later chapters, the causal modeling framework allows one to easily encode this sort of probabilistic relationship.

which to compare the numbers provided. What are some possibilities? The two most obvious comparisons are against another company with a similar patient base as Hubris, or general health statistics (for that region or the whole nation). It turns out that the latter statistics are commonly collected and reported by government agencies, professional groups, and academic researchers. Asking the interviewer for more information along these lines is rewarded with the following general statistic.

Doctors make up 0.29% of the general US population, and 1/10 of those are general practitioners.

What do these numbers imply for Hubris Health? Considering the population of 1.5 million enrolled employees, they should have $1{,}500{,}000 \times 0.0029 = 4{,}350$ doctors in total. Furthermore, 10%, or 435, of these should be general practitioners. Since you are told that they actually employ 500 general practitioners, they have this portion overstaffed based on US statistics. Also, based on these numbers, they should have $4{,}350 - 435 = 3{,}915$ specialists on the team to cover expected treatments, and they actually have only 1,500. They maintain 25% of their doctors as generalists rather than the 10% found across the whole medical profession.

This doesn't necessarily point to a problem, but it certainly demands investigating. Why might Hubris' doctors vary so much with what is expected from population averages? One possibility is that the spectrum of specialists is so broad that it doesn't make sense for them to keep too many of them in-house. They might, instead, rely on referrals to fill the gaps in specialists. As long as the number and cost of referrals is stable (or at least predictable), it can be incorporated when setting the fixed fee. It is only if these referrals are increasing unexpectedly or faster than Hubris' speed of fee adjustment, would a lack of specialists be causing Hubris' cost problem.

Another common explanation for these differences is that the people actually enrolled in Hubris' program are not representative of the population at large. If their enrolled employees tend to be older/younger, healthier/sicker, or more susceptible to certain specific health problems than the overall population, then having this skewed mix may be appropriate for them. Just as with the number of specialists, if the demographics of enrolled employees are changing, then the "appropriate adjustments" they are making by scaling back the number of doctors at the same 1:3 ratio may not suffice. They may actually need more doctors and/or a different mix to accommodate the current patients.

Although you are told that the referral fee has remained constant, the number of referrals has increased 5% per year. That is more than half as fast as the 8% increase in specialist treatments. These numbers both indicate that something is increasing the rate at which enrolled employees require specialist care. Furthermore, whatever the cause, it requires more specialists than Hubris currently employs to deal with. Up to this point, you have identified a problem (too few specialists) and a solution (hire more specialists), yet if you stopped there you would only get credit for pointing out an obvious, direct relationship. Why is that? It is true that the increase in referrals is increasing costs, but only in a way that can be (mostly) accommodated by fee increases, and fees are not a source of Hubris' problem. *Referrals* are not effecting *Profits*, which is the true objective.

The causal modeling approach entices you to pursue what the interview would prefer to see: a longer causal chain and interdependent elements. Take the lead and dig even deeper to identify the cause of increasing specialist treatments. The only option is that enrolled employees are getting sick more often in ways that require specialist care. However, there are many reasons for people to become sick more often: more hazardous working conditions, deteriorating environmental conditions, old age, poor eating habits, lack of exercise, and more. To analyze the cause of Hubris' increasing need of specialists you need more information about the enrolled employees over the past several years.

You could ask for information about all these characteristics and possible explanations separately. The advantage of that approach is that it reveals your ability to think through multiple hypotheses. The disadvantage is that you might get all that data, and then you have to sift through it to find what is useful. Remember that interviewers want you to make progress in a short time, so a simpler question is: Has any important characteristic of the enrolled population changed consistently over the past five years? The interviewer responds with the following new information.

The average age of the population enrolled with Hubris Health has increased 1.2 years per year on average over the past five years. The average income has also increased by 1.5% per year on average.

You should expect those two variables to be correlated with each other. Which one is more interesting for the analysis at hand? While it is true that there is a statistically significant correlation between wealth and health, that one is a negative relationship. A much better explanation is that the higher age distribution of the employee population is causing (at least partly) the rate of specialist treatments required to increase. That is your speculation. It is a good idea to check that with a request for more relevant data. When you ask for the breakdown of number and type of treatments per age group for the USA and/or Hubris (a specific question), you instead get the following coarse-grained fact relating age to doctor visitation:

People aged 65 and older visit general practitioners three times as often, and specialists five times as often, as people aged 18–65.

Because almost all employees enrolled in the Hubris Health plan are between 18 and 65, this brute fact will not allow you to calculate actual expected visitation numbers for Hubris' employee population. What you can infer from this information is that as people get older they use more health services and visit more doctors. The rate of both types of doctors increases, but specialist usage increases more. To incorporate these revelations into the model, add an *Average Age of Enrollees* node to the causal model and have it point to both types of treatments.

Identifying the increasing *Average Age of Enrollees* as a cause of the increasing *Specialist Treatments*, which in turn causes increasing *Referrals*, which causes increasing *Costs* is a solid analytical result demonstrating your ability to find chains of causes. You can do even better! If you end with increasing age as your root cause, then there is no direct intervention to fix the problem. Something is happening to increase the average age of the employees of the companies that use Hubris Health's services. It is important to again walk back the causal story and think about what might be causing such a change. The full model, including the revised financial part of the model, is presented in Figure 3.5. Is there something there that indicates where to look next?

3.5 BRING IT ALL TOGETHER

At this point, you wish to explain why the average age of employees of the companies utilizing Hubris' services is increasing, and what, if anything, Hubris can do about it. From the initial information you know that *Renewals* are decreasing. Might there be a connection? In the beginning of Section 3.2 you asked a series of questions, one of which was "Is there anything in common among the companies that are not renewing?" The purpose of this question was to flesh-out additional factors that might be influencing renewal rates. You did get an answer to that question from the customer satisfaction surveys. Now you are interested in finding out whether there is a link (direct or indirect and in which direction) between *Renewals* and *Average Age of Enrollees*. It seems you need more information about commonalities among the companies that are not renewing.

If Hubris Health were a real company, and you were really trying to solve this problem for them, then you could ask for company profiles for all clients for the past 10 years, put all that data in a spreadsheet (or have somebody build a database), and then look for patterns. In an interview you cannot do all that, so you need to ask a succinct question (indicating that you know what you are looking for) to get the information you need to continue with your analysis. You should ask the question similar to: "Are there any properties of the clients, such as company size, industry type, growth rate, or stock performance that correlate with their likelihood of renewal?" In response to this question you are given the following information regarding Hubris' past and current client companies.

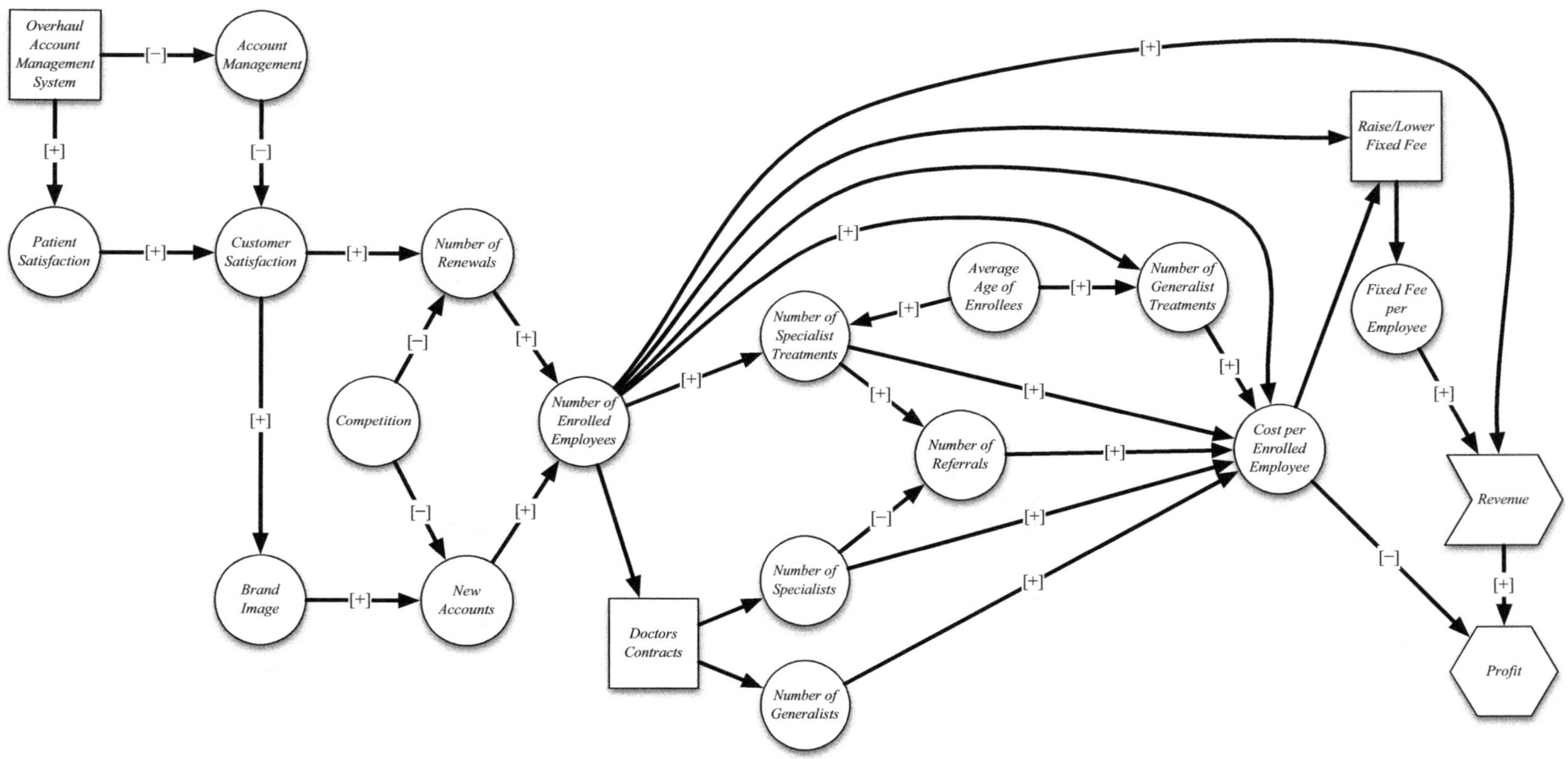

FIGURE 3.5 The complete model, including both modules analyzed in-depth and refined through digging deeper.

Ones that Got Away

Because we have been aware of this problem for some time, we have already performed some analysis of our client data. Below you will find a plot of the *Number of Clients by Industry* for the past five years. This data shows the total number; therefore it includes the effects of both fewer renewals and drops in new clients. Although some new clients did pick up our services during this period, they did not suffice to replace the clients that left. The decline in clients can be seen across the board.

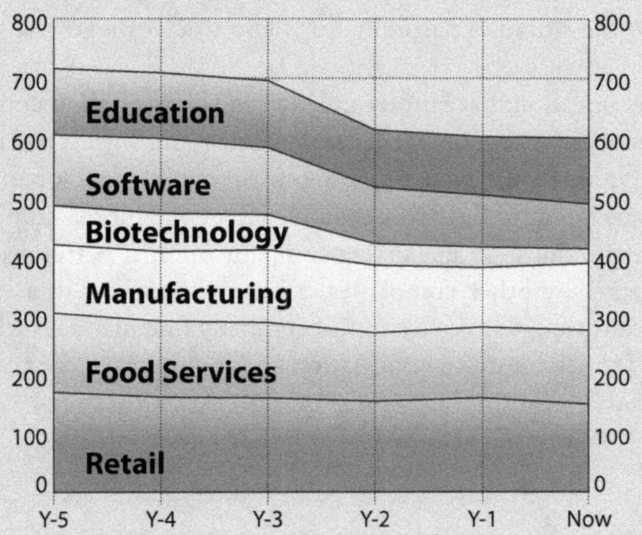

Number of clients by industry sector for the past five years.

The overall story is obvious from this plot: total client numbers have dropped from more than 700 to 600 over the past five years (an annual average reduction rate of 3.5%). But the details are more interesting than just that. Looking just at the top line (the total number) shows that these reductions were not at a constant rate throughout the period: there was a big drop in clients between two and three years ago, but they have been only slightly declining across other years. This bump may indicate that a significant event (with respect to client companies' health service plan purchases) happened during that year. Make a note to follow-up and look for further insights.

Because the graph is broken down by industry, this invites you to look beyond the overall numbers at the categories more closely. Although it looks like all industries got hit 2–3 years ago, some industries were clearly harder hit than others. Software and Biotechnology combined went from approximately 175 clients five years ago to less than 100 in the current year. The other sectors remained fairly consistent in the other years. In fact, Manufacturing and Education even seem to have increased over the past two years, just not enough to offset the losses from other industries.

What could explain this sudden drop in clients specific to these two industries? Looking again at your model thus far in Figure 3.5, the causes of *Renewals* are *Customer Satisfaction* and *Competition*. The causes of *New Accounts* are *Brand Image* and *Competition*. You have already received information about *Customer Satisfaction* and *Brand Image* in the second information box. Nothing there indicated a sudden change in the past five years. So using your causal model to reason through the situation naturally leads you to investigate the competition in more detail.

It is a good assumption that Hubris already keeps tabs on their competition, even though they seem quite confident in their position as the biggest and most profitable healthcare provider in the region. You can probably think of many ways that competition can change to have this negative effect on Hubris' numbers: a national provider moving to the area, an existing company offering better prices or services, heavy sales pushes by other companies, a new company with a whole different approach to healthcare, or even a copycat company that just does whatever Hubris does. You need to ask for more information about changes in the competition over the past five years.

Investigating the Competition

Hubris Health has maintained its enormous lead over the competition throughout the past five years, although it dropped from 67% to 62% of the market share in the region where it operates (among companies that service large corporate clients). There are four other significant players in the same market space, as well as some other minor ones that handle niche markets.

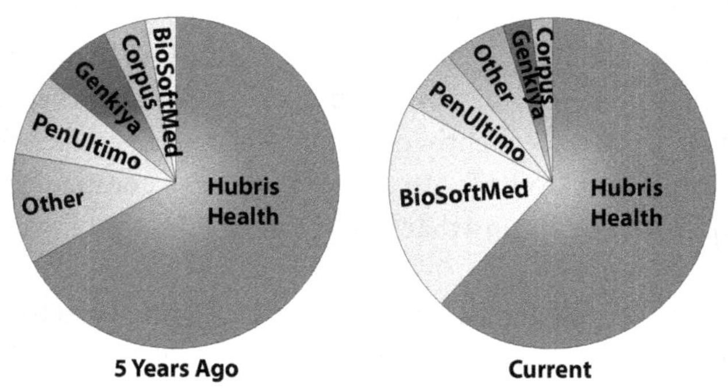

Regional market shares of health service companies.

The largest change is the growth of BioSoftMed, a company that caters mostly to high technology and research companies of any size. These sectors were originally dominated by large defense contractors, chemical, and genetic engineering firms. However, a few years ago, many of these companies broke up into smaller firms that specialize in one device, technology, or application. Some of these new small firms then became quickly successful and grew exponentially.

The story is now coming together. Hubris saw a big drop in clients around the same time that these large technology and biotech firms were breaking up into smaller firms. At that point, they may not have held enough employees to qualify for Hubris Health's services. Then, as these companies grew large again, they stayed with the health provider they used during the transition, which in many cases seems to have been BioSoftMed. But is that the whole story?

That does explain why renewals have been dropping. Furthermore, if the total number of large companies in the region has been dropping, then that would negatively influence new accounts as well. But neither of those directly explains the increasing average age of employees enrolled with Hubris. Looking again at the plot in the "Ones That Got Away" information box, the two industry sectors with the biggest losses are Biotechnology and Software Engineering. A plausible explanation for the observed increase in average age is that employees in Biotechnology and Engineering tend to be younger than in the other industries, and hence the loss of companies in these industries is pushing the average age of remaining enrollees higher.

Rather than just speculate, it would be better to confirm that speculation with supporting information. You have to ask for more information. Ideally you would like to link (1) the loss of Software and Biotechnology Industry clients to (2) the rise of BioSoftMed and (3) the increasing ages of enrolled employees. A natural thing to ask about is how the competition, and BioSoftMed in particular, operates.

From the BioSoftMed website:
BioSoftMed offers the ultimate convenience and flexibility in providing for the healthcare needs of your employees. We do not hire doctors directly, but instead contract for the services you need as required. Our administration is done almost entirely via our secure website: signing up, making appointments, checking accounts, making payments, ... everything your employees need to handle their medical needs without hassle. Of course there is always a person to talk to when you have questions. Our lower overhead allows us to price competitively while custom tailoring a solution to fit your company's needs.

You can decide how much coverage to offer your employees, and then they can take advantage of further care to fit their own needs through our discounted services. We believe in proactive healthcare. That is why we offer bonuses for health-club memberships, regular check-ups, and improving scores on key health measures (such as losing weight, lowering blood pressure, and reduced alcohol/tobacco consumption). Employees that sign their kids up with our service receive a free promotional WonderBot trainable toy robot.

We look forward to handling all your employee's health needs no matter if you are a small start-up or giant multinational. When you join BioSoftMed all your employees become a part of our family.

This competitor operates very differently from Hubris Health. They focus on flexibility, convenience, and providing incentives to reducing healthcare usage. Companies using BioSoftMed typically cover the fixed enrollment cost and annual fees as well as a specific annual allowance of medical expenses (which varies by employer and by employee status). This system allows employers to pick their own health coverage budget: employees that need more services pay from their own pocket, and ones that need fewer services simply cost the employer less.

Because BioSoftMed does not hire doctors in house, and relies on referrals, they pay more for actual treatments than Hubris. However, they provide incentives to reduce treatments and (as you have seen) they target companies such as biomedical research and software design that tend to have younger (and therefore probably healthier) employees. The features of their health services (online account management, flexibility, and bonuses for staying in good health) appeal to exactly this crowd of young, educated people.

In part, Hubris created this situation through its own policy to only accept large companies. When the large biomedical companies broke up into smaller companies, they were forced out of Hubris' system. Some of them were attracted to BioSoftMed's offerings, joined up, and stayed with them as they grew. Other companies, both large and small ones, saw people's satisfaction with the flexible, modern system and also joined up (especially among industries with like-minded employees). Up to now, BioSoftMed has taken share mostly from smaller competitors, only cutting into 5% of Hubris Health's market share. However, BioSoftMed has been growing in the small-to-mid-sized company market as well. It may only be a matter of years before many other clients switch away from Hubris pursuing more progressive healthcare options.

What these competition details reveal is that the competition's products and marketing have effected the mix of companies that sign on, and stay on, with Hubris health. The resulting change in client companies by industry generated changes in the average age of enrolled employees because some industries (especially high tech ones) employ a younger population. These relationships can be added to the model (as seen in Figure 3.6).

BioSoftMed does not want all of Hubris' customers, however, because its business plan only works when enrolled employees are mostly healthy and can be given incentives to stay that way. And it is exactly these services that appeal mostly to companies with young employees. So, one option is that Hubris could continue its current operations by specializing in care for older industries with older employees, although this would require a downsizing of its operation. A better recommendation would be for Hubris to adapt its health services by offering a variety of plans, including ones that will attract the youthful industries. Changing its product features is certainly within the control of Hubris' management. Doing so could not only increase patient/user satisfaction (and all the downstream effects of that your model shows), but also enhance Hubris' brand image as a forward thinking company that responds to changing market demands.

A node for *Features of Health Services* can be added to the model. Those features are determined by a decision node called *Diversify Product Features* controlled by Hubris. For example, instead of just a one-product-fits-all service, Hubris can offer that umbrella package as well as other customized options to attract lost companies and pursue new ones. The effects of competition, and this potential response to it, have been added to the model in Figure 3.6. Notice that these additions make a connection between the customer satisfaction and financial modules of the situation, thus building a complex web of causal relationships of all the elements brought out through your analysis.

3.6 PROVIDE SPECIFIC RECOMMENDATIONS

One could take the analysis further: push back the causes for more levels, collect more data, and repeat this cycle many times to gain a clearer and more in-depth understanding of the situation. In actual consulting analyses it typically becomes more detailed and a quantitative approach becomes necessary to make specific recommendations. For an interview situation, in which time and information is limited, the model depicted in Figure 3.6 is already more detailed than usual. How far you can take such an analysis depends both upon your facility with the tool as well as the time and information with which you are provided. The main point we wish to emphasize is that causal modeling is a tool that features scaleabity, modularity, generality, and formalism – all of which greatly increase your ability to handle a case on any topic, of any size, and at any level of detail.

At a certain point, no matter how much fun you are having analyzing the case, the interviewer will indicate that it is time to stop, discuss the findings, and make some recommendations based on your analysis. It is important to synthesize insights from each stage, pulling them together into a "big picture" of the situation. Referring to the causal model you constructed and recapping key points will help you make those recommendations specific.

3.6.1 Upgrade Account Management System

Based on the analysis on customer satisfaction, you recommend that Hubris Health improve the account management system for both clients and their employees. This is important because the current management system is possibly frustrating patients, leading to lower patient satisfaction. Because patient satisfaction is near

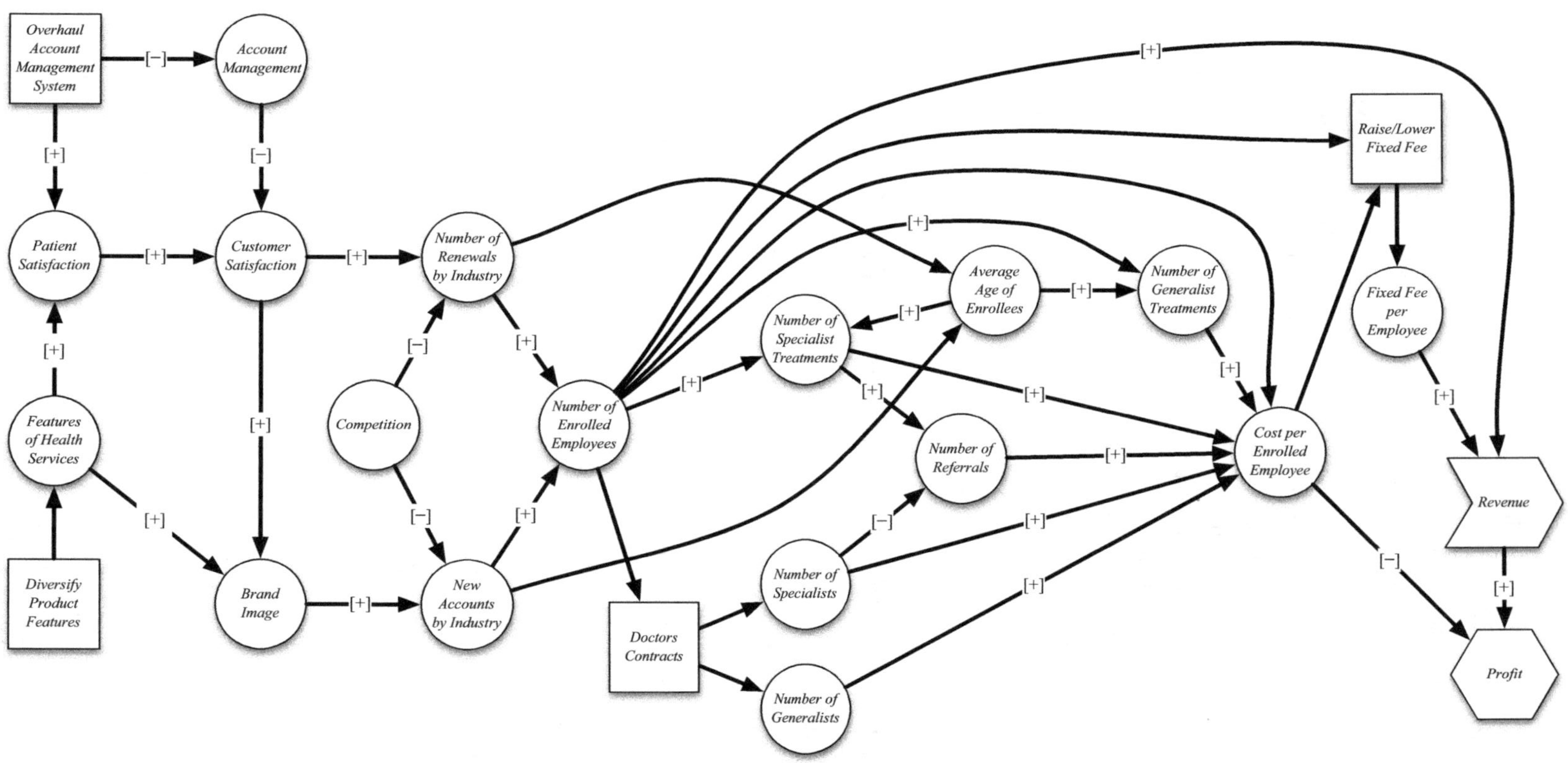

FIGURE 3.6 Adding the effects of product features and industry-specific competition to the model.

the root of the whole situation (as shown by your causal model), making an improvement here can have large downstream effects.

The initial response from Hubris Health is one of resistance: "We've been managing our accounts the same way for 20 years and the system works for us. We're by far the largest provider in the area. Why should we change the system that got us here?" You have to explain that the times are changing.

People have become more convenience-oriented over the years, and online access and management has become expected, not just a perk. The growth of BioSoftMed has demonstrated the appeal of these service features. Because Hubris services large corporations in retail, food services, education, etc. many enrolled employees are still young workers. It was revealed that employers are in fact sensitive to the preferences of workers regarding which health service company to use, so continuing with an outdated system could cut even further into their market share.

Of course even the younger generation wants to see real doctors face-to-face, and they still want a human to talk to if and when something goes wrong or is unclear. By streamlining certain processes Hubris Health can cut overhead costs and better satisfy patients and clients with the same move. Keep in mind that many clients' employees are satisfied with the Hubris system; the old-fashioned style and modern style can coexist to satisfy an even broader customer base.

3.6.2 Adjust to an Aging Population

The next major revelation is that the average age of the Hubris Health enrolled employee population is increasing. Because older people need more health services, they should increase the number of specialists, and research which specialties are requiring the most referrals. It might save Hubris Health money to hire some of those specialists in-house. They may also need more generalists doctors, but you would need more data on their usage rates in order to make that determination. Currently they are not incorporating the average age of employees into determining the mix of doctors to contract – they just use a trusted recipe to determine the mix. So your recommendation is that they *start* using the *Average Age of Enrollees* to inform *Doctors Contracts*. This arrow has been added in dotted black to Figure 3.7.

If an older population of enrolled employees is going to be Hubris' reality, then they should also increase their fixed fee. Whatever formula they use to determine how to cover costs has undercompensated for the actual changes. More treatments means more service staff, more doctor visits, more referrals and therefore more costs. Because this is new territory for Hubris, they may not have the expertise to cope with it as exactly as other fluctuations. You discovered that customers are not complaining about the fees, so they may be able to increase the fee without losing ground against the competition – especially if they become more competitive in other areas.

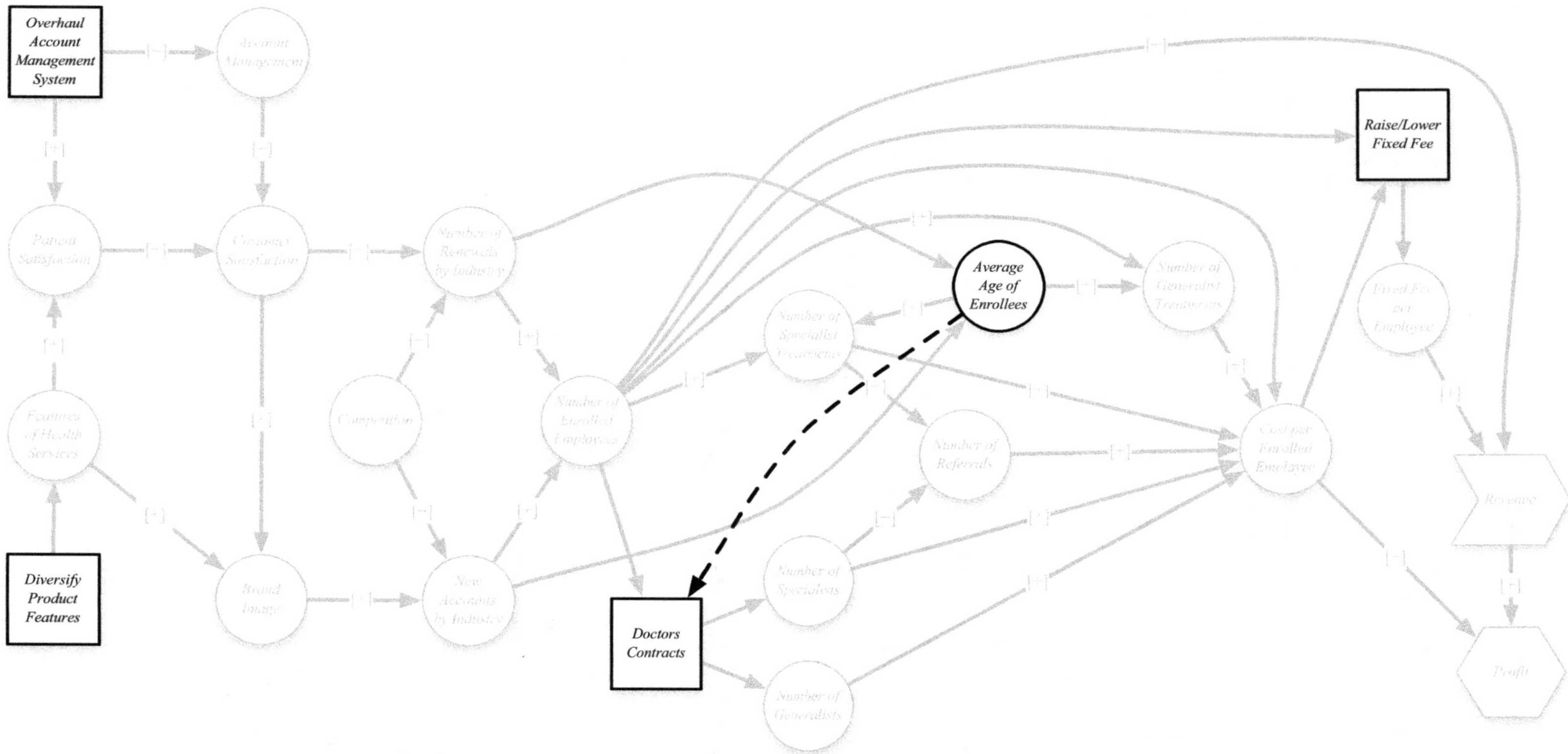

FIGURE 3.7 Highlighting the recommendations to Hubris within the model.

3.6.3 Appeal to a Younger Crowd

Employees themselves do not decide whether or not to use Hubris Health for their healthcare services, their employers do. Employers, as we have seen, are sensitive to their employees preferences on these matters, so it may be the case that there is something specific about Hubris' plan that is less appealing to younger people, and therefore to companies with higher numbers of young employees. We have already addressed automating and streamlining the account management system; what else can be done along these lines?

It turns out that young people are increasingly working in small companies, such as technology start-ups and niche market production industries. Because Hubris Health only pitches its plans to large companies, the easiest solution to correcting the average age balance is completely off Hubris' radar. This is an example of a lack of perspective on the part of the managers. There may not be anything wrong with Hubris' system for these smaller companies and their younger employees, there may just be more people (and especially more young people) in companies that are not eligible for Hubris' services.

A specific recommendation here, is to loosen the restriction on company size. Doing so will likely require other adjustments in operations: (1) assign staff members to multiple small companies instead of dedicated agents, (2) rely more on the new automated account management system they are going to build, and (3) reconsider the competition faced when expanding into these new markets. Though the management of Hubris health is despondent about the idea of using the upstart company BioSoftMed as a guide, they can probably learn a great deal about what product features and management tools are effective with this target customer base. Hubris can potentially leverage their in-house doctors, decades of experience, and large size to beat out the competition in the high tech and small-to-mid-size business markets if they can offer the appropriate product features.

3.6.4 Final Note

All the above specific recommendations to Hubris came from the structure of the causal model. In particular, Hubris can only change things that it controls, and those elements are all identified by decision nodes in the causal model. Figure 3.7 highlights the points where the above recommendations come into play; you will notice that every decision nodes implies a course correction for Hubris' business practices.

In addition to intervening at decision nodes, you can also make a specific recommendation about what information to use for decisions. In this case you recommend that Hubris base the number of doctors on the actual age distribution of enrolled employees rather than a standard formula based on historical performance. You might consider also adding an information link from *Competition* to *Diversify Product Features* to represent the advice that Hubris imitate BioSoftMed in its attempts to appeal to younger employees – if you thought that copying was a critical recommendation.

Using the model allows you tie all these together. Diversifying the product features and/or overhauling the management system alone may be enough to correct both the client and cost problems because these are root nodes. They both have downstream effects regarding the mix of client industries (through renewals and new accounts), and therefore the average age of enrollees, and hence the treatments required, and hence the costs. Changing the mix of doctors or the fixed fee, on the other hand, can be used to reduce costs, but will not likely effect the number or types of clients using Hubris Health. Because management time and other resources are always scarce, you can use the structure of the model to prioritize changes.

The logical, structured thinking process demonstrated here is the same process in actual consulting. The remainder of the book establishes the conceptual and technical foundation required to make use of the technique for whatever application may cross your path in an interview, consulting job, or other managerial problem. Applied models are typically more detailed and quantitative in nature. Chapters 6 and 8 provide examples of using software to analyze causal models and find optimal solutions to quantified problems.

KEY CONCEPTS: CHAPTER 3

- Begin with a situational assessment of given information, and develop the relationships among these factors, available decisions, and the objective (typically profits, but not always).
- "Walk back" through the causes in steps. What causes profit? Total revenues and total costs. What causes total revenues? Prices and quantities. What causes quantities? Price and demand. And so on.
- Make sure to identify strategic options (decision variables) in the system because ultimately the point of consulting is "What should Firm-X do?"
- Sign the links where possible. If you do not know the sign, or even whether a link is appropriate, ask for more information. Some potential links may be outside the scope of available information.
- Once you have a suitable number of causal steps in the model – typically 4 levels back from the objective, perhaps with some additional detail around important modules – make sure to think about links cross-cutting the modules that incorporate them into the wider scope.
- With the model in place, use it to ask and answer further questions. Investing time in filling out the model up-front, facilitate being able to drill down into a particular area as the case develops.
- If something comes up in a later stage of the case that is revealed to be an important omission on your part, just add the node(s) and keep on going!

- In the open-ended discussion occurring at the end of the interview, continue to use the model to structure your thinking. Since the case will typically have focused on one or two parts of the model, use this as an opportunity to opine on areas of the model that were not discussed earlier.
- Bottom line: Use the model! Think causally!

Chapter 4

Quantitative Causal Models

Qualitative causal models are useful either to structure your thinking at the start of a major analysis or in situations where time and information are limited (e.g., the consulting interview). However, the power of the methodology goes far beyond these qualitative settings. In this chapter, we provide the foundational material necessary to build and use quantitative causal models. This includes some basic probability theory and a few different ways to think about probabilities of states. We also provide a way to interpret qualitative causal models as quantitative ones, thus providing a segue from the quick and simple modeling technique to the inferential capabilities presented in the next chapter.

4.1 PROBABILITY BASICS

The fundamental difference between qualitative and quantitative causal models is the use of probabilities in capturing the changes in environmental nodes. Before we get to the actual implementation of this procedure, it will be helpful to review some basic facts about probabilities and also to elaborate the definitions and notational conventions we will use throughout the remainder of the book.

To put our discussion on a concrete footing, recall the Office Equipment Purchase example's non-monotonic relationships illustrated by Figure 2.3. To recap: You believe that the quality of the office equipment purchased affects the likelihood of different maintenance cost levels. There are three levels to product *Quality*: *low, medium,* and *high*. There are the same three possible outcomes for *Maintenance Costs*: *low, medium,* and *high*. To economize on notation, assign the numbers 1, 2 and 3 to *low, medium,* and *high*, respectively.

4.1.1 Variable States

As discussed earlier, a causal model describes certain types of relationships among a collection of variables. Each variable is represented as a node in the causal diagram, and arrows between a pair of nodes represents a particular kind of conditional effect one has on the other. We limit attention to discrete (rather than continuous) variables, so they can take only a finite number of values.[1] We often refer to the values that a variable can take as the set of possible **states** of that variable; and thus a change in a node's value can also be called a *change in its state*.

In the Office Equipment model we have two variables, each of which can be in one of three states. Let Q denote the quality of the office equipment being considered for purchase, and let M denote the maintenance cost outcome. Sets will be represented as lists within curly brackets like "{...}"; in this case we have $\{1, 2, 3\}$. To indicate set membership we use the standard symbol for "— is an element of —" which looks like this "$\in$". We combine these to indicate that a variable takes on one of a certain set of values. In this example, then, $Q \in \{1, 2, 3\}$ and also $M \in \{1, 2, 3\}$. As described above, $Q = 2$ means that the products purchased are of *medium* quality, while $M = 3$ means the maintenance costs are high. It is also correct to write $Q \in \{low,\ medium,\ high\}$, and this is sometimes used when there is no natural order to the values, or to clarify the state meanings.

> **Definition 4.1** *State: each of the possible values of a variable. Sometimes "the state" is used to refer to the outcome – the actual realized value of that variable.*

4.1.2 Events

Consider a balanced, six-sided die. Let D be the variable representing the outcome of a die toss. Then, $D \in \{1, 2, 3, 4, 5, 6\}$. As discussed above, each value in the set $\{1, 2, 3, 4, 5, 6\}$ is called a state of D. We refer to collections of states as **events**. Note that a single state is, itself, considered to be an event (though, a trivial one).

For example, symbolically, we would represent the statement, "the result of the die roll was even-numbered" as $D \in \{2, 4, 6\}$, where $\{2, 4, 6\}$ is the collection of states satisfying (or defined by) the event "the die roll was even-numbered." If we wish to indicate the probability of this event, we have a couple of options:

- The cumbersome version: $P(D = 2,\ \text{or}\ D = 4,\ \text{or}\ D = 6)$; or
- The compact version: $P(D \in \{2, 4, 6\})$.

1 Note that the causal modeling technique is not limited to discrete variables. We limit our exposition to discrete variables because doing so provides better intuitions for causal thinking and fosters calculations that can be done by hand (or with the help of a spreadsheet).

Since we like streamlined notation, we choose the second option and in cases where D is understood from the context we would just write $P(\{2,4,6\})$.[2] Of course, $P(\{2,4,6\}) = \frac{1}{2}$ for a fair die.

Reiterating, the notation $P(D \in \{2,4,6\})$ indicates the probability that the result of the die roll will be 2 *or* 4 *or* 6; i.e., one of the states listed. In probability notation, *we use commas only* (i.e., without "{...}") to mean "and". For example, $P(3,5)$ is the probability that the die roll is 3 and 5. Of course, the probability that the outcome of a single die roll is both 3 *and* 5 is zero (assuming it never tilts on an edge!), so $P(3,5) = 0$.

This convention is useful when dealing with multiple events (as well as multiple variables, as we discuss below). For example, suppose we wish to know the probability that the result of the die roll is odd *and* less than or equal to 3. For our system of just one die, these two events can be written $\{1,3,5\}$ and $\{1,2,3\}$, respectively. So, we could write $P(\{1,3,5\},\{1,2,3\})$ meaning the probability that both $D \in \{1,3,5\}$ *and* $D \in \{1,2,3\}$.

Of course, the only way for both $D \in \{1,3,5\}$ and $D \in \{1,2,3\}$ to be true is if $D \in \{1,3\}$ is true. We say that the event $\{1,3\}$ is the **intersection** of the two events $\{1,3,5\}$ and $\{1,2,3\}$ because it includes all and only the states shared by the original events. We are requiring that D be in this intersection: $D \in \{1,3\}$. To test your intuitions using this notation, it should be easy to see that $P(\{1,3,5\},\{1,2,3\}) = P(\{1,3\})$. That is, your relative certainty that $D \in \{1,3,5\}$ *and* $D \in \{1,2,3\}$ is equal to your relative certainty that $D = 1$ *or* $D = 3$.

Braces "{...}" indicate that at least one of the contained elements is true. This applies to events as well as states; e.g., $P(\{\{2,5\},\{1,4\}\})$ is the probability that d is either (2 *or* 5) *or* (1 *or* 4). The **union** of events is the collection of states appearing anywhere in those events; i.e., in this one or in that one or ... In these cases, we take the union of the events to write things as simply as possible. Here, $P(\{1,2,4,5\})$ is simpler than, and equal to, $P(\{\{2,5\},\{1,4\}\})$.

Now, suppose we wish to consider the outcome when we roll a die and flip a coin (C). For example, what is the probability that the coin toss equals Tails and the die roll is Even. Here, again, we have an "and" question – we wish to know the probability that both things are true. Therefore, we write $P(C = \textit{tails}, D \in \{2,4,6\})$. The notation $P(t, \{2,4,6\})$ is more compact and unambiguously represents the same probability, therefore it is preferred (and so you should become comfortable with this representation).

2 Careful readers will note that we could write the probability of a state as, e.g., $P(D \in \{2\})$ instead of $P(D = 2)$ or, even simpler, $P(2)$. The point of using notation is to ease our ability to make our discussion more precise. "$P(D \in \{2\})$" is technically correct, but overly cumbersome. However, sometimes the more cumbersome notation is necessary for clarity, such as when dealing with an event across multiple variables: $P(x \in \{2\}$, or $y \in \{2,4\}$, or $z \in \{6\})$.

Definition 4.2 *Event: a set of states for one or more variables; often defined as those states satisfying a condition of interest.*

Definition 4.3 *Intersection: an event combining other events using "and"; the resulting event only includes states shared by all of the original events.*

Definition 4.4 *Union: an event combining other events using "or"; the resulting event includes all the states in all of the original events.*

4.1.3 Probabilities

There are two different ways to think about probabilities. One is that the probability of some event is the objective *frequency* with which that event happens. For example, if one has a balanced coin, then saying, "the probability of heads = 1/2," means that if one were to flip the coin a large number of times, heads would come up half the time. The other interpretation is that the probability of something is a person's *subjective assessment* of its likelihood. For example, when a student who has never taken a class in causal modeling says, "the probability that I will get an A in this class is 90%." This may be expressing her subjective certainty about the prospect of this future event.[3]

Suppose C is the result of a coin toss. Then, C is a discrete variable with $C \in \{heads, tails\}$. The notation $P(C = tails)$ means "the probability that the outcome of a coin toss is tails." This is a number; e.g., if the coin is balanced then $P(C = tails) = \frac{1}{2}$. Stated differently: prior to the toss, *your degree of certainty that the coin will be tails is* $\frac{1}{2}$.

When the variable is clear from the context, we simplify things by writing, e.g., $P(tails) = \frac{1}{2}$ instead of $P(C = tails) = \frac{1}{2}$. When we want to make some general comment about the probability of a coin toss (i.e., without referencing any particular value) we write $P(C)$. As you will soon see, this form will be used to compare probabilities and to write out equations to calculate combined probabilities.

It is part of the definition of being a **probability** that it satisfies the following properties:

1 The probability that a variable is in one of its possible states is 1.
2 Probabilities are always positive; for any x, $P(x) \geq 0$.
3 For any events E_1 and E_2, the probability of their union is determined by

$$P(\{E_1, E_2\}) = P(E_1) + P(E_2) - P(E_1, E_2).$$

3 The frequency-based interpretation would be that the student is claiming that if she takes the class 100 times, then she would get an A 90 times – an unusual claim to make in this situation.

4 If E_1 and E_2 are events with no states in common (i.e., they are mutually exclusive), then

$$P(E_1, E_2) = 0.$$

5 If E_1 and E_2 are stochastically independent (more soon), then

$$P(E_1, E_2) = P(E_1) \times P(E_2).$$

From Item 1 we know that $P(x \in \{\text{heads,tails}\}) = 1$ and $P(D \in \{1, 2, 3, 4, 5, 6\}) = 1$. This is what it means when we say we are certain that the variable is in one of its possible states. This is the quantitative version of the mutually exclusive and collectively exhaustive requirement; i.e., states cannot be counted twice and we have not forgotten any states that could occur with some positive probability. As a corollary, we also know that if we add the probabilities across all the possible states then they must sum to 1.

Item 3 says that, in the balanced die example, the probability that the die roll equals either one of 1 or 3, or one of 1 or 4 is:

$$P(\{\{1,3\},\{1,4\}\}) = P(\{1,3\}) + P(\{1,4\}) - P(\{1\}) = \frac{2}{6} + \frac{2}{6} - \frac{1}{6} = \frac{1}{2}.$$

We can double check this. As discussed above in the definition of event unions, $P(\{\{1,3\},\{1,4\}\}) = P(\{1,3,4\})$. Since states are considered to be mutually exclusive events by themselves, application of Items 3 and 4 yields:

$$\begin{aligned} P(\{1,3,4\}) &= P(1) + P(3) + P(4) \\ &= \frac{1}{6} + \frac{1}{6} + \frac{1}{6} \\ &= \frac{3}{6} \\ &= \frac{1}{2}. \end{aligned}$$

Note that whenever events are mutually exclusive there are no overlapping states, and so the probably is just the sum of the individual event probabilities. That is, from Items 3 and 4 it is clear that when the states of E_1 are mutually exclusive of the states of E_2, $P(\{E_1, E_2\}) = P(E_1) + P(E_2)$.

Definition 4.5 *Frequency-based probability: the likelihood that an event happens is determined from the historically observed events.*

Definition 4.6 *Subjective probability: a likelihood assessment based on personal expectations (which may or may not be influenced by information).*

4.1.4 Conditional Probabilities

Given two variables x and y, the **conditional probability** $P(x|y)$ denotes "the probability of x given y." In our context, this means the probability of x when y is true *and nothing else matters in our assessment of* x (the relevance of this important caveat will become clear as we move on to causal models).

If the value of y is itself irrelevant with respect to the value of x, then $P(x|y) = P(x)$ and we say x is **stochastically independent** of y. Note that we have just introduced a new concept of independence between variables in addition to c-independence presented in section 2.3 – one based upon probabilities!

In causal models, we use **conditional probability tables** to elaborate the conditional dependency of one variable's values on another's. For example, suppose a major soft drink manufacturer notices that its competitor tends to offer price discounts following a major ad campaign. Let $Price \in \{low, high\}$ be the competitor price variable and $Advertising \in \{no, yes\}$ indicate whether the competitor advertised. The following table shows the conditional probabilities of *Price* given *Advertising* or $P(Price|Advertising)$.

Conditional Probability Table

	Price		
Advertising	*low*	*high*	**Sum**
no	0.5	0.5	**1.0**
yes	0.8	0.2	**1.0**

Throughout the book, we adopt the convention of putting the values of the conditioning variables in rows (*Advertising* here) and those of the focal variable in columns (*Price* here). The table says that there is a 50:50 chance of *low* prices when there is *no* advertising, but that likelihood increases to 80% if there has been advertising. Therefore the conditional probabilities of *high* prices are $P(high|no) = 0.5$ and $P(high|yes) = 0.2$.

Note that *each row sums to 1*. This is a necessary feature of conditional probabilities. That is, for each state of the conditioning variable, the sum over all the states of the focal variable must equal 1. This is just Item 1 in the list of properties shared by all probabilities – conditional probabilities are still probabilities after all.

> **Definition 4.7** *Conditional probability: the probability of one event occurring given that some other event has happened.*
>
> **Definition 4.8** *Stochastic independence: the probability of one event is not influenced by another event.*

Definition 4.9 *Conditional probability table (CPT): a convenient tabular presentation of the probability of each state of the focal variable for each state of the conditioning variable.*

4.1.5 Joint Probabilities

Suppose we wish to know the probability that some variable x equals some particular value *and* some other variable y equals some particular value; i.e., the **joint probability** of those two events which we've already seen written as $P(x,y)$. It turns out that a **fundamental rule** in probability is:

$$P(x,y) = P(x|y)P(y). \tag{4.1}$$

Applying this rule to a joint probability like $P(x,y)$ is called **factoring**. Continuing with the price and advertising example, suppose you wish to know the probability that next month your competitor will advertise and discount prices. That is, you wish to know $P(yes, low)$. As equation 4.1 indicates, the preceding table provides part of the answer (the $P(Price|Advertising)$ part). However, we still need to know $P(Advertising)$. Assume $P(yes) = 0.3$. By implication (because probabilities over states have to sum to 1), $P(no) = 0.7$. Now we can proceed:

$$P(yes, low) = P(low|yes)P(yes) = 0.8 \times 0.3 = 0.24.$$

We continue like this for all the other combinations of *Advertising* and *Price* to construct a full table of joint probabilities. To do this you simply multiply the entries in the conditional probability table by the probabilities of the rows; in this case $P(no)$ for the top row and $P(yes)$ for the bottom row. In our example, this leads to:

Joint Probability Table

	Price		
Advertising	*low*	*high*	**Sum**
no	0.35	0.35	**0.7**
yes	0.24	0.06	**0.3**
Sum	**0.59**	**0.41**	1.0

By summing all the column entries across a row in a joint probability table, we obtain the **marginal probabilities** for *Advertising*. That is, $P(no) = 0.7$ and $P(yes) = 0.3$. Similarly, summing a column down the rows delivers the marginal probabilities of *Price*: *P*(*low*) = 0.59 and *P*(*high*) = 0.41. Note that the marginal probabilities for each variable sum to 1.

Definition 4.10 *Joint probability: the probability that a particular combination of variable states occurs.*

Definition 4.11 *Marginal probability: the probability that a single variable takes a particular one of its states.*

4.1.6 System States

The information in the preceding joint probability table can also be organized into a **system-level joint probability table**. Such a table lists all possible combinations of the variable states in columns on the left and for each combination the probability of that outcome on the right. In our advertising example, it looks like this:

System-Level Joint Probabilities

Advertising	*Price*	Probability
no	*low*	0.35
no	*high*	0.35
yes	*low*	0.24
yes	*high*	0.06
	Sum	**1.00**

The preceding table is informative in that each row corresponds to one possible combination of states for every variable in the system, and we can do this for any number of variables in the system (whereas the other tables are limited to showing two at a time). Moreover, each row is a mutually exclusive combination of the system's states because the rows of the table elaborate every possible combination that can be taken by all the variables in the system. We refer to each such combination as a **state of the system**.

The column on the right shows the probability distribution over states of the system. Since the rows are mutually exclusive, we can compute marginal probabilities by adding all the rows in which a focal value appears. For example, the probability that *Price* is *low* is the sum of rows 1 and 3 (0.35 + 0.24 = 0.59).

More generally, when we refer to a "state" with respect to a *set* of variables, we mean a combination of states, one for each of the variables in the set. Thus, we can talk about the state of the entire system (all the variables in a causal model), the state of a node's set of parent variables, and so on.

4.1.7 Bayes' Rule

Often, we wish to update an assessment about the state of one variable based upon new information or evidence on the state of another. To do this, we use Bayes' Rule:

$$P(x|y) = \frac{P(y|x)P(x)}{P(y)}. \tag{4.2}$$

Notice that equation 4.2 is simply a rearrangement of the terms in equation 4.1.[4] $P(x)$ is referred to as the **prior probability** (or sometimes simply as the "prior") of x because it is the value that is getting updated. $P(x|y)$ is likewise called the **posterior probability** of x *given* y.

In our previous example, suppose you are told *price* is *high*. What is the probability that there was an advertising campaign? The prior probability of there being an advertising campaign is $P(yes) = 0.3$ (refer to the preceding joint probability table). The probability that *Price* is *high* given *Advertising* is *yes*, $P(high|yes)$, is 0.2 (from the conditional probability table). Overall, the marginal probability that *Price* is *high*, $P(high)$, is 0.41. Therefore,

$$P(yes|high) = \frac{P(high|yes)P(yes)}{P(high)} = \frac{0.2 \times 0.30}{0.41} = 0.14.$$

In other words, without any knowledge about your competitor's price reduction, the likelihood that there was advertising was 30%. However, once told that their prices remained high, Bayes' rule provides a way to update your assessment. The new, or posterior, assessment that there was advertising, conditional upon this information, is only 14%. That your expectation regarding an advertisement campaign would decrease in this case is probably intuitive. The added detail of these quantitative models will refine those intuitions and also provide insights into cases too complicated for intuitions to be reliable.

4.2 QUANTIFYING A QUALITATIVE MODEL

The qualitative causal modeling technique presented in Chapter 2 can be interpreted in a purely quantitative way; i.e., it is really just a low resolution version of a quantitative model. In this section we show you how to add a little more meat to a qualitative model so you can do more analysis with it. It also acts as a segue from the basics of probability just presented to their application in causal models.

For each variable (node) in the model, we attach a conditional probability table in which the rows capture the possible states of all its parents and the

---

4 From equation (4.1) we know both $P(x, y) = P(x|y)P(y)$ and also $P(x, y) = P(y, x) = P(y|x)P(x)$...so $P(x|y)P(y) = P(y|x)P(x)$. To get Bayes' Rule for $P(x|y)$, just divide both sides by $P(y)$. Note that this definition of $P(x|y)$ only makes sense if $P(y) \neq 0$.

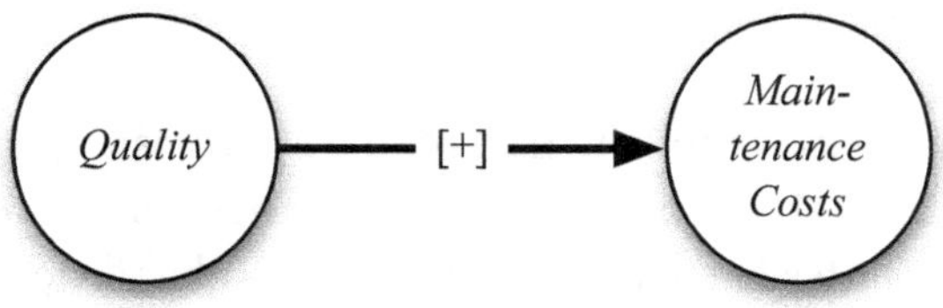

FIGURE 4.1 A simple causal model.

columns correspond to the states of the node itself. The entries in the table are the conditional probabilities of the node's states given the possible states of the parent variables. The positive "[+]", negative "[−]", and non-monotonic relationships specified for each arrow gain an interpretation in terms of conditional probabilities. Here we must add something to the models beyond what qualitative models require: *if a node is a root node (i.e., it has no parents), its table simply presents the prior probabilities on the node states.*

We start with converting a simple example from section 2.2 with just two variables: *Quality* and *Maintenance Costs*; both having three states: *low*, *medium*, and *high*. There is a causal link from *Quality* to *Maintenance Costs* marked by a [+] to indicate the relationship is positive (see Figure 4.1).

Recall from Chapter 2 that the way to interpret the [+] on the arrow is that this is a monotonic relationship and, holding everything else constant, finding out that the value of *Quality* is *low* makes your certainty that *Maintenance Costs* is *low* increase, and simultaneously decreases your certainty that *Maintenance Costs* is *medium* or *high*. And so on for the other states of the variables. The conditional relationships are quickly made clear by representing them in a table such as this:

	Maintenance Costs		
Quality	*low*	*medium*	*high*
low	+	−	−
medium	−	+	−
high	−	−	+

Now we are going to quantify these relationships. We want to do this in a way that makes as few assumptions as possible and satisfied the requirements for probabilities presented earlier. Because these +s and −s are stand-ins for conditional probabilities, we know that each row must sum to 1. Imagine that we had a row of all +s or all −s. The way to interpret this is that given information that the parent node's state corresponds to this row, we cannot change our certainty regarding the state of the child. Another way to say this is that each state of the child is equally likely, and this is best translated as equal probabilities across the row. If s is the number of states available for the column variable, then $\frac{1}{s}$ is the probability of each one – in this case it would be $\frac{1}{3}$ because *Maintenance Costs* has three possible states.

Looking at the table above for the monotonic relationship, each row has exactly one + and two – signs. To quantify this case, we therefore need to make one state more likely and two states less likely. Because the probabilities still have to sum to 1, we know that the cell with + must be greater than $\frac{1}{3}$ and the cells with – must be less than $\frac{1}{3}$. There are many ways to assign these values (an infinite number of ways, actually). One extreme is to make the values 1 for + and 0 for –; the other extreme is to make the difference as tiny as possible. The minimal assumption, however, is to break up the probability mass into equal blocks and distribute the blocks as necessary to satisfy the +s and –s.

For one + and two –s we need four blocks; the + gets two blocks and the – each get one block. Since *s* (the number of blocks) equals 4 in this case, and $\frac{1}{4} = 0.25$, the + is assigned a probability of 0.50 and the –s are each assigned a probability of 0.25. Applying these translation rules to the table presented above, we produce the following quantitative conditional probability table:

	Maintenance Costs		
Quality	*low*	*medium*	*high*
low	0.50	0.25	0.25
medium	0.25	0.50	0.25
high	0.25	0.25	0.50

We may instead have a non-monotonic relationship between the variables that involves a row with two + signs. In order to satisfy the conditions of making the probability of + greater than just the average, and – less than the average, and the sum equals 1, *and* none of the values less than zero, we need to divide the full probability into five blocks. This way, we can again give more blocks to the + cells than to the – cells. Five blocks implies that each block translates into 20%; the +s become 0.40 and the –s become 0.20. Using these translations we can fill in a quantified conditional probability table for the non-monotonic relationship presented for these variables in Figure 2.3 and recreated below.

	Maintenance Costs		
Quality	*low*	*medium*	*high*
low	+	–	–
medium	+	–	+
high	–	+	+

	Maintenance Costs		
Quality	*low*	*medium*	*high*
low	0.50	0.25	0.25
medium	0.40	0.20	0.40
high	0.20	0.40	0.40

This quantification technique can be applied to any number of states by creating enough blocks so that each + has one more block than the – signs. We can also use this technique to assign probabilities to the root nodes. If you have no reason to believe one value is greater than another, just assign equal values. If you do, then generate the initial probability table for each root node with a + or − in each column indicating how likely that state is compared with the other states. This single-row table can then be quantified using probability blocks in exactly the same manner just presented. Though probability tables for root nodes were not necessary for the kind of reasoning we did in Chapter 2, they are necessary for the quantitative analyses presented throughout the rest of this book.

4.2.1 More Refined Approximations

Expanding the vocabulary of symbols allows this qualitative/quantitative hybrid technique to produce more refined (but still rough) approximations to building quasi-formal causal models. By introducing symbols such as "++", "−−", "+++", etc. we can break the probability blocks into smaller units. This is useful when building models as part of a discussion to indicate "when we observe this conditioning value, the probability of this state goes up, but the probability of this other state *really* goes up."

Keep in mind that these are probabilities *relative* to the other states: ++ means one more block of probability than +, and +++ means two more blocks than +. If you assign a row with −, +, ++, and +++ then the cell with + will have a less-than-the-average probability. That assignment is also equivalent to −−, −, +, ++, which produces more intuitive probability assignments. Furthermore, it may be useful for you to also add the symbol "⊘" to indicate "the probability in this cell is equal to exactly 0" (an important feature of causal models which will show up later in the book).

Though these refinements can make qualitative models more precise and more useful, they are no substitute for building quantitative causal models from data. Unfortunately, real-world data is not always available. When relevant data is not available – perhaps due to time pressure, the cost of obtaining the data, or the simple fact that the data does not exist – other methods of quantification must be taken. Often, a solid substitute is to solicit expert opinion to help quantify the model. If that is not possible, then the analyst has no choice but to base the probability tables on his or her own best subjective judgment.

This quantification approach provides a way to combine those variables for which you do have data with those for which you don't. By converting the qualitative aspects into quantitative models this way, you are essentially just guessing the probabilities based on your experience or the experience of experts, but if the qualitative model captures informative relationships among the variables, then conversions made by the above approach are the best guesses. Once you have a quantitative model then you can apply all the operations presented throughout the rest of this book.

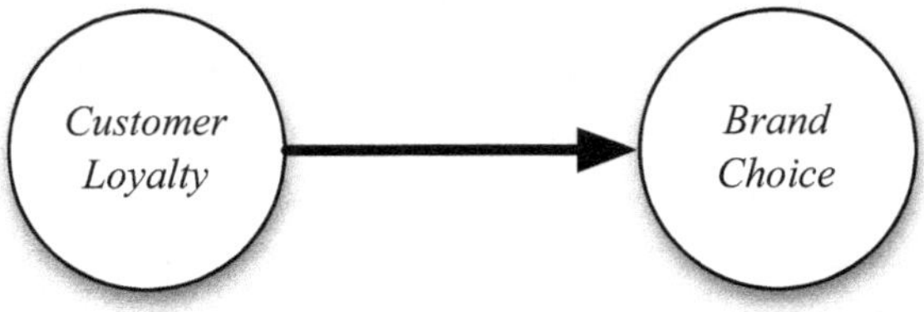

FIGURE 4.2 Consumer choice revisited and simplified.

4.3 WORKING WITH QUANTITATIVE MODELS

To demonstrate how to build a causal model from data, we return to the advertising example initially discussed in section 1.2 for Feather Touch toilet tissue. To start things simply, we will restrict the initial number of variable values to two: *Customer Loyalty* (*L*) and *Brand Choice* (*B*). *Loyalty* can be either *low* or *high* while *Brand Choice* is either *no* or *yes* (*yes* means customers choose to buy the Feather Touch brand). The simplified model is shown in Figure 4.2.

There are only two nodes in this version: a parent (root) node and its one child node. The model therefore needs a simple probability table for *Customer Loyalty* and a conditional probability table for the probabilities of *Brand Choice*, given each state of *Customer Loyalty*. The first step is to outline the problem by creating all the probability tables we need to fill in from the data.

Typically, it is easiest to begin with the root nodes. Our *Customer Loyalty* variable only has two possible states, *low* and *high*, and its table needs a column for each one. Because it does not depend on anything (by definition of it being a root node), there is only one row. That completes the structure of the marginal probability table for *Customer Loyalty* as shown here:

Customer Loyalty	
low	*high*
$P(low)$	$P(high)$

All variable nodes that are not root nodes need a conditional probability table. Each row corresponds to a set of input states, and there is a column for each state of the node in question. The cells are filled with the appropriate conditional probabilities.

Continuing with the Feather Touch example, our simplified causal diagram says that *Customer Loyalty* (and nothing else) causes *Brand Choice*. Again, the meaning of "causes" here is that changes in *Loyalty* tend to change the likelihoods of the states associated with *Brand Choice*. Since both *Loyalty* and *Brand Choice* take two values, the template for the conditional probability table for *Brand Choice* looks like this:

	Brand Choice	
Loyalty	*no*	*yes*
low	$P(no\|low)$	$P(yes\|low)$
high	$P(no\|high)$	$P(yes\|high)$

Now we need to fill in those values.

4.3.1 Probabilities from Count Data

The premise in this example is that the consultants collected data from 1,000 tissue paper buyers, thereby allowing us to make precise estimates of the probabilities needed for our quantitative causal model. To fill in the preceding tables from the data, we count the number of times loyal consumers choose to buy Feather Touch tissue as well as the number of times they do not, and so on. Then, assuming the data is representative, we use it to estimate the probabilities. Suppose the data breaks out as follows:

	Count Data		
	$\#B = no$	$\#B = yes$	**Total**
$\# L = low$	652	48	700
$\# L = high$	58	242	300
Total	710	290	1,000

We start by filling in the root nodes. Reviewing the data, we discover that there were 300 loyal Feather Touch buyers out of 1,000 surveyed, so $P(L = high) = 0.30$. This means that 30% of the people choosing to buy toilet tissue are predisposed toward the Feather Touch brand. That is the number revealed from the data, and by using it in the model we are assuming that it is representative for the population at large.[5] The other number is $\frac{700}{1000} = 0.70$, which not coincidentally equals $1 - 0.30$. Therefore, the probability table attached to the *Customer Loyalty* node is simply:

Customer Loyalty	
low	*high*
0.70	0.30

Remember that the probability values in any row of a conditional probability table must sum to 1. The marginal probability tables for root nodes can be thought of

5 At this point, those of you with statistical training may start to worry about confidence levels, variance, sample bias, and other such considerations. Although such concerns naturally do apply to using data for causal models, we leave those issues alone for now to focus on teaching you the causal modeling technique.

as conditional probability tables, in which this node's values do not depend on another node. The point is that the values across the single row of a marginal probability table must also sum to 1 (they are probabilities, after all).

Conditional probabilities are a bit more involved, but still straightforward. **In order to get the conditional probabilities of *Brand Choice* given *Customer Loyalty*, divide each of the row amounts by the row total**. When you do this, **each row** of the table becomes its own probability distribution over changes in the states of the child variable within that row.

As the parent variable changes state, the probabilities over its children also change. This captures the sense in which changes in the parent cause changes in its children; specifically, the changes that are captured are the changes in our **certainty** regarding the states of the child node given changes in the state of the parent node.

To calculate $P(no|low)$, we therefore need two numbers from the count data: the count of people who have *low Loyalty* and did *not* buy Feather Touch tissue, and the total number of people with *low Loyalty*. Going back to the Count Data table we see that the first number is 652, and the row total is 700. Thus, the probability that somebody will *not* choose to buy Feather Touch given that they have *low Loyalty* is $\frac{652}{700} = 0.931$. We continue like this to fill in the rest of the table, and the resulting conditional probability table is:

	Brand Choice	
Loyalty	*no*	*yes*
low	0.931	0.069
high	0.193	0.807

If you perform all those calculations correctly, the rows each add to 1. Up to this point we have built the structure of the causal model, identified what probability tables we need for it, and analyzed count data to determine the appropriate probabilities to fill in the tables. The last step is to bring it all together into a complete quantitative causal model as shown in Figure 4.3.

This model completely encodes our knowledge of the situation both in terms of the pertinent causal relationship as well as a probabilistic quantification of this relationship. Notice that we have no more use for the "+" notation on the arrow or in the probability table because we have now fully quantified the "+" effect. Given a person has a *high* degree of loyalty, that person is much more likely to purchase Feather Touch tissue than to not purchase it. And when we change our consideration from a disloyal to loyal consumer, our expectation of that person choosing the Feather Touch brand increases from 6.9% to 80.7%.

	Customer Loyalty	
	low	*high*
$P(L)$	0.70	0.30

Marginal Probabilities

$P(C \mid L)$	Brand Choice	
Loyalty	*no*	*yes*
low	0.931	0.069
high	0.193	0.807

Conditional Probabilities

FIGURE 4.3 Consumer choice, quantified.

4.3.2 Joint Probability Tables

As discussed earlier, we are often interested in the joint probabilities; sometimes as a step toward calculating conditional probabilities and sometimes to list the probability of being in each possible system state. For example, we may wish to know the probability that *Loyalty* = *low* **and** *Brand Choice* = *no*, that is $P(low, no)$ (or equivalently $P(no, low)$ because the order doesn't matter for joint probabilities – but it does for conditional probabilities).

If you recall that fundamental rule of probability captured in equation 4.1, it tells us that $P(low, no) = P(no|low)P(low)$. We just calculated all the conditional probabilities above, so **to get the joint probabilities, multiply the entries in each row of the conditional probability table by the marginal probability for that row.** The rows of a child correspond to the columns of the parent(s), so we are using $P(Loyalty = low) = 0.7$ for the top row, and $P(Loyalty = high) = 0.3$ for the bottom row. The system-level joint probability table is shown below. Now we can see that the probability of *Loyalty* = *low* **and** *Brand Choice* = *no* is 65.2%. Remember that in a joint probability table, the sum of *all* the entries must equal one – though there may be a small discrepancy if the individual values are rounded.

System-Level Joint Probabilities

Customer Loyalty	*Brand Choice*	Probability
low	*no*	0.652
low	*yes*	0.048
high	*no*	0.058
high	*yes*	0.242
	Sum	**1.00**

4.3.3 The Complete Advertising Model

The simplified two-node version of the advertising model example is nice for teaching the basic mechanics of dealing with probabilities for causal models, however actual causal model will be much more complicated. Now we present the full advertising model of Figure 1.2, along with the actual conditional probability tables in Figure 4.4. These tables were used to generate the data upon which the statistical analysis discussion in that chapter is based. The *Profit* node has been removed because it is a deterministic function of the values of other variables, rather than a probabilistic one. It must be modeled as a functional or payoff node, and we will discuss payoffs shortly (but not now).

Armed with a quantified causal model, there is much we can do in the way of interesting analysis. One thing to notice is that the probabilities computed in the preceding discussion are based upon the two-variable model in which *Customer Loyalty* alone causes *Brand Choice*. In the full model, *Brand Choice* has two causes: *Customer Loyalty* and whether or not *Coupons* are being offered.

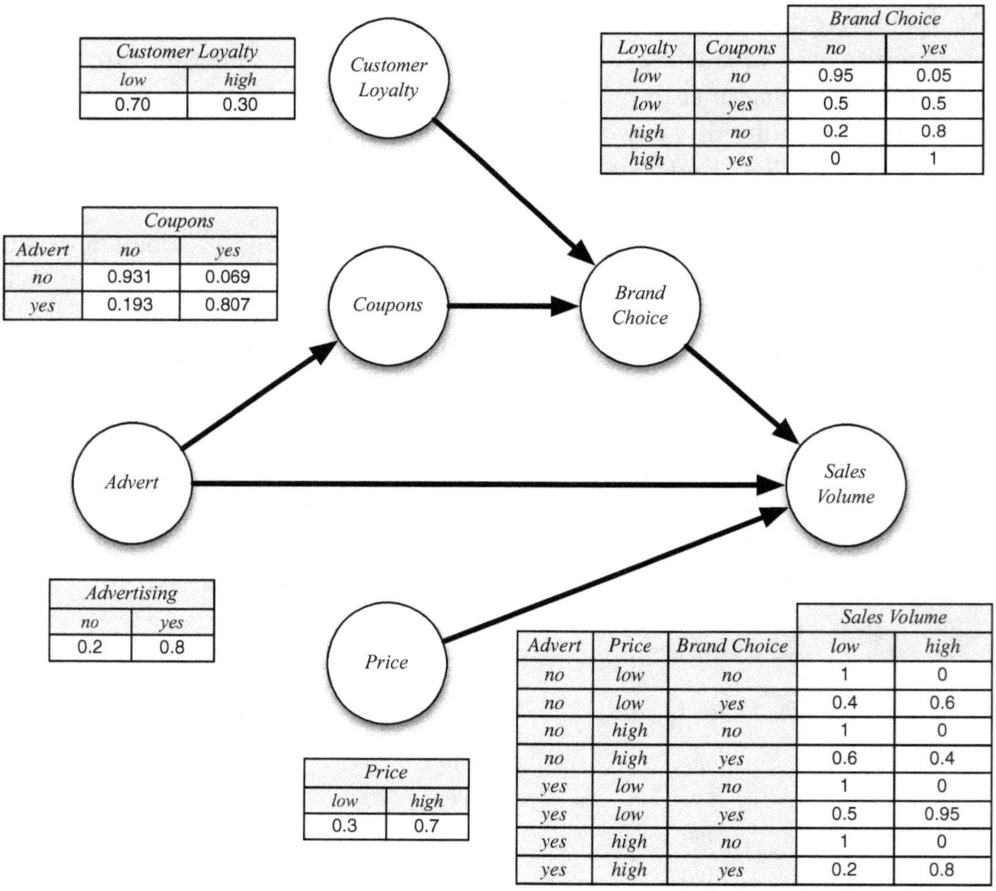

Customer Loyalty	
low	*high*
0.70	0.30

		Brand Choice	
Loyalty	*Coupons*	*no*	*yes*
low	*no*	0.95	0.05
low	*yes*	0.5	0.5
high	*no*	0.2	0.8
high	*yes*	0	1

	Coupons	
Advert	*no*	*yes*
no	0.931	0.069
yes	0.193	0.807

Advertising	
no	*yes*
0.2	0.8

Price	
low	*high*
0.3	0.7

			Sales Volume	
Advert	*Price*	*Brand Choice*	*low*	*high*
no	*low*	*no*	1	0
no	*low*	*yes*	0.4	0.6
no	*high*	*no*	1	0
no	*high*	*yes*	0.6	0.4
yes	*low*	*no*	1	0
yes	*low*	*yes*	0.5	0.95
yes	*high*	*no*	1	0
yes	*high*	*yes*	0.2	0.8

FIGURE 4.4 Advertising model with conditional probability tables.

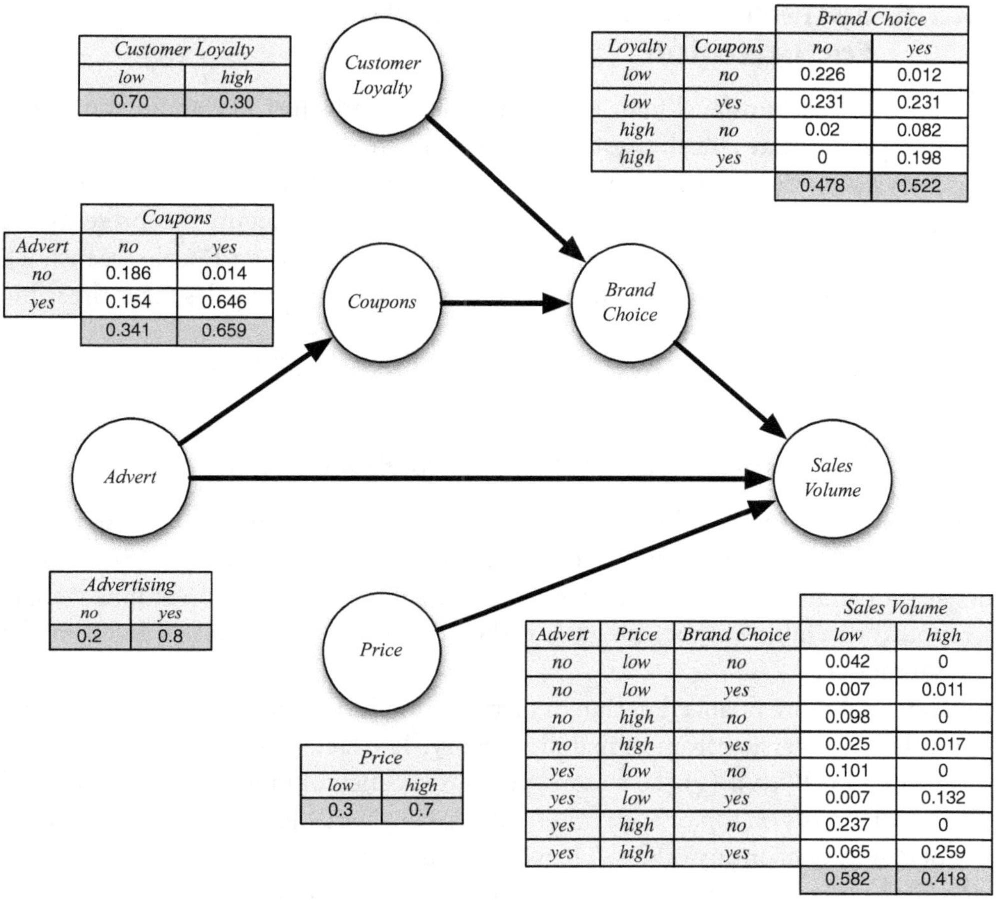

Customer Loyalty	
low	high
0.70	0.30

	Coupons	
Advert	no	yes
no	0.186	0.014
yes	0.154	0.646
	0.341	0.659

Advertising	
no	yes
0.2	0.8

Price	
low	high
0.3	0.7

		Brand Choice	
Loyalty	Coupons	no	yes
low	no	0.226	0.012
low	yes	0.231	0.231
high	no	0.02	0.082
high	yes	0	0.198
		0.478	0.522

			Sales Volume	
Advert	Price	Brand Choice	low	high
no	low	no	0.042	0
no	low	yes	0.007	0.011
no	high	no	0.098	0
no	high	yes	0.025	0.017
yes	low	no	0.101	0
yes	low	yes	0.007	0.132
yes	high	no	0.237	0
yes	high	yes	0.065	0.259
			0.582	0.418

FIGURE 4.5 Advertising model with joint and marginal probabilities.

The complete conditional probability table for *Brand Choice* is the one shown in the figure.

Figure 4.5 illustrates the same model with the *joint probability tables* (along with the marginal probabilities) for each variable instead of the conditional probability tables. **Summing down the columns of the joint probability table yields the marginal probability for each state of the variable at that node.** For example, in this model the overall (*aka* marginal) probability that Feather Touch is currently running a coupon campaign is 65.9%. Similarly, the marginal probability that the *Sales Volume* is *high* is 41.8%.

There are two ways of interpreting this. The first is from the situational assessment perspective: without any other information or evidence regarding the states of the various variables in the system, our relative certainty that sales are high is 41.8%. The second is from the managerial intervention perspective: if senior management sits back and does nothing, the firm can be expected to hit a high level of sales 41.8% of the time.

4.3.4 System-Level Joint Distribution & Factorization

Creating the system-level joint probability table works just like above, but there are many more states for this system. Since there are six variables and each variable has two states, there are $2^6 = 64$ states at the system-level (each row is one of 64 unique outcomes that might be observed in this system). Figure 4.6 presents the joint probabilities of every possible state of the system (as an Excel screenshot). The probability for a row is the likelihood that each of the variables takes the value specified for it in that row.

The row probabilities correspond to the joint probability for the intersection of each variable's value:

$$P(\textit{Loyalty}, \textit{Coupons}, \textit{Advertising}, \textit{Brand Choice}, \textit{Price}, \textit{Unit Sales}),$$

which we will now abbreviate to $P(L, C, A, B, P, U)$. As it turns out, that fundamental rule of probability from equation 4.1 can be applied to the system-level probabilities, and it is known as **factorization**. With factorization, we can compute the joint probabilities using the conditional probabilities given in the causal model.

There is a more nuanced notion of factorization for causal models, but we begin with the basic version from probability theory. The factorization for two variables was: $P(x, y) = P(x|y)P(y)$. To factorize the intersection of more variables you build the list of factors on top of each other. Choose one variable as the "base" and make it not conditional on any others. Then choose another variable (any other variable), and make it conditional on your base variable. Next choose a third variable, and **make it conditional on the intersection of the previous two variables**. Continue on like this until every variable appears once on the left side of a pipe character, "|" (except the base variable which has no "|"). Applied to the advertising example, one break-down we could get is this:

$$P(L, C, A, B, P, U) =$$
$$P(L|C, A, B, P, U)P(C|A, B, P, U)P(A|B, P, U)P(B|P, U)P(P|U)P(U)$$

You already know that the order of multiplication doesn't matter for joint probabilities, and that $P(x, y)$ can be written as either $P(x|y)P(y)$ or $P(y|x)P(x)$. This equivalence extends to any number of variables, so the above factorization is just one of many possible factorizations. We could also write any of the following:

$$P(U|C, A, B, P, L)P(C|A, B, P, L)P(A|B, P, L)P(B|P, L)P(P|L)P(L)$$

$$P(B|C, A, P, U, L)P(C|A, P, U, L)P(A|P, U, L)P(P|U, L)P(U|L)P(L)$$

$$P(B|C, L, P, U, A)P(C|L, P, U, A)P(L|P, U, A)P(P|U, A)P(U|A)P(A)$$

...

and so on. If we do not know the joint probabilities, but we do know the conditional probabilities that correspond to one of the possible orderings of the variables, then we can compute the joint probabilities doing the appropriate multiplication.

For causal models, we can go even further. Recall that if the value of x does not depend on y, then $P(x|y)$ just equals $P(x)$. Likewise, if the value of x is conditional on z, but not on y, then $P(x|y,z) = P(x|z)$. We can use this fact of probabilities, and careful attention to which variables depend on which others, to compress the full factorization of the advertising example. The result is:

$$P(L, C, A, B, P, U) = P(U|A, B, P)P(P)P(B|L, C)P(L)P(C|A)P(A). \qquad (4.3)$$

You can see the connection between the probabilities (including the conditional probabilities) in that equation and the causal structure represented in the diagram. Root nodes are captured with simple (non-conditional) probabilities. The variables of nodes with parents are conditional on the values of all its parent nodes. This is not a special feature of causal models – it is universally true and falls directly out of the laws of probability – but causal modeling does make good use of this property.

> **Definition 4.12** *Factorization: expressing a joint probability as a multiple of conditional and simple marginal probabilities.*

4.3.5 Marginalization

The detailed system-level probabilities in the table of Figure 4.6 can be used to compute the marginal probabilities of any variable or combination of variables. This is done by simply **summing up the rows that correspond to the states of interest**. For example, summing the first 32 rows – those corresponding to a lack of *Customer Loyalty* – results in 0.7. This number matches, not coincidentally, with $P(L = low)$ from that node's marginal probability table. If we add up all the rows for which *Advertising* = *no* and *Coupons* = *no*, we obtain 0.16. This is the value of $P(A = no, C = no)$ i.e., the probability of that combination being the case, ignoring the states of other variables. The process of adding up the rows to get the probability of a combination of states of interest is called **marginalization**.

Of course, for real world applications, the number of variables and states can be quite large. So large, in fact, as to make it infeasible to compute a joint probability table for the entire system. Using conditional probability tables, however, is typically feasible and computationally efficient.

> **Definition 4.13** *Marginalization: determining the probability of a specific combination of states.*

Example 4.1: Semiconductor Manufacturing

Loyalty	*Coupons*	*Advertising*	*Price*	*Choice*	*Volume*	**Prob**
0	0	0	0	0	0	0.0371
0	0	0	0	0	1	0.0000
0	0	0	0	1	0	0.0008
0	0	0	0	1	1	0.0012
0	0	0	1	0	0	0.0867
0	0	0	1	0	1	0.0000
0	0	0	1	1	0	0.0027
0	0	0	1	1	1	0.0018
0	0	1	0	0	0	0.0308
0	0	1	0	0	1	0.0000
0	0	1	0	1	0	0.0001
0	0	1	0	1	1	0.0015
0	0	1	1	0	0	0.0719
0	0	1	1	0	1	0.0000
0	0	1	1	1	0	0.0008
0	0	1	1	1	1	0.0030
0	1	0	0	0	0	0.0014
0	1	0	0	0	1	0.0000
0	1	0	0	1	0	0.0006
0	1	0	0	1	1	0.0009
0	1	0	1	0	0	0.0034
0	1	0	1	0	1	0.0000
0	1	0	1	1	0	0.0020
0	1	0	1	1	1	0.0014
0	1	1	0	0	0	0.0678
0	1	1	0	0	1	0.0000
0	1	1	0	1	0	0.0034
0	1	1	0	1	1	0.0644
0	1	1	1	0	0	0.1582
0	1	1	1	0	1	0.0000
0	1	1	1	1	0	0.0316
0	1	1	1	1	1	0.1265
1	0	0	0	0	0	0.0034
1	0	0	0	0	1	0.0000
1	0	0	0	1	0	0.0054
1	0	0	0	1	1	0.0080
1	0	0	1	0	0	0.0078
1	0	0	1	0	1	0.0000
1	0	0	1	1	0	0.0188
1	0	0	1	1	1	0.0125
1	0	1	0	0	0	0.0028
1	0	1	0	0	1	0.0000
1	0	1	0	1	0	0.0006
1	0	1	0	1	1	0.0106
1	0	1	1	0	0	0.0065
1	0	1	1	0	1	0.0000
1	0	1	1	1	0	0.0052
1	0	1	1	1	1	0.0208
1	1	0	0	0	0	0.0000
1	1	0	0	0	1	0.0000
1	1	0	0	1	0	0.0005
1	1	0	0	1	1	0.0007
1	1	0	1	0	0	0.0000
1	1	0	1	0	1	0.0000
1	1	0	1	1	0	0.0017
1	1	0	1	1	1	0.0012
1	1	1	0	0	0	0.0000
1	1	1	0	0	1	0.0000
1	1	1	0	1	0	0.0029
1	1	1	0	1	1	0.0552
1	1	1	1	0	0	0.0000
1	1	1	1	0	1	0.0000
1	1	1	1	1	0	0.0271
1	1	1	1	1	1	0.1085
						1.0000

FIGURE 4.6 System-level joint probability distribution.

We will now reinforce the material just presented with a straightforward example of a semiconductor manufacturing process. These manufacturing processes are conducted in "clean-rooms" – areas in which the air is highly filtered and personnel must wear special garb designed to prevent contamination of the product. This is because even the tiniest particles floating in the air can ruin an entire run of semiconductor chips. Such sensitivity also requires stringent quality control and testing.

Suppose the manufacturing *Process* can be *clean* or *dirty* on any given day. Depending upon whether the facility is dirty, contamination is either more or less likely. If wafers become contaminated, then the entire *Batch* is *bad*, otherwise it is *good*. It is impossible, however, to tell whether the chips are actually bad just by looking at them. Therefore, a subset of the chips are subjected to a *test* to determine whether or not to keep them. If the result of the test is *pass* then they are all kept, but if the test result is *fail* then they dispose of the entire batch! The testing process, like all testing processes, is prone to Type I (false positive) and Type II (false negative) errors.

Now, without explaining how we came to know it, suppose the probability distribution over all states of the system are as elaborated in Table 4.1. For example, $P(clean, good, pass) = 0.8379$. This means that the probability that, on a given day, the manufacturing process is *clean* and the batch is *good* and the test result is *pass* is 83.79%.

Table 4.1
System-Level Joint Probabilities

Process	Batch	Test	Probability
dirty	*bad*	*fail*	0.0255
dirty	*bad*	*pass*	0.0045
dirty	*good*	*fail*	0.0014
dirty	*good*	*pass*	0.0686
clean	*bad*	*fail*	0.0383
clean	*bad*	*pass*	0.0068
clean	*good*	*fail*	0.0171
clean	*good*	*pass*	0.8379

To calculate marginal probabilities, add up the numbers in the rows corresponding to the property of interest. For example, if you wish to know the probability that a test result indicates a corrupted batch, just add up the probability values of the rows in which *Test* = *fail*:

$$P(fail) = 0.0255 + 0.0014 + 0.0383 + 0.0171 = 0.0823,$$

or, 8.23%. This works for any marginal probability. For example, it should be easy for you to confirm that $P(bad) = 0.0751$, or 7.51%.

Now, suppose you wish to know the conditional probability that the batch is bad given a failing test result. This is also an easy, very mechanical, calculation when you have that joint probability table. Recall Bayes' Rule from equation 4.2:

$$P(bad|fail) = \frac{P(fail|bad)P(bad)}{P(fail)}.$$

Also recall that, from the fundamental rule in equation (4.1),

$$P(fail, bad) = P(fail|bad)P(bad).$$

Since the numerator of Bayes' Rule matches the right side of the fundamental rule, we make a replacement with the left side to derive another way of writing Bayes' Rule:

$$P(bad|fail) = \frac{P(fail, bad)}{P(fail)}.$$

In other words, the probability that the batch is actually bad given a failed test result is equal to the probability that the test failed *and* the batch is bad divided by the probability that the test failed.

Using the table, $P(fail, bad)$ is found by adding up the probabilities of the rows in which both *fail* and *bad* occur:

$$P(fail, bad) = 0.0255 + 0.0383 = 0.0638.$$

Therefore,

$$P(bad|fail) = \frac{0.0638}{0.0823} = 0.7752,$$

or, 77.52%. In other words, knowing that the state of the *Test* variable is *fail*, leads to a substantial increase in the certainty that the batch of semiconductors is actually bad – from 7.5% to 77.5% certainty.

4.4 THE MOVE TO CAUSAL MODELS

The system-level joint probability distribution (JPD) provides the most basic probability information possible. If one knows the JPD, then any probability value – conditional, marginal, or joint – can be calculated. In the preceding example, we assumed that the JPD shown in Table 4.1 was known prior to starting the analysis. From that, we were able to compute all the quantities of interest. Yet, this raises the question: **where does one obtain the information to fill in such a distribution?** This is an important question because, as most managers will attest, the primary problem in most business analyses is not a lack of analytical tools but, instead, a lack of reliable information upon which to apply those tools.

One option, as suggested in the previous chapter, is the frequentist approach to probabilities. In a semiconductor manufacturing setting, the quality control manager may be able to collect and maintain accurate records on the key variables. If so, Table 4.1 could be constructed by counting how many times each system state occurred and then converting those counts into probabilities the way we did for the raw count data earlier.

There are typically two obstacles to the frequentist approach. First, when dealing with real world cases, both the number of variables and the number of variable states is prohibitively large. It does not take long before the number of rows in the JPD table blows up to an unmanageable size. As a result, collecting a statistically reliable database may be overly time consuming and costly. Second, some variables may actually be impossible to measure. In the semiconductor example, it was assumed that testing was required precisely because the true batch quality could not be observed. Yet, if the true batch quality cannot be observed, then it is impossible to collect data on the true batch quality to estimate the system-level probability distribution in the first place.

In these situations, causal models come to the rescue. Instead of needing to fill in the whole JPD, **causal models only require the node-level conditional probability tables.** Thus, the amount of information required is smaller and much more manageable. Furthermore, computations in causal models can be done node by node; which when combined with the implications of causal independence results in additional efficiencies.

A second benefit of a causal model over the JPD is that **soliciting the information required for a causal model is typically much easier because people tend to find causal thinking more natural than probabilistic thinking.** For example, whereas an expert may be qualified to provide estimates of the conditional probabilities at a particular node, he or she may have little understanding of the system as a whole; e.g., in our semiconductor example, certain engineers may know the Type I and Type II error characteristics of the testing procedure but know little about the key cleanliness characteristics of the actual manufacturing process. Even if explicit data or expert opinions are not available, the causal structure invites managers to call upon their own experience and understanding of the situation in a familiar yet rigorous way.

Example 4.2: Semiconductor Manufacturing

Assume that the true causal relationships in our semiconductor example are those shown in the model depicted in Figure 5.3. It is worth noting that this model incorporates a basic, serial connection structure (which was covered in section 2.3.1 for qualitative models). The model indicates that, 90% of the time the manufacturing process meets a satisfactory level of cleanliness. When the *Process* is *clean*, semiconductor *Batch Quality* is *good* 95% of the time. However, when the manufacturing *Process* is *dirty*, 30% of the batches are *bad* quality. The conditional probabilities for the *Test* node indicate that the test makes Type I (false positive) errors 2% of the time and Type II (false negative) errors 15% of the time.

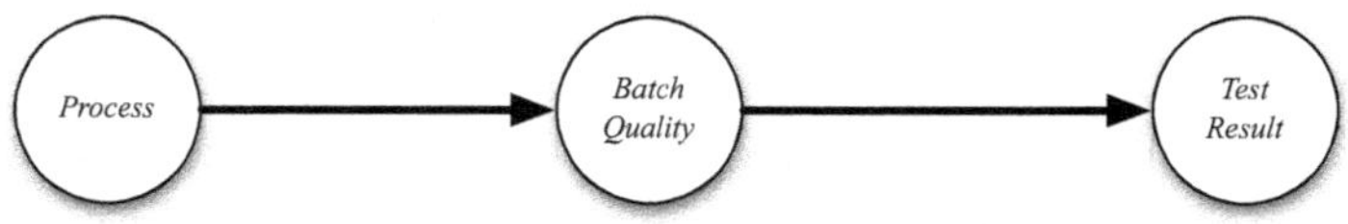

Process Cleanliness	
clean	*dirty*
0.9	0.1

	Batch Quality	
Process	*good*	*bad*
clean	0.95	0.05
dirty	0.7	0.3

	Test Result	
Batch	*pass*	*fail*
good	0.98	0.02
bad	0.15	0.85

FIGURE 4.7 Causal model of the semiconductor quality example.

Suppose we wish to compute the JPD. The information presented in Figure 5.3 is sufficient to this task. First, to economize on notation, we abbreviate *Process* cleanliness as P, *Batch* quality as B, and the *Test* result as T. We *again* use the fundamental rule from 4.1 to factorize the joint probability distribution. From it we know that the following list elaborates *all* of the possible factorizations of $P(P, B, T)$:

$$
\begin{aligned}
P(P,B,T) &= P(T|P,B)P(B|P)P(P) \\
&= P(B|P,T)P(T|P)P(P) \\
&= P(T|B,P)P(P|B)P(B) \\
&= P(P|B,T)P(T|B)P(B) \\
&= P(P|T,B)P(B|T)P(T) \\
&= P(B|T,P)P(P|T)P(T).
\end{aligned}
$$

This is not especially helpful given the model in Figure 5.3 because the model does not provide any of the final terms. For example, looking at the first row, the model provides $P(P)$ and $P(B|P)$ but not $P(T|P,B)$. It does provide $P(T|B)$, but this is not the information required to use the factorization to determine the joint probability.

We now make good use of that feature of probability that allowed us to compress the factorization of the advertising model to section 4.3. It also allows us to overcome this information deficiency to show that the JPD for a causal system can be computed directly from its conditional probabilities. Specifically, **the probability of a system state is the product of the conditional probabilities of the variables *given the states of their parents***. When applied to our current example, this means that:

$$P(P, B, T) = P(P)P(B|P)P(T|B).$$

No further information is required.

Pause for a moment to reflect upon this. $P(T|P, B) = P(T|B)$ means that T is probabilistically independent of P given B. Recalling our definition of c-independence from definition 16, we know that T is also causally independent of P given B. Does this relationship – between a qualitative notion of causal independence defined in terms of a causal diagram and the quantitative notion of probabilistic independence – hold more generally? Surprisingly, **the relationship between c-independence and probabilistic independence is quite robust.**[6] In all of the models considered in this book, computing joint probabilities via factorization works. In fact, Table 4.1 was originally computed directly from the causal model in Figure 5.3.

KEY CONCEPTS: CHAPTER 4

- This chapter introduces some new notation:
  1. A variable's states are written $V \in \{x_1, x_2, x_3, x_4, \ldots\}$.
  2. An *event* is a subset of states for a variable; it is also indicated with braces, e.g., $\{4, 5, 6\}$.
  3. When the variable, say V, is known from the context, $P(\{1, 3, 5\})$ means the probability that V takes on one of the states listed: $V = 1$ or $V = 2$ or $V = 3$.
  4. When multiple variables are involved, we separate the arguments with commas. For example, $P(tails, \{1, 3, 5\})$.
  5. The joint probability of x and y is written $P(x, y)$.
  6. The conditional probability of x given y is written $P(x|y)$.
- All probabilities share certain properties by definition:
  1. Probabilities across all states sum to 1 for each variable.
  2. Probabilities are never negative.

6 There are special cases in which the factorization implied by a quantitative causal model does not hold. However, these cases are rare and weird. In fact, if the model does not generate a probability distribution consistent with the graph, it is usually taken to mean there is something wrong with the model!

3 Event probabilities can be calculated from the probabilities of component events/states.

- One can convert a qualitative model into a rough quantitative one by utilizing approximate relative probabilities.
- The fundamental rule of probability is: $P(x, y) = p(x|y)p(y)$. One can derive Bayes' Rule from this fact:

$$P(x|y) = \frac{P(y|x)P(x)}{P(y)}.$$

- A *system state* is a list of all system values, one for every variable in the system.
- To quantify the model, add conditional probability tables to each node. Conditional probabilities of a root node is just its marginal probabilities.
- The probabilities of states of the system are computed by multiplying the probabilities of all the variables conditional on their parent nodes.

CHAPTER 4 REVIEW QUESTIONS

4.1 Which of the following is an expression of Bayes' Rule? ****Check all that apply****

(a) $P(x|y) = \frac{P(x)P(y)}{P(y|x)}$
(b) $P(x|y) = \frac{P(y|x)P(x)}{P(y)}$
(c) $P(x|y) = \frac{P(x, y)}{P(y)}$
(d) $P(x|y) = \frac{P(y|x)P(y)}{P(x)}$

4.2 Which of the following is **not** a basic rule of probability? ****Check all that apply****

(a) The sum of the probabilities over all possible states of a variable equals one.
(b) Probabilities are all greater than or equal to zero.
(c) If E_1 and E_2 have no states in common then $P(\{E_1, E_2\}) = P(E_1)P(E_2)$.
(d) If E_1 and E_2 do have states in common then $P(\{E_1, E_2\}) = P(E_1) + P(E_2) - P(E_1, E_2)$.

4.3 A magic die has six sides and looks like a normal die. After rolling the die many, many, many times you determine that $P(D \in \{1, 2, 4, 6\}) = 1$. Which of the following statements **must** be true for this die?

(a) $P(3) + P(5) = 0$
(b) $P(5) = 0$
(c) $P(\{2, 3\}) \leq P(\{2, 4\})$
(d) $P(\{1, 2\}) < P(\{4, 6\})$
(e) $P(6) > P(5)$

4.4 As part of a quality control project for a manufacturer of jet engine parts, you are investigating issues with their best selling item. Your client produces a line of ceramic turbine blades which, due to molecular bond variation in the drying process, can have a heat tolerance of either 1,000, 1,100, 1,200, or 1,300°C. Also, because of shrinkage in the firing process there are microscopic size variations in the resulting units: $-0.02\mu m$, $-0.01\mu m$, $0.00\mu m$, $+0.01\mu m$, or $+0.02\mu m$ from the design specification. Using compact mathematical notation, how would you represent each of the following?

(a) The probability that the heat tolerance is above 1,100°C and the size is exactly on spec.
(b) The probability that one of the extreme cases of both variables is observed.
(c) The probability that the heat tolerance is 1,100 or 1,200 and the size is within $0.01\mu m$ of the correct size.

4.5 UltraBoards Incorporated produces snowboard decks for many brands. The boards can be categorized into three *Quality* levels: *Kids, Hobbyist,* or *Extreme*. The boards are painted a solid color before being sent to the individual brand shops for custom effects and decals. The *Colors* are pulled from either *pastel, earthtone,* or *vivid* color swatches. After years of producing boards, UltraBoards has collected data on the numbers of orders of each type and put them in the following table.

	pastel	*earthtone*	*vivid*	*Total*
Kids	300	50	150	500
Hobbyists	300	650	150	1100
Extreme	50	50	400	500

Given this information, answer the following questions:

(a) What are the conditional probabilities of each *Color* given a level of *Quality*?
(b) One of the product representatives tells you that an energy drink company is getting some custom boards made for a promotional snowboard racing event. What is the probability that they will order brown and gray (i.e., *earthtoned*) snowboards?
(c) A new ski resort is opening up, and the rental shop plans to put together a range of snowboard packages for its expected customers. They tell you that the distribution of customers they expect is 30% *kids*, 60% *hobbyists*, and 10% *extreme* sports fanatics. They want you to use your past data to send them appropriate numbers of each *Color* type for each user group (which corresponds to a *Quality* level). If they plan to order

1,000 boards in total, how many of each *Quality* and *Color* combination should they purchase?

(d) A friend (who is not a child) tells you that he plans to visit this new ski resort and he really prefers brightly colored snowboards. Knowing that the color of board they give you at the shop is basically random within each category, would you recommend that he/she ask for a *hobbyist* or *extreme* board? Explain your reasoning.

4.6 Consider the conditional probability table below for the *Number of Patents* a company holds given the *Level of Research and Development* at that firm. If these are the real conditional probability values underlying the causal structure of system behavior, what **qualitative** relationship(s) would be revealed by this process? Specifically, what should appear instead of the question mark on the link?

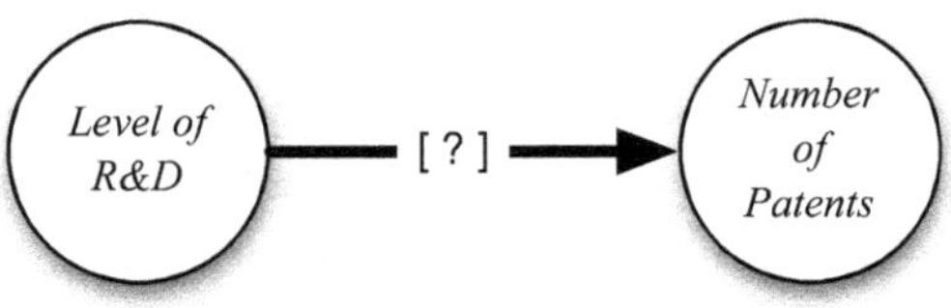

	Number of Patents		
Level of R&D	*few*	*moderate*	*many*
low	0.82	0.17	0.01
medium	0.02	0.97	0.01
high	0.03	0.08	0.89

4.7 A regional coffee chain wants to optimize its marketing budget, equipment maintenance, and staffing by determining the likelihood of different types of drinks being purchased on specific days.

Temperature	*Weekend*	*Drink Type*	Probability
cold	*no*	*iced*	0.02
cold	*no*	*hot*	0.33
cold	*yes*	*iced*	0.01
cold	*yes*	*hot*	0.10
nice	*no*	*iced*	0.08
nice	*no*	*hot*	0.03
nice	*yes*	*iced*	0.06
nice	*yes*	*hot*	
warm	*no*	*iced*	
warm	*no*	*hot*	0.07
warm	*yes*	*iced*	0.10
warm	*yes*	*hot*	0.04

Use the joint probability table above to answer the following questions:

(a) There are two missing entries in the probability table. You are told that the probability of being a *warm* day in the region is 31%. Use this information to determine the probability that somebody buys a *hot* beverage on a *nice Weekend* day.
(b) What is the conditional probability that somebody orders an *iced* beverage when *Weekend* = *yes*?

4.8 Which of the following is **not** a factorization of $P(a, b, c)$? **Check all that apply**

(a) $P(a)P(b|a)P(c|a, b)$
(b) $P(a)P(b|a)P(b|a, b)$
(c) $P(c)P(b|c)P(a|c, b)$
(d) $P(b)P(a|b)P(c|a, b)$
(e) $P(c)P(c|a)P(c|a, b)$

4.9 Given the causal model depicted in the figure below, write the system-level probability equation capturing the conditional relationships of each variable in the model. [Hint: use abbreviations for the variables.]

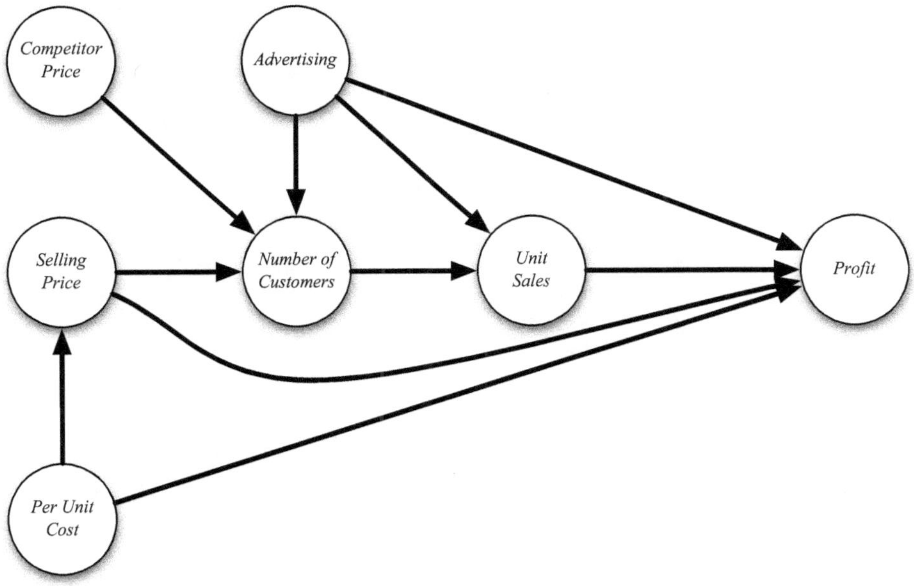

4.10 A laptop's price increases with faster processor speed and a lower weight, but collecting a list of laptop *Prices* (*low*, *high*) for various units by *Speed* (*slow*, *fast*) and *Weight* (*light*, *heavy*) doesn't immediately reveal the conditional relationships.

Speed	*Weight*	*Price*	Probability
slow	*light*	*low*	0.12
slow	*light*	*high*	0.08
slow	*heavy*	*low*	0.20
slow	*heavy*	*high*	0.13
fast	*light*	*low*	0.05
fast	*light*	*high*	0.25
fast	*heavy*	*low*	0.05
fast	*heavy*	*high*	0.12

Given the system-level probability table above for these variables, answer the following questions:

(a) What is the probability that a laptop is inexpensive, given that it is heavy?
(b) What is the probability that a laptop is fast, given that it is expensive?
(c) What is the probability that a fast and light laptop is also cheap?

Chapter 5

Situational Analysis

One of the primary uses of quantitative causal models is as "information oracles" to support decision making. Once a situation is described to a manager's satisfaction in terms of a quantitative causal model, new information about the true states of a particular set of variables can be used to make inferences about the unknown states of other variables.

A classic example is the quality control (QC) manager of any manufacturing facility, such as the semiconductor wafer and chip example from the last chapter. The QC manager's job is to diagnose quality problems that arise during the production process. Wafer fabrication is a sequential process: planarization, cleaning, etching, diffusion, chemical coating, lithography, oxidation, implantation, sputtering, grinding, polishing, vapor deposition, testing, and spinning. Quality problems can arise at each step. Suppose semiconductor chips can only be tested at the end of the entire process. If chips fail the test, how should a QC manager decide which steps in the process are most likely causing the problem?

Another common example in business is the use of focus groups in marketing. A marketing manager may wish to estimate the size of the consumer segment for a new product. Rather than launch a new product in a test market (which is expensive and may lead to bad publicity in the event of failure), firms often hire market research firms to conduct focus groups. Essentially, a small, representative sample of the consumer population is brought in to examine the new product and their responses are analyzed in detail. Given the marketing manager's prior information about the target market and evidence of product demand as provided by the focus group, how should the manager update his beliefs?

In strategy, one way to neutralize a competitor's performance advantage is to imitate the sources of that advantage. Suppose a firm's advantage is due to its informal organization – that is, the actual flow of productive activities across functional areas (e.g., the conception-to-market process involving R&D,

manufacturing and marketing) versus the reporting relationships written down in the official organizational chart.[1] This source of competitive advantage may be difficult to imitate precisely because it is informal and, hence, difficult to discern from outside the organization. Just the same, even though a competitor's informal organization is invisible, various activities arising though that organization may be observed (e.g., customer service offerings, distribution channel features, specific dimensions of the marketing campaign, alliance partners, patent history, and so on). With a causal model, such evidence might be used to infer the hidden organizational structure and, hence, to imitate the advantage.[2]

In this chapter, we use quantitative causal models to examine basic updating of beliefs regarding states of unknown variables given evidence on the states of other variables. We elaborate a general updating procedure for each of the basic 3-node structures already discussed in section 2.3: serial, diverging, and converging. Our objective here is to build your intuitions regarding the Bayesian updating technique so you understand changes in beliefs across a causal structure as well as to illustrate the benefits of using causal models for this purpose.

In the last chapter, we described how to calculate marginal probabilities and update beliefs about unknown variable states based upon new information about the states of other variables using a known system-level joint probability distribution (JPD). We also explained how to derive a JPD from the conditional probability tables in a causal model. In this chapter, we explain how to calculate marginal probabilities and update them in response to new information directly from the causal model (i.e., without having to calculate the entire JPD).

We restrict our attention to the basic 3-node components of causal models: serial (causal chain), divergent (multiple effects), and convergent (multiple causes). Our goals are to illustrate how Bayes' Rule works in this context and to build intuition about evidence-based inference under causality. The procedures described below will, without much adjustment, cover a wide range of causal structures, including chains, trees and polytrees. That being said, calculations can become fairly involved in the general case. The good news is that, at of the time of this writing, there are already several excellent software programs available (both open- and closed-source) designed to automate the process. Our purpose is to equip readers with a substantive understanding of the quantitative concepts underlying these kinds of analyses, not train them to hand-substitute available software.

1 Southwest Airlines has been an oft-cited example of a firm with such an advantage. See, e.g., Porter (2002).

2 This possibility is analyzed in Ryall (2009).

5.1 MARGINAL FROM CONDITIONAL PROBABILITIES

Recall that the marginal probability is the unconditional likelihood that a variable is in a particular state. Looking at the root node in Figure 5.3 it is easy to see that the probability that *Process* = *clean* is 90%. Not as obvious from the diagram is the probability that the quality of the *Batch* = *dirty*. Still, the information contained within a quantitative causal model does allow us to determine the initial marginal probabilities for all variables in the model – without computing the entire JPD.

For each of the three basic structures, the determination of marginal probabilities is straightforward. If a node is a root node, then its marginal probabilities are simply the probability values in its table – these are already the probabilities for each state of that variable. **For a variable that is influenced by other variables, the marginal probability of one of its states is the linear combination of the probabilities in that state's column with the influencing marginal probabilities of the rows.** The linear combination is computed by multiplying the conditional probabilities in the column corresponding to that state by the marginal probabilities of each row as determined by the influencing variable, and then summing the results down the column. For example, if a node has one parent, then the rows of its conditional probability table correspond to the columns of the parent. In this case, the marginal probability of a row is equal to the marginal probability of the parent state that corresponds to that row.

5.1.1 Serial Connections

That may sound complicated, but it is rather simple when you see it in practice. In Figure 5.1, the marginal probabilities of each variable state are shown at the bottom of the corresponding column of their table (gray boxes). For example, we see that our relative certainty that a *Batch* is *good* is 92.5% – meaning that, on average, a batch of semiconductors is good 92.5% of the time. Similarly, the process is clean, on average, 90% of the time.

Those of you with linear algebra experience will recognize that linear combination of probabilities as matrix multiplication. That marginal probability table for *Process* can also be written as the vector [0.90 0.10][3] or exactly one column ($N \times 1$ Matrix). These are sometimes referred to as a "row vector" or "column vector", respectively. The conditional probability table for *Batch* can likewise be written as the 2×2 matrix

$$\begin{bmatrix} 0.95 & 0.50 \\ 0.70 & 0.30 \end{bmatrix}.$$

3 A vector for our purposes is just a matrix with either exactly one row ($1 \times M$ matrix) or exactly one column ($N \times 1$ matrix).

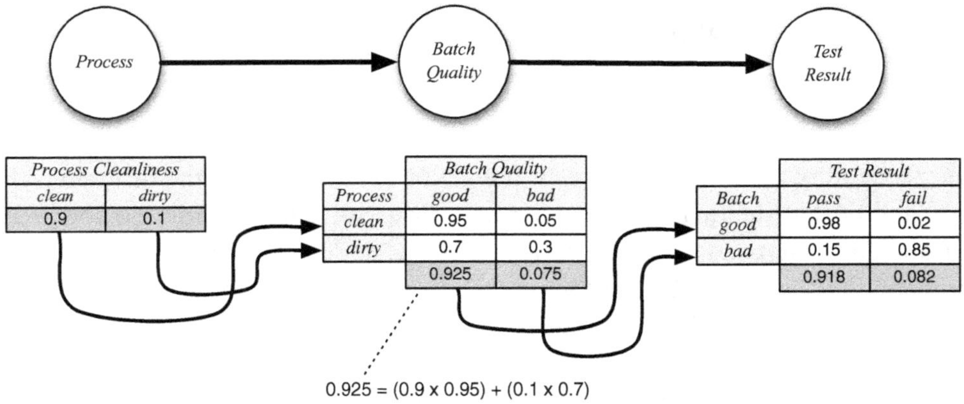

FIGURE 5.1 Marginal probabilities for serial semiconductor example.

The general pattern for multiplying a 1 by 2 vector with a 2 by 2 matrix is:

$$\begin{bmatrix} x & y \end{bmatrix} \begin{bmatrix} a & b \\ c & d \end{bmatrix} = \left[(x \times a) + (y \times c) \quad (x \times b) + (y \times d) \right].$$

Using this, the above calculation of marginal probabilities for the states of *Batch* can be written as the matrix multiplication

$$[0.90 \;\; 0.10] \begin{bmatrix} 0.95 & 0.50 \\ 0.70 & 0.30 \end{bmatrix} =$$

$$[(0.90 \times 0.95) + (0.10 \times 0.70) \;\; (0.90 \times 0.50) + (0.10 \times 0.30)] = [0.925 \;\; 0.075].$$

To determine the marginal probabilities for *Test*, you just need to follow the same multiplication procedure, this time using the marginal probabilities calculated for *Batch* as the vector, and the conditional probabilities of *Test* given *Batch* as the matrix. The result of this calculation is also shown in Figure 5.1. What this means for you is that the order in which you perform these operations matters. Again, start with the root nodes, and multiply down the causal chain until you have all the marginal probabilities.

5.1.2 Diverging Connections

The calculation for a diverging connection follows exactly the same procedure as the calculation for serial. In a divergent triple there is one root node and two children. So the marginal probabilities of the parent's states are multiplied by the conditional probability tables of **both** child nodes. It works just like the marginalization of *Batch* from *Process* in Figure 5.1 for both (or really all) child nodes in a divergent connection.

5.1.3 Converging Connections

The last type of basic structure is a converging connection, one in which there exists a node with multiple parents. Because there are multiple parents, the rows of the conditional probability tables must capture every possible combination of input states to that node – the probabilities of the child node's states are conditional on both parents' states.

The number of rows you need for a convergent connection depends on the number of variables and the number of states for each variable. Let s_1 be the number of states of some variable V_1, s_2 is the number of states for variable V_2, and so on to variable V_n. The number of rows you need is $s_1 \times s_2 \times \ldots \times s_n$. For example, if each variable has two states, and you have six variables, then you need $2 \times 2 \times 2 \times 2 \times 2 \times 2 = 2^6 = 64$ rows. The order in which you list them doesn't matter, but be consistent.

Tip: Large Tables Here is a tip to help you organize these tables as they become larger. For the first variable, go through each of the states with one row each until you fill up all the rows you calculated you need. Then for the second variable, put the first state next to one of each state of the first variable, the second state next to each of the states of the first, and repeat until you fill the table. And repeat this process until you have covered all the variables and completely filled in the possible value combinations of all incoming variables. This is exactly the same combinatorial exercise necessary for the large system-level joint probability tables in section 4.3.4, and hence you can see a result of this kind of clean ordering in Figure 4.6

We start with a simple converging triplet, so there are exactly two parent variables, and each one has two possible states. The conditional probability table will therefore need four rows. Figure 5.2 shows a sample model with this structure. Both parent nodes simply have marginal probabilities; i.e., the probabilities for each state happening absent any other information. To determine the marginal probabilities for *Batch*, we need to combine the probabilities of all the cases of the other variables in which *Batch* = *good*. This is another linear combination, but a slightly more complicated one.

Remember the factorization $P(B,P,E) = P(B|P,E) \times P(P) \times P(E)$ (don't forget that because P and E are independent, $P(P|E) = P(P)$). This tells us that to determine the joint probability of the event {*Batch* = *good*, *Process* = *clean*, *Error* = *false*} we need to calculate:

$$P(good|clean,false) \times P(clean) \times P(false).$$

The marginal probability of *Batch* = *good* is the sum of the joint probabilities of all the cases in which *Batch* = *good*, so you just perform this same multiplication for

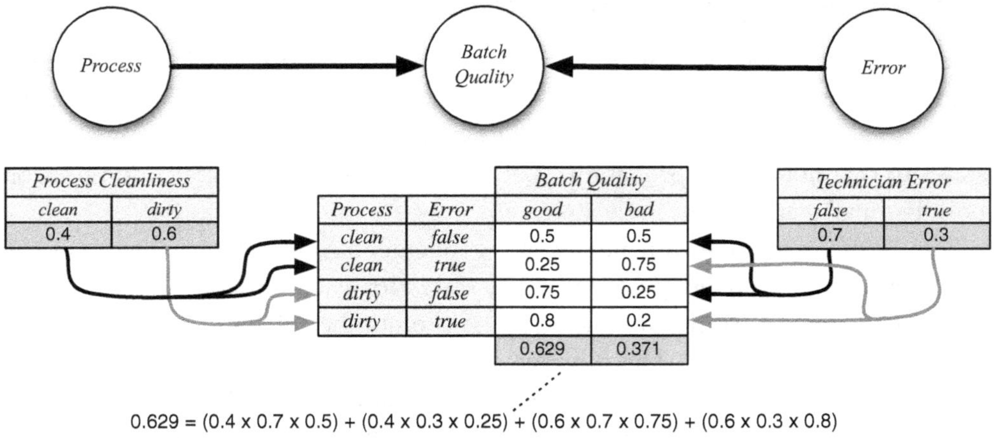

FIGURE 5.2 Marginal probabilities for convergent semiconductor example.

all combinations of states of *Process* and *Error* – of which there are four. You can see this calculation, and the result, in Figure 5.2.

5.2 EVIDENCE & INFERENCE IN CAUSAL MODELS

Marginalization is the process by which you ascertain the probabilities of specific states in the absence of new information or intervention. We now describe how to update the certainties of the states of a variable based upon new knowledge about the states of other variables. Probabilities calculated from the unadjusted values in a causal model are referred to as *prior beliefs* (e.g., according to our running example, your belief that the *Batch* quality = *good* is 92.5% ...prior to receiving any information beyond that encoded by the model). If you observe the actual states of some subset of the variables in the model, then you must update your prior beliefs for the remaining, as yet unobserved, variables. Beliefs that have been updated to reflect new information are referred to as *posterior beliefs*. When the actual state of a variable is observed, we say you have obtained *evidence* with respect to it. Inferring posteriors from priors in response to evidence is called *evidence-based reasoning*.

As we proceed, think of each node in a causal model as an independent information processor. When a node's true state is observed, it updates its own marginals and "sends messages" up and down the causal chain. Upstream and downstream variables receive the message, update their own marginals, and send new messages to the next variables in the chain. Keep in mind that *c-independence implies probabilistic independence*.[4] **Thus, observing the state of one variable that is c-independent of another variable results in no change in certainty regarding the state of the latter.** This is a labor-saving feature of causal models.

4 Again, one can construct exceptions to this rule, but we will not encounter any in this book.

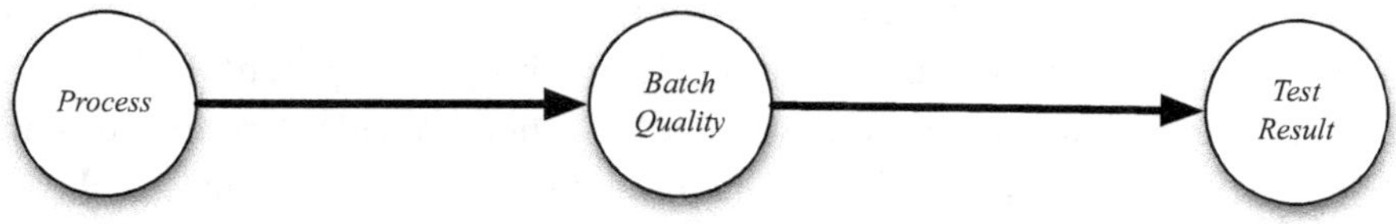

Process Cleanliness	
clean	*dirty*
0.9	0.1

	Batch Quality	
Process	*good*	*bad*
clean	0.95	0.05
dirty	0.7	0.3

	Test Result	
Batch	*pass*	*fail*
good	0.98	0.02
bad	0.15	0.85

FIGURE 5.3 Semiconductor example of a serial connection.

As mentioned above, think of each node as a self-contained information processor: it has a current condition captured with its prior probabilities, it takes input from other nodes, it changes its condition based on its previous condition and the input, and then it sends output to other nodes. The input and output need not follow the arrow directions of the causal diagram; recall from Chapter 1 that causal *reasoning* can go in both directions up and down the causal chain. But the directions of the arrows does effect which calculation is appropriate for new information coming "upstream" or "downstream".

5.2.1 Serial Connection

To illustrate updating in a serial connection, we continue with the semiconductor model of Figure 4.7, presented again here for reference. The procedure for all evidence-based reasoning uses a few key components: (1) the original probability tables in the causal model; (2) upstream and downstream likelihood ratios; (3) upstream and downstream messages; and (4) evidence.

Just like the marginalization process above, the description of the method is a lot more cumbersome and involved than the technique itself. It's like reading a description of how to ride a bike versus actually doing it. Once you have some experience with the technique, it becomes clear – perhaps even intuitive. As we proceed, keep in mind that performing the calculations in practice is actually much simpler than the verbal instructions required to explain them!

Setup

We have created a spreadsheet in Microsoft Excel that reflects the results of all the setup instructions contained here, and the updating calculations presented next. As we describe each element of the setup, it will probably be helpful for you to refer to Figure 5.4.

1 Under the probability portions of each probability table, add a row for each of the following new elements (explained below): L^+, L^-, x, *posterior beliefs,*

and *evidence*. L^+ is used for downstream messages (from parents to children), while L^- is used for upstream messages (from children to parents). The "x" row is for an intermediate calculation; you will also need to calculate the sum of the entries in the x row. The row for posterior beliefs initially contains the prior beliefs for each state of that variable; i.e., each state's marginal probability. The evidence row (we use "e" to abbreviate evidence) is used to indicate the states of variables when they are observed. These rows are labeled in this order in Figure 5.4.

2 To begin, leave the L^+, x, and beliefs rows blank. Set all the values in the L^- and evidence rows to 1 *for all nodes*. If all entries in the evidence row are 1, then that means no evidence has been received regarding the state of this node.

3 Starting with the root node, set the L^+ (downstream message) values equal to the values in the probability table. Remember that because it is a root node, the probabilities here are not conditional on another variable, so you can think of it as a conditional probability table (CPT) with just one row. The probabilities in a root node's table are both conditional probabilities (but not conditional on anything) and marginal probabilities (because it is the probability of each state). Thinking of it this way will help you later to see the general rules for setting up these calculations.

4 Each column in a parent node's L^+ corresponds to a row in its child's conditional probability table. Multiply the root node L^+ values across the cells in the corresponding row of the child's CPT, then sum down the columns. This yields L^+ values for the child of the root node. Thus, we can think of L^+ as a message sent from the parent node (the root node in this case) that is then processed upon receipt by its child. You may also notice that this calculation is the same as marginalization, a linear combination of the parent's L^+ with the child's conditional probabilities.

For example, the first entry in the L^+ row for B in Figure 5.4 is 0.925. This is equal to $(0.9 \times 0.95) + (0.1 \times 0.7)$ (the same calculation shown in Figure 5.1). The second value in that row is $(0.9 \times 0.05) + (0.1 \times 0.3) = 0.075$. Those marginal probabilities are the L^+ of *Batch* and the downstream message from B to T (written as "M(B-to-T)" in the Excel screenshot). *Test*'s L^+ row is calculated the same way: multiply the first row of T's conditional probability table by the first item of the message, and its second row by the second item (and you would continue for every item if there were more than two).

5 The numbers we calculated for the L^+ rows were indeed the marginal probabilities, but that is only true for the setup; when we show you how to update beliefs to account for new evidence the L^+ row will capture different information. But we will still want to know the actual marginal probabilities. To get the marginal probabilities for each node, start by multiplying each column's L^+ and L^- values and writing the product in the x row. Then sum across the

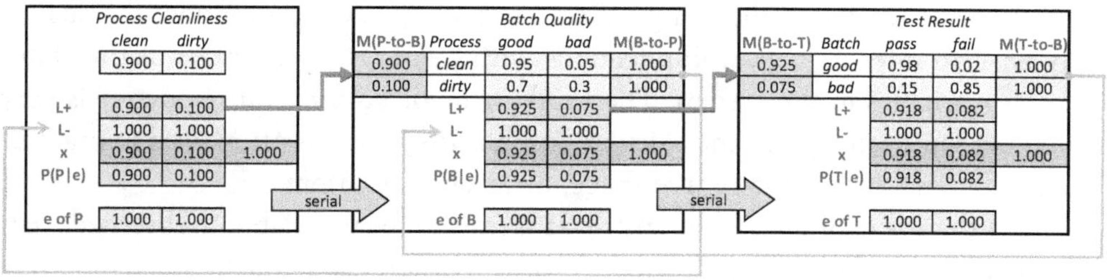

FIGURE 5.4 Initial setup, serial connection.

columns and record that $\sum x$ number to the right of the x row (as seen in Figure 5.4).

As you already know, probabilities must sum to 1, but when updating probabilities based on evidence we are dealing with likelihood values that aren't exactly probabilities. To convert them back into probabilities, we need to normalize them by calculating **likelihood ratios**. The way to do this is easy: divide each column's x value by the sum of the x values. The result is that node's marginal probabilities (the $P(P|e)$, $P(B|e)$, and $P(T|e)$ rows in Figure 5.4). Without any evidence the L^+, x, and marginal probability values will be identical.

Updating

In a causal model *evidence* refers to new information about the true state of a variable (information that arises after the initial model has been constructed). When evidence arises, we must update our beliefs in conformance with Bayes' Rule – i.e., compute conditional probabilities using the observed value as the conditioning variable. For example, having observed *Process* = *clean*, how should we update our belief that *Test* = *pass*? That is, we wish to know $P(Test = pass|Process = clean)$.

If we happen to have the joint probability table on hand, we can always go ahead and do the arithmetic to get $P(Test = pass|Process = clean)$ or $P(Process = clean|Test = pass)$ or anything else. Once again, however, this approach becomes prohibitively cumbersome when dealing with large causal systems. Instead, we can use the structure of the causal model to do the updating via local, node-level calculations, thereby saving a lot of computation.

Note: Evidence can be "hard" in the sense that the actual value of a variable is observed. Evidence can also be "soft" in the sense that it instead provides information requiring you to update your certainties about the states of some variables (that is, you still don't *know* the actual value, but you have refined probabilities over the states). In this book, we will only consider evidence in the form of actual observations; i.e., in which the true states of some variable or variables are revealed to the analyst. Based upon this new information, the analyst may wish to infer its logical implications with respect to certainties of the states of other variables in the system. The thinking involved for "soft" evidence is very much the same, and so are the calculations, but they are less often used in practice and so we omit them to conserve space, time, and attention.

When the value of a variable upstream from the focal variable is observed, that information must be "**passed down the chain**." This procedure is the same as setting up the initial marginals:

1 When the true state of a variable is revealed, set the evidence row (bottom row in Figure 5.4) for that node equal to 1 in the column corresponding to the true state and 0 in the other columns.
2 Whenever there is evidence for the actual state of a node, both the L^- and L^+ rows of that node are set equal to the evidence row.
3 The L^+ row is the message sent down to the node's child. As in the initial setup, the L^+ of the receiving node (not the one with evidence, the one directly downstream) becomes the linear combination of the incoming message and its conditional probabilities. Each element of the message is multiplied by the elements in the corresponding rows of the child's conditional probability table and summed down the columns. These sums become the new L^+ row of the child.
4 With L^+ so updated, the marginal probability of the downstream node is recalculated according to the likelihood ratio approach presented above. For each column i, find $x_i = L_i^+ \times L_i^-$, and then the marginal probability for column i is $x_i / \sum x$. If the downstream message is the only message, then the x and marginal probability values will still be the same as the L^+ values.
5 The updated L^+ for this node is then used as the new message to send downstream to the next child – and this whole process is repeated there.

When the value of a variable downstream from our variable of interest is observed, that information must be "**passed up the chain**". We are observing the state of the effect variable, and using it to reassess the certainties of the states of one of its direct or indirect causes (e.g. using the observation of the *Test* to update our certainty about the *Batch* quality or *Process* cleanliness). The procedure is as follows:

1 As above, when the true state of a variable is revealed, set the evidence (bottom row marked with e) row for that node equal to 1 in the column corresponding to the true state and 0 everywhere else. Both the L^+ and L^- rows for this variable are changed to match this evidence. These new values are used to recompute this variable's marginal probability which, of course, must now match the evidence we received.

2 Next, the L^- row of the updated node is used to compute a message for the parent node. The message is a linear combination of the values in the L^- row (which will be zeros and a one) with the node's conditional probabilities. That is, each column of L^- is multiplied by the corresponding column value from the conditional probability table, and then the rows are summed across. Because we are using only "hard" evidence, the message contains a list of the conditional probabilities of this node from the column matching the known state of this node. The parent node replaces its L^- with this message.
3 With the parent's L^- so updated, the marginal probability of the parent is recalculated using likelihood ratios. Just as above, for each column i, find $x_i = L_i^+ \times L_i^-$, and then the marginal probability for column i is $x_i / \sum x$.
4 Continuing the procedure, the updated L^- row for the parent is then used to compute a message for **its parent** using the same linear combination approach from step 2. But this time L^- will probably **not** be zeros and a one. Assume L^- is the list $\begin{bmatrix} x & y \end{bmatrix}$, and the conditional probabilities of this node are $\begin{bmatrix} a & b \\ c & d \end{bmatrix}$, then the message to its parent node is the list $\begin{bmatrix} (x \times a) + (y \times b) & (x \times c) + (y \times d) \end{bmatrix}$. Note that this is not matrix multiplication (as we saw in marginalization). That message becomes the new L^- of this nodes parent, and the process continues up the chain.

Example 5.3: Evidence Flowing Upstream

Passing information upstream is a bit more complicated (or at least less familiar) than the downstream calculations, so we now provide a walk through of the calculations using our running semiconductor processor example.

1 Assume that we observe the result of the test, and *Test* = *pass*. Because we are now certain that *Test* = *pass*, we enter a 1 into the corresponding bottom *e* cell in Figure 5.5 (setting *fail* = 0). Both the L^+ and L^- rows for *Test* are changed to be identical to this evidence: $\begin{bmatrix} 1.0 & 0.0 \end{bmatrix}$. *Test*'s new marginal probabilities must now put a full weight of 1 on *Test* = *pass*, because we *know* that's the value.
2 Next, the L^- row is used to compute a message for *Batch* by multiplying the values in the L^- row times the corresponding values in each row of the conditional probability table and summing them across. Since *pass* has weight 1, the calculation is easy:

$$\begin{bmatrix} (1 \times 0.98) + (0 \times 0.02) & (1 \times 0.15) + (0 \times 0.85) \end{bmatrix} = \begin{bmatrix} 0.98 & 0.15 \end{bmatrix}$$

These row sums are the message $M(T\text{-to-}B)$ sent up the chain to *Batch*. *Batch* replaces its own L^- with $M(T\text{-to-}B)$.

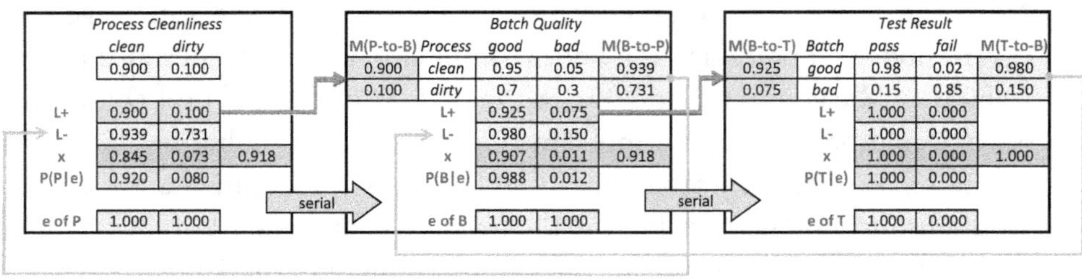

FIGURE 5.5 Updating procedure when the true state of *Test* is discovered to be *pass*.

3 With *Batch*'s L^- updated to $\begin{bmatrix}0.98 & 0.15\end{bmatrix}$ by the message, the marginal probability of *Batch* is recalculated. *Batch*'s L^+ is $\begin{bmatrix}0.925 & 0.075\end{bmatrix}$, so we set the x values to the piecewise multiplication of those lists:

$$\begin{bmatrix}0.98 \times 0.925 & 0.15 \times 0.075\end{bmatrix} = \begin{bmatrix}0.907 & 0.011\end{bmatrix}.$$

To normalize these likelihoods we find their sum: $0.907 + 0.011 = 0.918$. The new marginal probabilities for *Batch* are:

$$\begin{bmatrix}0.907/0.918 & 0.011/0.918\end{bmatrix} = \begin{bmatrix}0.988 & 0.012\end{bmatrix}.$$

As we see, knowing that *Test* = *pass* results in a revision of our certainty that *Batch* quality is *good* from 92.5% to 98.8%.

4 Continuing the procedure, the updated L^- row for *Batch* is then used to compute a message for *Process*. The details of the calculations follow the same steps as above: the message becomes *Process*'s new L^-, which is multiplied by the L^+ to get new xs, which are normalized by their sum to get the new marginal probabilities. You can see the results of each of the calculations in Figure 5.5. What we can infer through this upstream flow is that knowing *Test* = *pass* implies that our certainty in *Process* being *clean* should change from 90.0% to 92.0%.

In this model, new information regarding the *Test* is propagated all the way up the chain, changing our certainties regarding the states of both *Batch* quality and *Process* cleanliness. Now we have seen both downstream and upstream information flows. As we will see later, it is possible to have both at the same time, but note that upstream and downstream messages do not interact by altering each other. Evidence flowing upstream operates through the L^- rows, and evidence flowing downstream operates through the L^+ rows. Evidence at the root node (*Process*) only flows down the chain, and evidence at the leaf node (*Test*) only flows up the chain. Knowledge about the true state of a non-root/non-leaf node (e.g., *Batch*

	Process Cleanliness	
Batch	*clean*	*dirty*
good	0.924	0.076
bad	0.6	0.4

Batch Quality	
good	*bad*
0.925	0.075

	Test Result	
Batch	*pass*	*fail*
good	0.98	0.02
bad	0.15	0.85

FIGURE 5.6 Semiconductor example in which *Batch* is a common cause.

quality in our example) sends messages in both directions. Its updated L^+ row gets sent to its child and its updated L^- row gets sent to its parent.

5.2.2 Divergent Connection

For the following discussion, we take the now-familiar semiconductor example and make one major change. In this version, assume that *Batch* quality is purely random (i.e., whether or not the *Batch* is *good* is due to causes not included in the model) and that, when a *Batch* is *bad*, it has a greater likelihood of releasing contaminants into the manufacturing environment (i.e., *Batch* quality now causes *Process* cleanliness). *Test* results are still caused by *Batch* quality. The full model is depicted in Figure 5.6.

Setup

The initial setup of a divergent triplet is quite similar to the serial case. Here, the root node is *Batch*, and the relationship that *Process* used to have with *Batch* is now the relationship that *Batch* has to both *Process* and *Test*.

1 The values in the root node's L^+ rows are set equal to the probabilities in its table.
2 The values in all L^- rows (i.e., for all variables) are set equal to 1, just as before.
3 The initial messages from the root node to both its children are the values in its L^+ row. Multiply these message values across the cells in the corresponding row of each child's conditional probability table, then sum down the columns (same calculation as marginalization). This yields L^+ values for both children of the root node, exactly as in the serial case, except now there are two children instead of one.
4 For each child node, multiply each column value in the L^+ row by the corresponding column in the L^- row to get the x values, and then normalize (dividing by the sum across the x row). This completes the procedure to obtain the initial marginal beliefs.

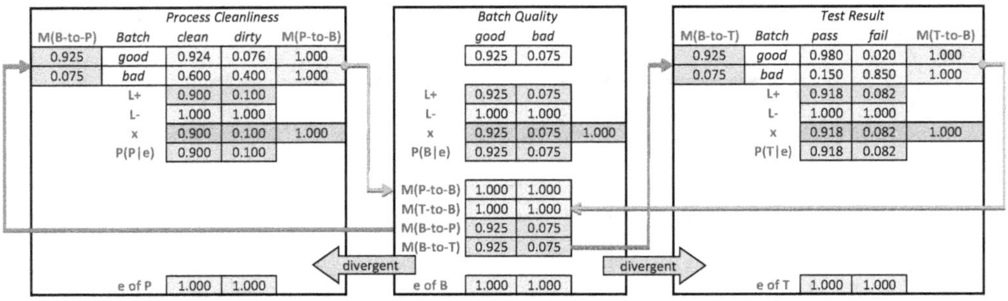

FIGURE 5.7 Initial setup for an example divergent connection.

What is different about this case is that the common cause node (*Batch*) must be able to **synthesize information** from, and to, both of its children. This requires us to be a bit more careful about which messages get sent to and received from which child. The initial setup is shown in Figure 5.7. We have made room at the bottom of the table for *Batch* to keep track of separate messages to and from each child.

Updating

Let us first consider what happens when we observe the state of the common cause (*Batch* in our example).

1 First, place a 1 in the column of the evidence row matching the state that was observed, and 0s in all the other columns of that row.
2 Both the L^- and L^+ rows are set equal to the evidence row.
3 The L^+ row is sent as the message to both children.
4 Both children take a linear combination of the message and their conditional probability tables to calculate their new L^+ values.
5 Recompute marginal probabilities for both children using the same normalized likelihood ratio approach as before and the process is complete.

That is so simple (basically it is just marginalization again) that we don't bother showing a diagram for it. The trickier part is what happens when we observe the state of a child. The procedure is as follows (see Figure 5.8):

1 For the observed child, a 1 is placed in the appropriate column of its evidence row (for the observed state) and 0s elsewhere. Both the L^- and L^+ rows of this node are set equal to the evidence row.
2 A new upstream message is computed for the common cause by multiplying the L^- row by each row of the conditional probability table and summing the rows across. Because we are only considering "hard" evidence, the message will contain the probabilities in the column corresponding to the observed state. In our semiconductor example, the message from *Test* to *Batch* is $\begin{bmatrix} 0.980 & 0.150 \end{bmatrix}$.

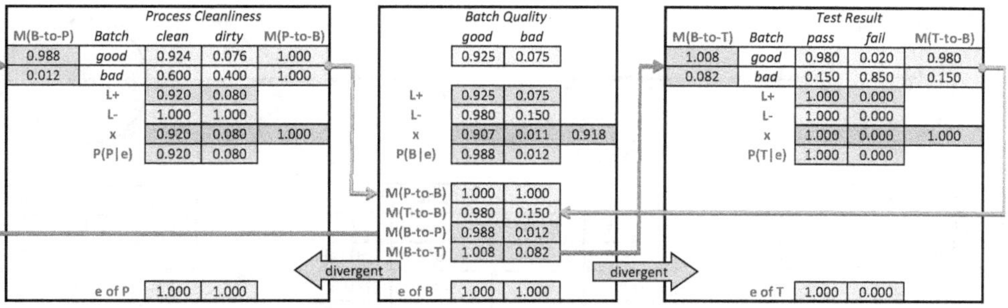

FIGURE 5.8 Updating a divergent connection upon discovery that *Test* = *pass*.

3 At the common cause node, L^- is computed as **the product of the messages from its children**. In this case, there is no change in *Process* and all L^- values are initially 1. As a result *Batch*'s L^- row will be equal to the message from *Test* (because you are multiplying the entries by 1).

4 For the common cause node, multiplying L^- by L^+ and normalizing to get likelihood rations yields the new marginal probabilities over the states of *Batch* (row marked as $P(B|e)$).

5 Now, **messages are sent from the common cause to both children**. The outgoing messages from the common cause are based upon the updated marginals (row marked as $P(B|e)$). For each outgoing message, divide the updated marginal probabilities by that node's incoming message to determine the outgoing message. In this case the incoming message from *Process* is all 1s. The message to the child with the known state (*Test* in this case) is ignored by that child because its state is already known with certainty.

6 The incoming message to the child having an unknown state is multiplied through the conditional probability table in the usual way to get that child's new L^+ row. As per usual, multiply that new L^+ by L^- and normalize to obtain the updated certainties on the child (the row marked $P(P|e)$ in our example).

5.2.3 Convergent Connection

Finally, consider the model of semiconductor quality shown in Figure 5.9 that relates the quality of a batch to both the cleanliness of the production process and the errors of technicians. In this case, we have multiple causes of a common effect. The two causes (*Process* cleanliness and technician *Error*) are **both** root nodes with simple probability tables, and the third node (*Batch* quality) is conditional on the states of those variables.

Setup

The initial setup is easy in the sense that two of the variables require no fuss – the root node marginals are simply equal to the table values. The slight complication

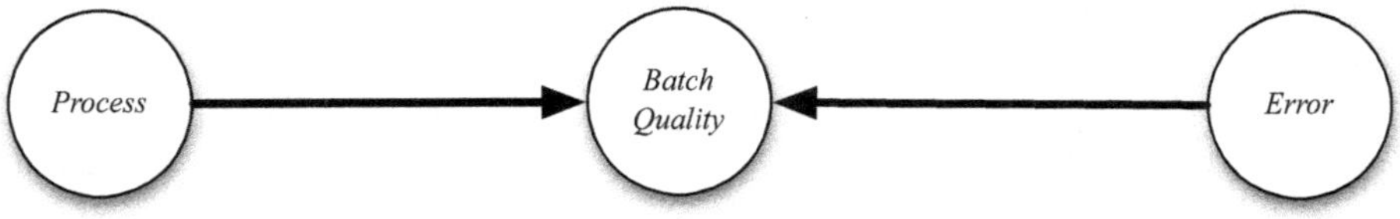

Process Cleanliness	
clean	*dirty*
0.8	0.2

		Batch Quality	
Process	*Error*	*good*	*bad*
clean	*false*	0.95	0.05
clean	*true*	0.8	0.2
dirty	*false*	0.85	0.15
dirty	*true*	0.7	0.3

Technician Error	
false	*true*
0.9	0.1

FIGURE 5.9 Semiconductor example of a convergent connection.

in the setup is that multiple parents result in a conditional probability table at the child enumerating **all possible combinations** of parent states (at *Batch*). Though our example with two input variables having two states each is very manageable, at this point we refer you to our tip for building large tables in section 5.1.3 because it will certainly come up as you use this technique later.

We showed you how to get marginal probabilities for a convergent connection in section 5.1.3, and we'll use the same calculation to find L^+ here.

1 The new L^+ for the common effect is the linear combination of the probabilities from both parents of the corresponding states and each state of the child node. For example, the conditional probability values in the row for *Process* = *clean* and *Error* = *no* are multiplied by the product *P*(*Process* = *clean*) × *P*(*Error* = *no*) (or, $0.8 \times 0.9 = 0.72$). Do this for every row, and sum down the columns to derive L^+ for *Batch*.
2 As before, L^- is set to all 1s.
3 Multiply L^+ by L^- to get an x value for each column and normalize them by the sum of the xs to get the marginal probabilities for the common effect (row marked as $P(B|e)$).

The initial setup is shown in Figure 5.10.

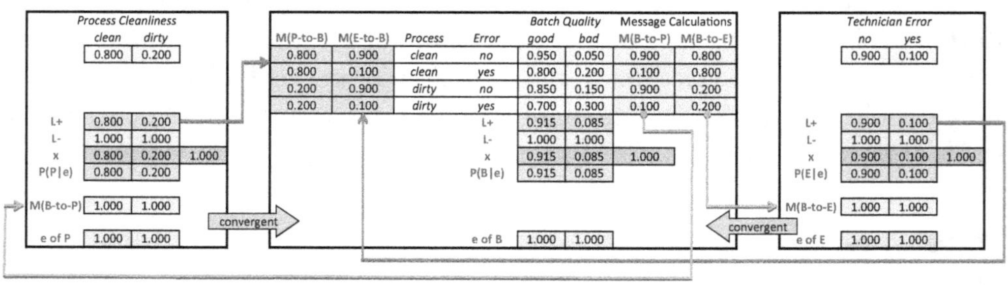

FIGURE 5.10 Initial setup of the semiconductor quality example with multiple, converging causes.

Updating

If one or both of the causes is observed, then updating from two causes to the common effect is easy. The L^+ messages are sent, the rows of the child's conditional probability table are multiplied by the updated probabilities in the same way as in the setup, a new L^+ row is calculated from the column sums, and the marginal probabilities are computed by multiplying L^+ by L^- and normalizing. There is nothing unusual except that **the rows of the child's table are multiplied by the product of the parent probabilities.**

Making inferences from effect to its causes is slightly more involved (though still intuitive as you will see), and so we will demonstrate the technique using our running example. The reason for the complication is tied to c-independence as explained in section 2.3.3: knowing the true state of the mutual child allows information to pass from one parent to the other. Thus, the message sent up to each parent must carry the information sent down by the *other* parent. Suppose we discover that a batch of semiconductors is bad (i.e., *Batch* = *bad*). What are our updated certainties with respect to the states of *Process* cleanliness and technician *Error*?

1 In the evidence row, set the value in the column for *Batch* = *bad* to 1 and the column for *Batch* = *good* to 0. Set the L^+ and L^- rows for *Batch* equal to this row.

2 Generate messages for both parents using the L^- (because the messages will both flow upstream) **and** the message received by the other parent. To calculate the message, you will need an entry for each row of *Batch*'s CPT. To demonstrate, for the message from *Batch* to *Process*, the first element corresponds to the case when *Process* = *clean* and *Error* = *no* with probabilities $\begin{bmatrix} 0.95 & 0.05 \end{bmatrix}$. Each of these is multiplied by the corresponding L^- entry $\begin{bmatrix} 0 & 1 \end{bmatrix}$ *and that whole row is also multiplied by the corresponding element of the message from Error* (in this case 0.90).

   So for the first row, take the product $(0.95 \times 0 \times 0.9) + (0.05 \times 1 \times 0.9) = 0.045$. For the next row (*Process* = *clean*, *Error* = *yes*), take the product $(0.8 \times 0 \times 0.1) + (0.2 \times 1 \times 0.9) = 0.02$. This must be done for every row to complete the first part of the message calculation. The next step for a message from *Batch* to *Process* is to add the values for rows representing the same state of *Process* and use these as the message to *Process*. For example, the message entry for *Process* = *clean* is calculated as: $0.045 + 0.020 = 0.065$. Repeat this whole process to calculate a message for *Error*.

3 The messages thus calculated are sent up to the parents and used as the parents' L^- rows. These are multiplied by their L^+ rows and normalized to get the updated marginal probabilities.

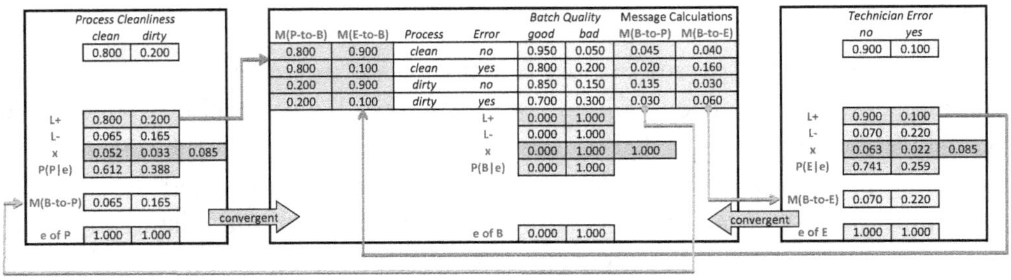

FIGURE 5.11 Updating beliefs when *Batch* = *bad*.

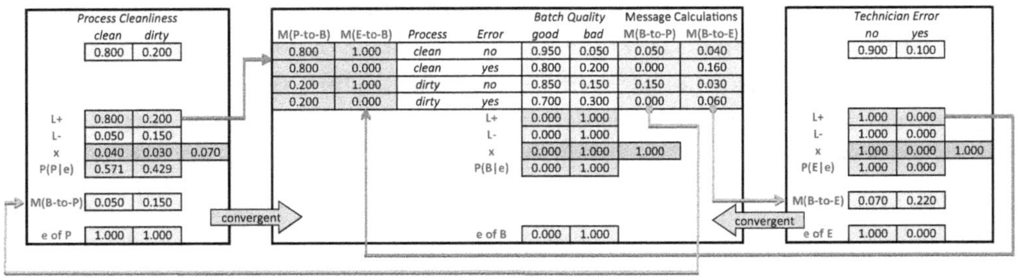

FIGURE 5.12 Updating beliefs when *Batch* = *bad* **and** *Error* = *no*.

The calculations for the case in which *Batch* = *bad* are shown in Figure 5.11. As we would expect, discovering a bad batch of semiconductors substantially increases our certainty that either there was a technician error or a dirty processing facility. For example, our certainty that there was technician error goes from 10% before the evidence of a bad batch to 25.9% after.

Suppose we discover further that the problem was definitely not technician error. That is, assume we know both *Batch* = *bad* **and** *Error* = *no*. Knowledge that *Error* = *no* triggers a downstream message to *Batch* as before. The message from *Batch* to *Process* is computed in the same way as just described. However, notice that because the *Error* message components get multiplied in our calculation of the message from *Batch* to *Process*, our beliefs about the state of *Process* are affected by our knowledge of *Error*. The calculations for this case are shown in Figure 5.12. Notice that our certainty that the facility was not clean increases even further.

KEY CONCEPTS: CHAPTER 5

- In a causal model, evidence-based inference can be conducted using local computations; i.e. calculations at each node based only on its inputs and outputs.
- If interventions are not involved, then one can infer the state of a variable given knowledge of another variable (or variables) by using a conditional probability; e.g.,

$P(x=0|y=1)$ is the appropriate measure of certainty that $x=0$ when it is revealed that $y=1$.

- Rather than compiling the whole joint probability table, it is possible to compute any marginal probability from the causal model's conditional probabilities.
- The marginal and conditional probabilities can then be used together in order to calculate any needed joint probability.
- Different structures require slightly different calculations for inference upstream and downstream. The exact algorithms for each type of structure were explained in detail.
- Evidence regarding the true state of a variable affects the marginal probabilities of the states of other variables, and an intervention to set a variable to a particular state counts as "evidence" for inference purposes.
- For each of the three basic node structures we elaborated the setup and updating procedures in detail. Though these calculations quickly become burdensome to do by hand for large models, it is important that you be clear on the mechanics of belief revision in Bayesian networks if you are planning to use the technique.

CHAPTER 5 REVIEW QUESTIONS

5.1 The upper management of GoodyGoods Inc., a large retail goods firm, has hired you as a consultant to help them improve their workflow and (hopefully) their profits. The company allocates operational tasks to the appropriate semi-autonomous divisions. The Manufacturing division is responsible for receiving *Estimated Sales* figures from the Research division and producing an appropriate number of *Units* to sell. To simplify things, let's assume that the sales forecast can be either *low* or *high* and the number of *Units* will be *few*, *moderate*, or *many*. The number of *Units* determines the *Profit* level, which can be *low*, *medium*, or *high*. The causal structure, as well as the conditional probability tables for each variable, are presented below.

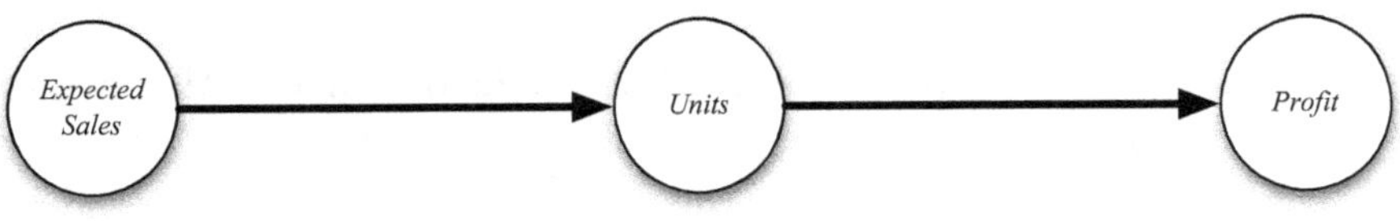

Expected Sales	
low	*high*
0.4	0.6

	Units		
Expected Sales	*few*	*moderate*	*many*
low	0.35	0.6	0.05
high	0.05	0.45	0.5

	Profit		
Units	*low*	*medium*	*high*
few	0.45	0.45	0.1
moderate	0.15	0.5	0.35
many	0.1	0.3	0.6

Use this causal model to answer the following questions:

(a) What is the marginal probability that *few Units* will be produced?
(b) What is the marginal probability that *Profit* is *high*?

(c) What is the probability of each profit level, given that you receive information that the *Estimated Sales* report said *low*?

(d) The Production manager later tells you that they produced *many Units*; what is your updated estimate that the *Profit* is *high*?

(e) After you receive this unit quantity information from the Production manager, you then find out that the *Estimated Sales* were actually *high*. How does this new information affect your assessment of *Profit* level? Why or why not?

5.2 Continuing your consulting for GoodyGoods Inc., they tell you that another component of their process is that they must store the units they produce in a warehouse. Because there is a long lead time in leasing warehouse space, the Facilities management division must reserve a block of *Warehouse* space using the *Estimated Sales* figures from the Research division (rather than based on the number of units actually produced). As before, the *Estimated Sales* is either *low* or *high*, and the *Warehouse* can be either *small* or *large*. The number of *Units* is also still as described above (*few*, *moderate*, or *many*), although in this step we are not considering the effect on profit.

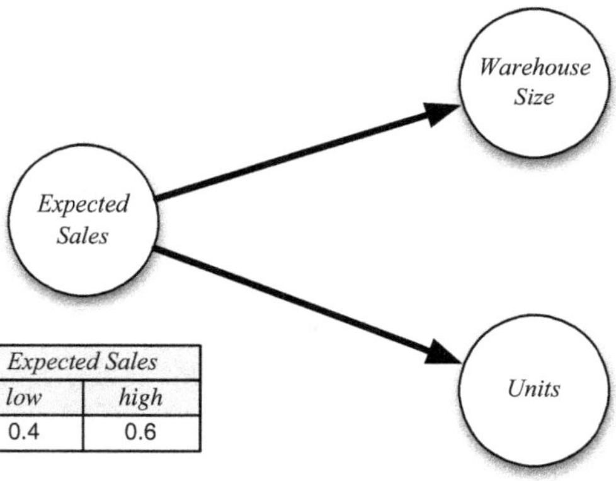

	Warehouse Size	
Expected Sales	*small*	*large*
low	0.8	0.2
high	0.15	0.85

Expected Sales	
low	*high*
0.4	0.6

	Units		
Expected Sales	*few*	*moderate*	*many*
low	0.35	0.6	0.05
high	0.05	0.45	0.5

Use this causal model to answer the following questions:

(a) Given the conditional probabilities provided, what is the marginal probability that a *large Warehouse* has been leased?

(b) You are trying to determine the number of *Units* being produced, but the Production manager hasn't returned your call or e-mail. The Facilities manager does respond to your queries and tells you that they reserved a *large Warehouse*. Given this information, what is the probability that *many Units* are being produced?

(c) Let's say that when the Facilities manager gave you the above information, you already knew that the *Estimated Sales* report projected a *high* level. What is the probability that *many Units* are being produced under this scenario?

5.3 It turns out that the number of units being produced and sold does not alone determine profit, it only determines gross revenue (as I'm sure you've already noted). Producing the units and leasing a warehouse have associated costs, and these costs take away from profit. Before tackling the estimate of profits, you need to better understand the determiners of *Combined Unit Costs*. The *Combined Unit Costs* can be *low*, *medium*, or *high* and the variable incorporates both fixed expenses from the warehouse and variable costs from unit production.

Warehouse Size	
small	*large*
0.41	0.59

Units		
few	*moderate*	*many*
0.17	0.51	0.32

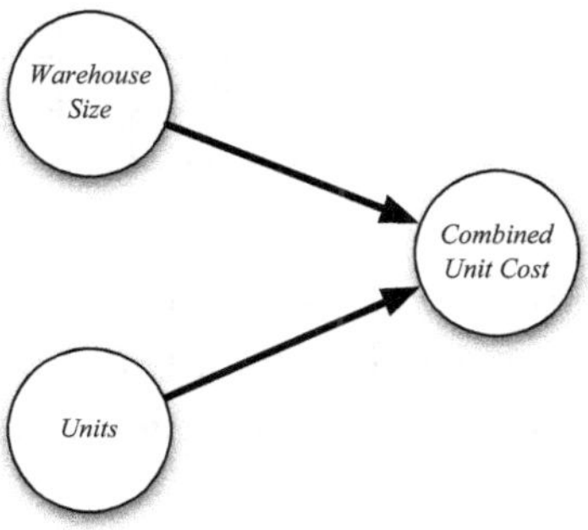

		Combined Unit Cost		
Warehouse	*Units*	*low*	*medium*	*high*
small	*low*	0.1	0.85	0.05
small	*high*	0.8	0.15	0.05
small	*low*	0.03	0.07	0.9
small	*high*	0.07	0.08	0.85
small	*low*	0.09	0.82	0.09
small	*high*	0.75	0.13	0.12

Use this causal model to answer the following questions:

(a) What is the marginal probability that the *Combined Unit Costs* are *low*?

(b) What is the probability that the *Combined Unit Costs* are *low* given information that *many Units* were produced.

(c) The accounting department tells you that the *Combined Unit Costs* came within the *medium* level. If you don't know the actual *Warehouse* size, what is your estimate that a *moderate* number of *Units* were produced?

(d) You then find out that the *Warehouse* being leased is *large*. What is your updated probability distribution over the number of *Units* produced?

5.4 The following table details the $L+$ and $L-$ values for a node capturing the *Number of Units Sold* (U) with three states (*low*, *medium*, or *high*). Calculate the probabilities for each state $P(U=i)$ implied by the $L+$ and $L-$ figures.

Number of Units Sold

	$i = none$	$i = low$	$i = medium$	$i = high$
$L+$	0.65	0.03	0.25	0.07
$L-$	0.80	0.50	0.70	0.30

5.5 Consider the causal diagram presented below and answer the following questions:

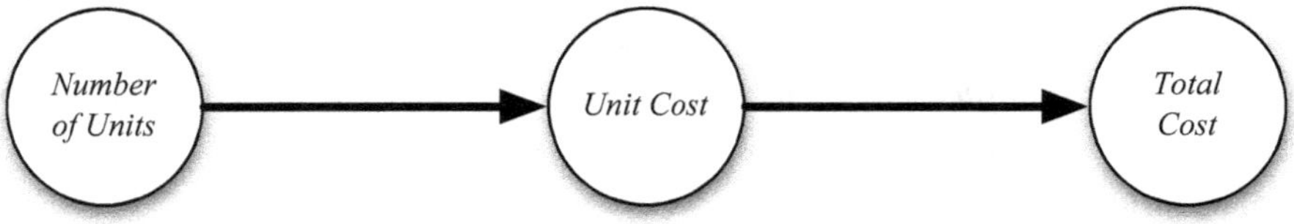

Number of Units	
low	*high*
0.25	0.75

	Unit Cost	
Number of Units	*low*	*high*
low	0.8	0.2
high	0.55	0.45

	Total Cost	
Unit Cost	*low*	*high*
low	0.6	0.4
high	0.22	0.78

(a) What are the $L+$ values for nodes *Units, Unit Cost,* and *Total Cost?*
(b) What are the $L-$ values for nodes *Units, Unit Cost,* and *Total Cost?*
(c) What are the marginal probabilities for nodes *Units, Unit Cost,* and *Total Cost?*
(d) If you discover that the true value of *Unit Cost* = *low,* what are the $L+$ and $L-$ values of nodes *Units, Unit Cost,* and *Total Cost?*
(e) When *Unit Cost* = *low,* what is the probability that *Total Cost* = *high?*

5.6 The following diagram depicts the relationship of a piece of machinery's *Energy Use* to its *Fuel Cost* and *Heat Output*. Answer the following questions based on the information contained in that causal model:

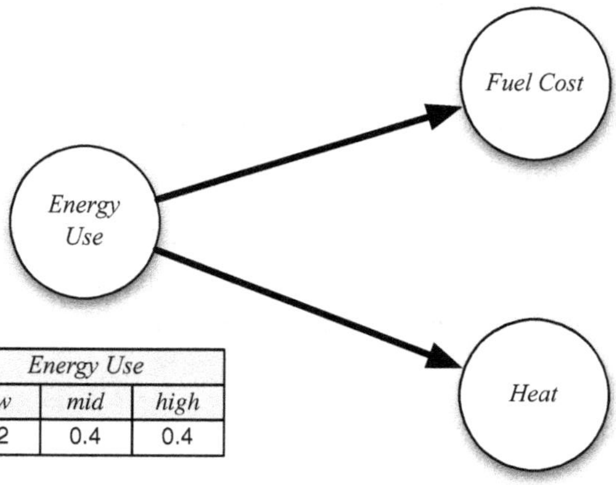

	Fuel Cost	
Energy Use	*cheap*	*expensive*
low	0.75	0.25
mid	0.5	0.5
high	0.1	0.9

Energy Use		
low	*mid*	*high*
0.2	0.4	0.4

	Heat	
Energy Use	*cold*	*hot*
low	0.85	0.15
mid	0.45	0.55
high	0.25	0.75

(a) What are the $L+$ values for nodes *Energy Use, Fuel Cost,* and *Heat Output?*

(b) What are the $L-$ values for nodes *Energy Use, Fuel Cost,* and *Heat Output*?

(c) If you discover that the true value of *Energy Use* = *mid*, what are the $L+$ and $L-$ values of nodes *Energy Use, Fuel Cost,* and *Heat Output*?

(d) When *Energy Use* = *mid*, what is the probability that *Heat Output* = *hot*?

(e) **Instead** of an observation of the *Energy Use*, assume you discover that the true value of *Fuel Cost* = *cheap*. Now what is your estimate that *Heat Output* = *hot*?

5.7 The following diagram depicts the relationship of the *Inflation Rate* on *Housing Prices* via the *Interest Rate*. Answer the following questions based on the information contained in that causal model:

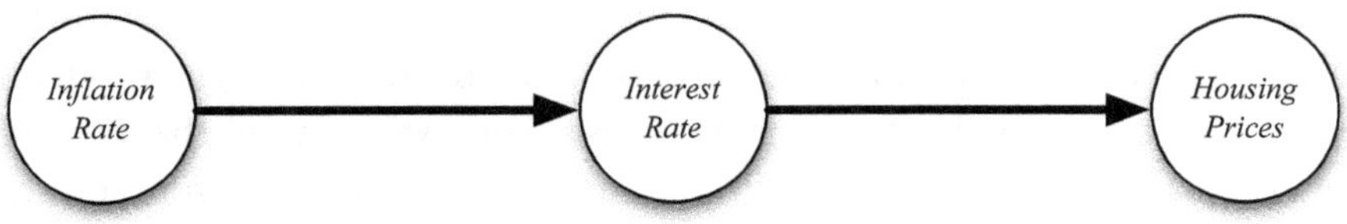

Inflation Rate		
low	*med*	*high*
0.15	0.5	0.35

	Interest Rate		
Inflation	*low*	*medium*	*high*
low	0.85	0.1	0.05
medium	0.2	0.6	0.2
high	0.1	0.2	0.7

	Housing Prices	
Interest	*cheap*	*expensive*
low	0.85	0.15
mid	0.4	0.6
high	0.25	0.75

(a) What is the marginal probability that the *Housing Prices* are *cheap*?

(b) If you observe that the true state of *Interest Rate* is *low*, what is your updated belief that *Housing Prices* are *cheap*?

(c) If you next observe that the true state of *Inflation Rate* is *med*, what is your updated belief that *Housing Prices* are *cheap*?

(d) Now assume that you haven't observed anything about the *Interest Rate* or the *Inflation Rate*. You look at the local housing listings and observe that the *Housing Prices* are *expensive*. Given this *Housing Prices* information, what is your updated belief that the *Interest Rate* is *high*?

Chapter 6

Application

Modeling Business Financials

In this chapter, we illustrate an important business application of situational analysis – understanding the drivers of a firm's financial performance. Financial forecasting is a ubiquitous activity in business. For example, at the firm level, managers forecast financial statements for annual budgets and long-horizon strategic plans. These typically include *Profit and Loss* (P&L) statements, cash flow, and balance sheet projections. At the division level, managers track and forecast revenues and costs every month over the course of a fiscal year. Even down to the project level, approval of capital expenditures, new product launches, marketing plans, strategic alliances, and capacity expansions are all contingent upon favorable financials.

Before getting started, we note that we now begin to incorporate using screenshots from causal modeling software in some of our illustrations. Once causal has more than a small number of nodes, each with a very limited number of states, the calculations required to conduct quantitative causal analysis quickly become too cumbersome to do by hand. For any serious application, specialized analytical software will be required. The good news is that there are many options available in market, from open source (free) to very sophisticated commercial-grade software. At the time of this writing, we examined three full-featured, commercial programs: Analytica, BayesiaLab, and Hugin.

The screenshots that follow are taken from the Mac version of BayesiaLab (because this is the software we use in our courses). The user interfaces across all these programs work similarly – the user constructs the qualitative part of the model clicking on node types and arrows, then enters conditional probability tables, functional relations, etc. This is also true of the open source options. Note that each one will *look* different. The interfaces, dialog boxes, and diagram displays vary in appearance across software packages, and you will see that in the diagram screenshots here, though they are each easy to interpret given your exposure so far.

To make things concrete, consider the task of creating a firm's annual financial forecast. In a large corporation, this work normally falls to a Financial Analyst (generally, someone with an advanced business degree). The first step in the process is to develop detailed assumptions from which to generate projections of revenues, costs, and capital expenditures. To do this properly, the Analyst must sit down with all the functional area managers and solicit their expert judgments on the relevant details pertaining their areas. Typically, Sales provides detailed assumptions about customers, pricing, and unit sales by product category; Marketing elaborates their upcoming marketing programs including price promotions, trade shows, media buys, consumer research and new product development; Manufacturing provides information on capital expenditures, process-related R&D programs and, in conjunction with Purchasing, projections of cost-of-goods-sold; all departments provide forecast headcount requirements along with job titles and descriptions of responsibilities and, in conjunction with Human Resources, projected direct and indirect compensation levels. In large businesses, just pulling this information together, making sure everyone's assumptions are consistent with one other, and constructing the implied financial projections is a big job – one that characteristically requires multiple iterations to complete.

Essentially, the Analyst must build a detailed *financial model* of the business and then populate it with base-case assumptions provided by area experts. The result is an important tool for senior managers. With the model in hand, managers can, not only review the plan and decide whether it meets with the strategic goals of their firm, but they can also explore the effects of varying assumptions to better understand their implications. Businesses operate in dynamic, complex environments. Many of the assumptions upon which a plan is built may be obsolete within a short time of its approval. Thus, the financial model is a decision tool that can be used both to optimize resource allocations under the base-case assumptions as well as to think through the appropriate responses as actual events play out.

To illustrate the power of situational analysis using quantitative causal models for planning and forecasting, let us explore an extended numerical example of a publishing firm forecasting its revenues for the coming year. Slimtree Publishing Inc., a purely fictional firm, produces novels for sale through retail outlets and textbooks for sale through student bookstores. First, we examine a standard, spreadsheet style analysis of the kind commonly practiced in business. Then, we show how expanding the setup to a quantitative causal model provides a much more nuanced and valuable decision tool for managers.[1]

1 The following example will illustrate two well-known issues that arise in statistical inference: the Prosecutor's Fallacy, and Simpson's Paradox. For a very nice analysis of these (and other) issues in non-business settings using quantitative causal models, see Conrady and Louffe (2011).

Slimtree Publishing Inc.
Forecast Profit and Loss Statement

	Units	$/unit	Total
Gross Margin			
Retail			
Sales	1,000,000	$ 15.00	$ 15,000,000
Variable Cost		10.00	10,000,000
Retail Margin		$ 5.00	5,000,000
College			
Sales	250,000	$ 85.00	21,250,000
Variable Cost		40.00	10,000,000
College Margin		$ 45.00	11,250,000
Total Margin			**16,250,000**
Fixed Operating Expenses			
Marketing Programs			1,200,000
General			500,000
Sales			600,000
Manufacturing			1,000,000
Total			**3,300,000**
Profit Before Interest & Tax			**$ 12,950,000**

FIGURE 6.1 Slimtree Publishing Inc. P&L spreadsheet.

6.1 THE SPREADSHEET APPROACH

By far, the most common approach to financial modeling is the use of digital spreadsheets. For example, suppose the Sales manager says that retail sales in the coming year should total 1,000,000 units at an average price of $15 per unit. She also estimates that textbook sales will total 250,000 units at an average price of $85 per unit. She mentions that these units and prices take into consideration the $1.2 million (m) retail promotional and advertising plan developed by Marketing. Sales overhead, primarily for sales persons' fixed salaries, are expected to total $0.6m. The Manufacturing manager says that the per unit variable cost will average $10 for retail books and $40 for textbooks. Manufacturing overhead is estimated at $1.0m. Last, general administrative expenses are forecast to come in at $0.5m. These figures are assembled into a P&L spreadsheet in Figure 6.1.

Once the spreadsheet is setup, analysts typically check the robustness of their forecasts using sensitivity analysis and contingency evaluation. The objective of sensitivity analysis is to understand how variations in key assumption variables affect the significant objective variables. For example, the Analyst might examine how a ±15% variation in market prices affect firm profit. Contingency planning involves identifying the most significant uncertainties and then quantifying their effects. For example, the Analyst might be asked to quantify the effect on profit of a six-month delay to a planned product launch.

Managers are concerned about the robustness of the forecast. In particular, they worry that the retail sales forecast may be soft and that manufacturing may not be able to deliver processing improvements that underpin the variable unit cost on the retail side. Using the spreadsheet, managers ask the Analyst to run the numbers, first with 10.0% lower units sold on the retail side, and then with a 10.0% increase in the variable cost on the retail side. According to the spreadsheet, the drop in retail sales results in a −3.1% change in total margin and a −3.9% change in profit. The increase in cost results in a −20.0% change in margin and a −7.7% change in profit. Based upon this analysis, if managers had to pick a single focal issue, the variable cost risk is the greater concern.

This is fine, as far as it goes. However, it is worth noting that implicit in the information provided by the various managers are a large number of implicit assumptions. For example, the sales forecast includes unspoken assumptions about customer demand, competitive pricing, the range of products on offer in the marketplace, and the effectiveness of the sales force and marketing plan. Note also that the spreadsheet analysis imposes deterministic assumptions (common in these models). The Sales manager has not provided any explicit information about the uncertainties underlying her numbers. Yet, the real world is made up of complex webs of direct and indirect influence relationships. These influence relationships are themselves subtle – changes in the state of one variable may shift the *likelihoods* with which the states of other variables occur. Little, if any, of this complexity is picked up via standard spreadsheet approaches.

6.2 BUILDING A CAUSAL MODEL

The easiest place to start building the causal analysis is upon the familiar ground of the spreadsheet. That is, begin with the standard relationships required to construct the P&L and use them to build the core of the causal model. Then, use causal and probabilistic thinking to expand and refine it.

The first step is shown in Figure 6.2. This graph is a qualitative causal model that illustrates the relationships implied by the P&L in our example. Each root node is given a deterministic value equal to its value in the spreadsheet. For example, *Retail Units* = \$1.0m. The children of the root nodes are defined as functions of their parents in the obvious way. For example, *Retail Margin* is the following function of its parents: $RM = RU \times (RP - RVC)$; *College Margin* is: $CM = CU \times (CP - CVC)$; and *Operating Expenses* is the sum of its parents. *Profit* is shown as a utility node,

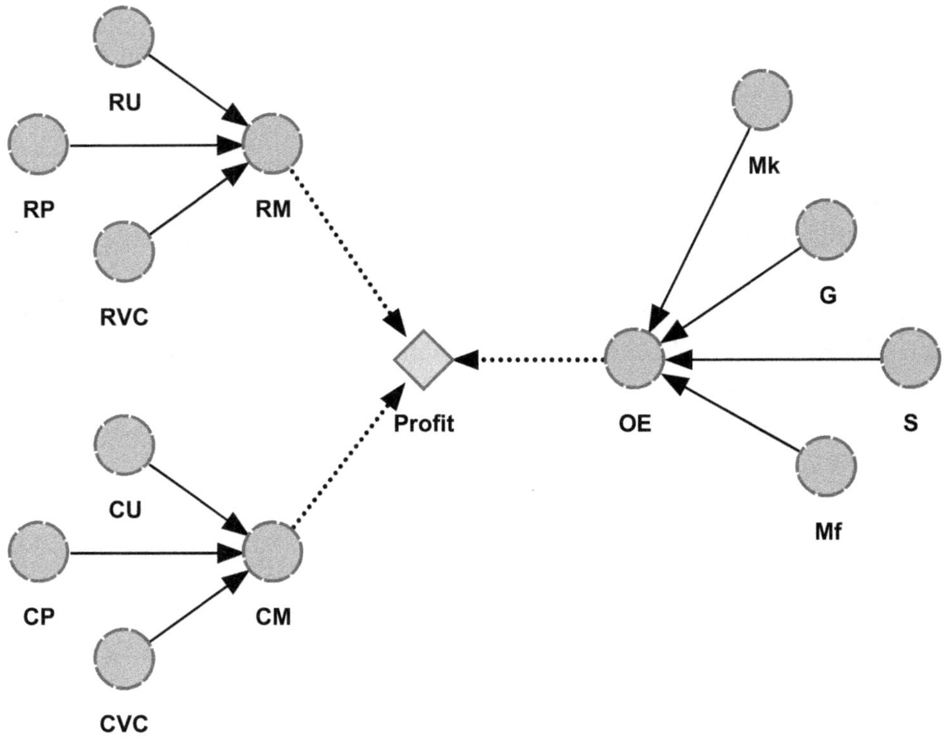

RU	*Retail Units*	CU	*College Units*	Mk	*Marketing*
RP	*Retail Price*	CP	*College Price*	G	*General*
RVC	*Retail Variable Cost*	CVC	*College Variable Cost*	S	*Sales*
RM	*Retail Margin*	CM	*College Margin*	Mf	*Manufacturing*
				OE	*Operating Expenses*

FIGURE 6.2 Initial causal diagram for the determinants of *Profit*.

the value of which is given by: $RM + CM - OE$. Doing this creates a causal model that is exactly equivalent to the deterministic spreadsheet model.[2]

Of course, simply building the causal analog of a deterministic spreadsheet is not particularly interesting. Rather, the causal model invites us to extend the spreadsheet in two important directions: (1) adding probabilistic information; and (2) "walking back" the initial model (as was discussed in Chapter 1) by thinking about causes of the P&L items, causes of those causes, and so on.

Let us begin by adding probabilities. The easiest way to get started is to identify reasonable value ranges for the root node variables and fill in uniform probabilities across the states. Later, the area experts (e.g., the Marketing manager) will be asked

2 As noted earlier, the diagrams drawn by the BayesiaLab software differ in appearance from the ones we generated for the text. Here, the environmental variables are gray circles instead of white, and the objective node is a rhombus ("diamond" shaped) rather than the hexagon we use elsewhere. You probably also noticed that these aesthetic changes did not cause you any confusion.

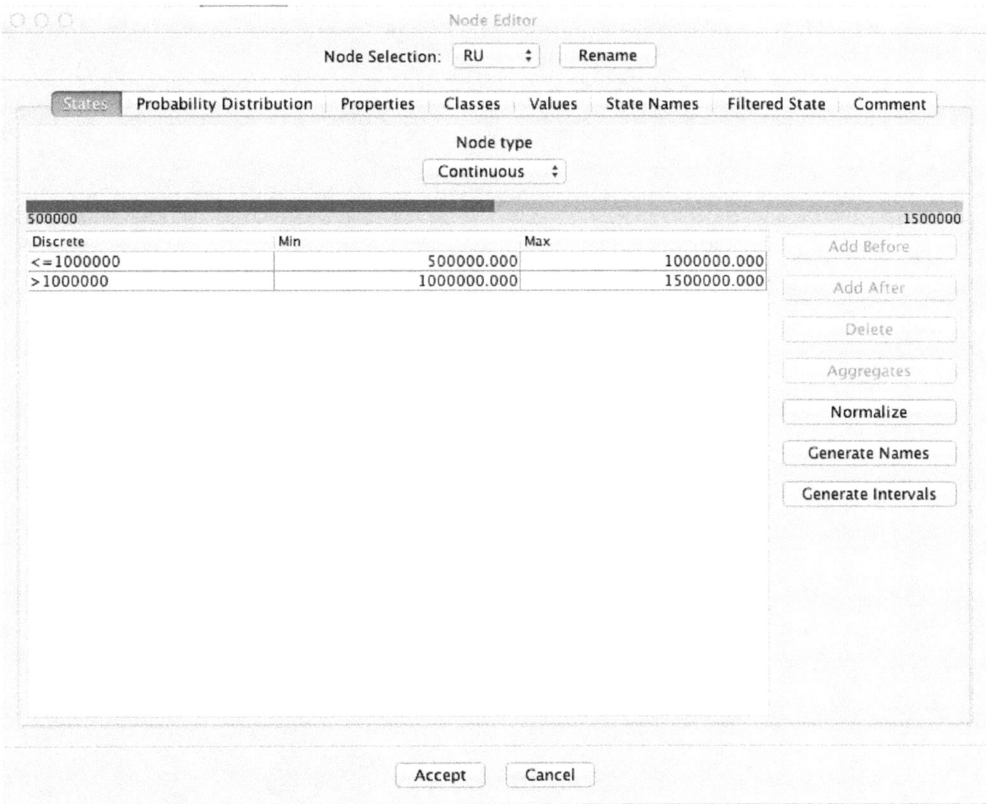

FIGURE 6.3 Variable ranges for *RU* (retail units sold).

to refine these. First, however, we simply want to generate a probabilistic version of the deterministic model. For purposes of this example, we break each root node into two states: a *high* and a *low* state and place 50% likelihood on each.

Figure 6.3 illustrates, for example, breaking the retail units sold variable (*RU*) into two possible states; a *low* state (corresponding to a range of 500,000 to 1,000,000 units) and a *high* state (corresponding to 1,000,000 to 1,500,000 units). Figure 6.4 shows the window in which the conditional probability table is entered for *RU*. Here, we have entered 50% probabilities for each of the two states. In this case, the program essentially takes the midpoint of each range (*low* = 750,000 units and *high* = 1,250,000 units), and multiplies each by 0.5 to get an expected value of 1,000,000 (which ties to the deterministic value in the original spreadsheet).

While we are here, this is a good place to mention a modeling technique that we have found helpful. Most causal modeling software provides the ability to add "function" nodes. These allow the user to create functions of nodes that have numeric values associated with their states. In our example, *RM*, *CM*, *OE* and *Profit* can all be defined as simple functions of their parent nodes. This is an efficient way to proceed when the modeler is certain that the relationships are correct and

FIGURE 6.4 Equal probabilities on the two *RU* states.

unlikely to change. Alternatively, however, one can use regular probability nodes for this purpose, where each state is associated with a numeric range. Then, the state implied by a combination of parent values is the one that corresponds to the range in which the function value falls. This approach allows the modeler to retain flexibility should she discover the need to add a new parent (direct influence) whose states are not associated with any numeric value. This flexibility is particularly helpful when one is in the process of *developing* the causal diagram (i.e., determining what the key variables are and how they are related).[3]

To see what we mean, see Figures 6.5 through 6.7. The *RM* variable is broken into eight states, each corresponding to a range of numeric values (Figure 6.5). By defining the function as shown in the upper box of the window in Figure 6.6, the values of the parent node states are automatically associated with the states of *RM*. Based upon the numerical mapping, the program automatically generates the conditional probability table shown in Figure 6.7. The purpose of the function definition in Figure 6.6 is only to associate parent states with *RM* states which, in

3 This is one of the reasons that, at the time of this writing, we favored the BayesiaLab software – it actually *requires* models to be built this way.

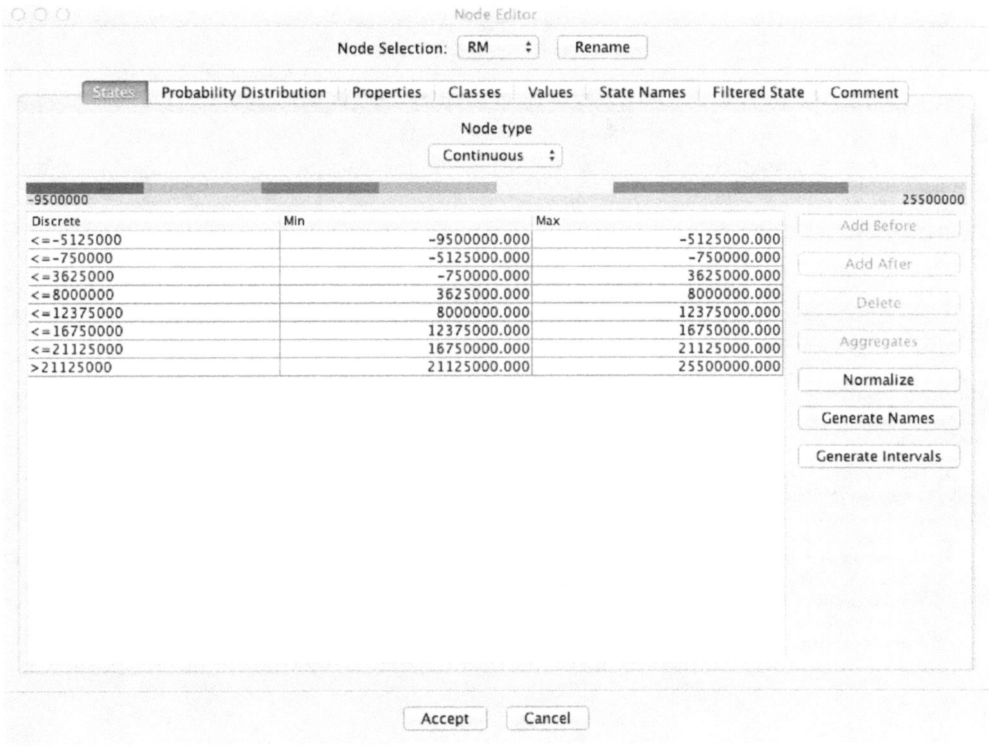

FIGURE 6.5 *RM* states correspond to variable ranges.

this case, can be done numerically because all the variable states are associated with real numbers. However, this approach leaves the door open, later down the road, to add a non-numeric cause. To do so, one would have to reformulate *RM*'s conditional probability table (CPT), which would certainly require some additional effort. Even so, no changes would have to be made to the children of *RM* (provided the *RM* states stayed the same). Had *RM* been originally defined as a (non-probabilistic) function node, adding a non-numeric cause would require much more revision: the node would have to be converted to a probability type, the CPT would have to be computed, and the CPTs of *RM*'s children would likely have to be updated as well.

Figure 6.8 shows the summary output for the causal model with the addition of the probabilistic information. Here, we see the probability distributions for the two-state root nodes. Notice that, having defined *RM*, *CM* and *OE* as probability nodes, the program also provides the implicit distributions over the states of those variables as well. Since *Profit* is defined as an objective node, the program reports its expected value of $12,887,566, which is a 0.5% difference from the deterministic value of $12,950,000 seen on the spreadsheet. Since this difference is the equivalent of a minor rounding error, we can be confident that the core part of the model is appropriately constructed and working properly.

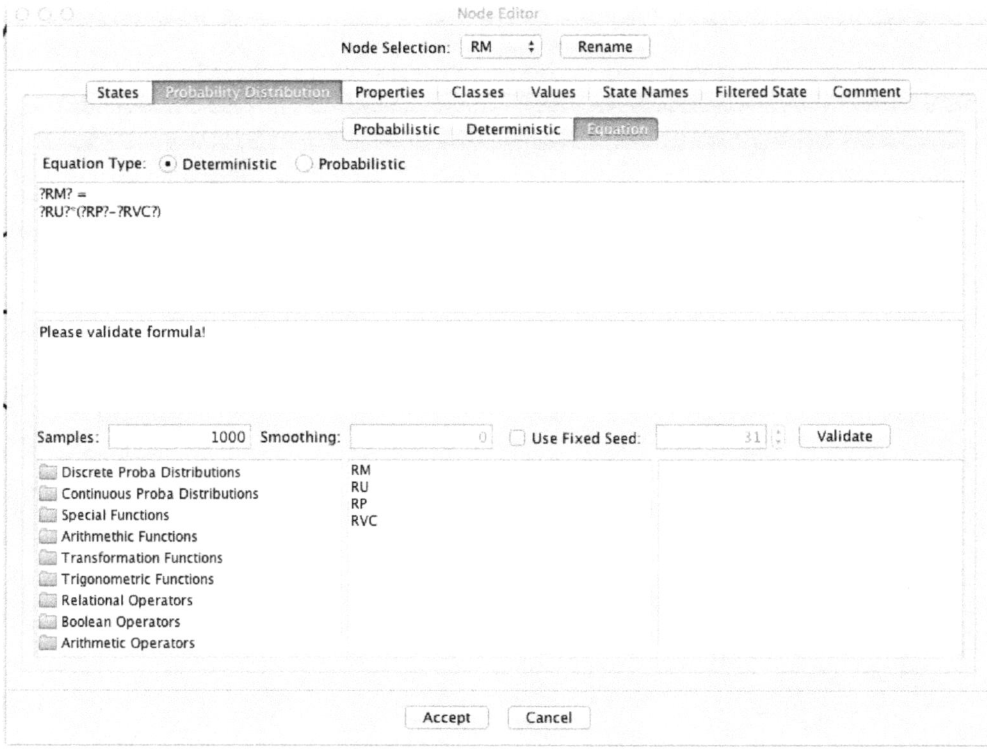

FIGURE 6.6 Parent states associated with *RM* states via a function.

RU	RP	RVC	<=-5125000	<=-750000	<=3625000	<=8000000	<=12375000	<=16750000	<=21125000	>21125000
<=1000000	<=15	<=10	0.000	0.000	51.500	47.500	1.000	0.000	0.000	0.000
		>10	0.000	32.800	66.600	0.600	0.000	0.000	0.000	0.000
	>15	<=10	0.000	0.000	1.500	61.300	35.900	1.300	0.000	0.000
		>10	0.000	0.000	50.500	48.300	1.200	0.000	0.000	0.000
>1000000	<=15	<=10	0.000	0.000	16.900	58.300	22.900	1.900	0.000	0.000
		>10	1.800	35.400	53.400	9.400	0.000	0.000	0.000	0.000
	>15	<=10	0.000	0.000	0.000	4.500	47.400	39.900	8.100	0.100
		>10	0.000	0.000	17.000	57.000	25.000	1.000	0.000	0.000

FIGURE 6.7 CPT automatically generated for *RM*.

Notice that, even with this basic model, rudimentary situational analysis is possible. For example, suppose managers discover that operating expenses are expected to fall in the range 3,612,500–3,768,750 with certainty; e.g., actual operating expenses come in at $3.7m. What inferences can be made about what caused this to happen? The answer is illustrated in Figure 6.9, in which the value of *OE* has been set to the state as shown to indicate the observation. We see that our updated beliefs on the causes of operating expenses reflect greater likelihoods each of those four variables were in their *high* states. The biggest change was with *Manufacturing Overhead* (*Mf*) – the prior was a 50:50 weight on both states. After observing the actual *Operating Expenses*, we see that the likelihood that *Mf* was in its *high* state (the 1.0m–1.5m range) increased to 89%.

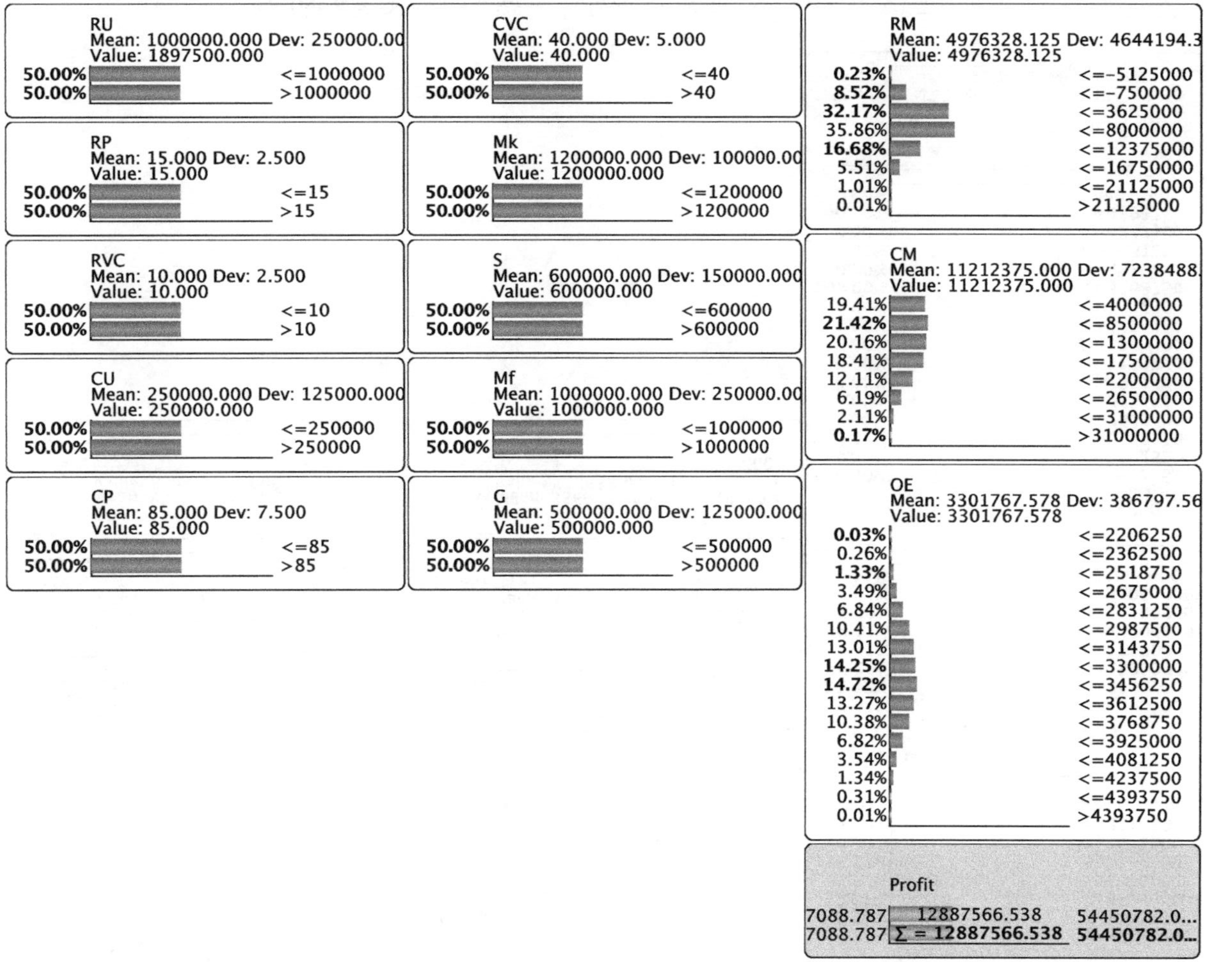

FIGURE 6.8 Summary information for the initial causal model.

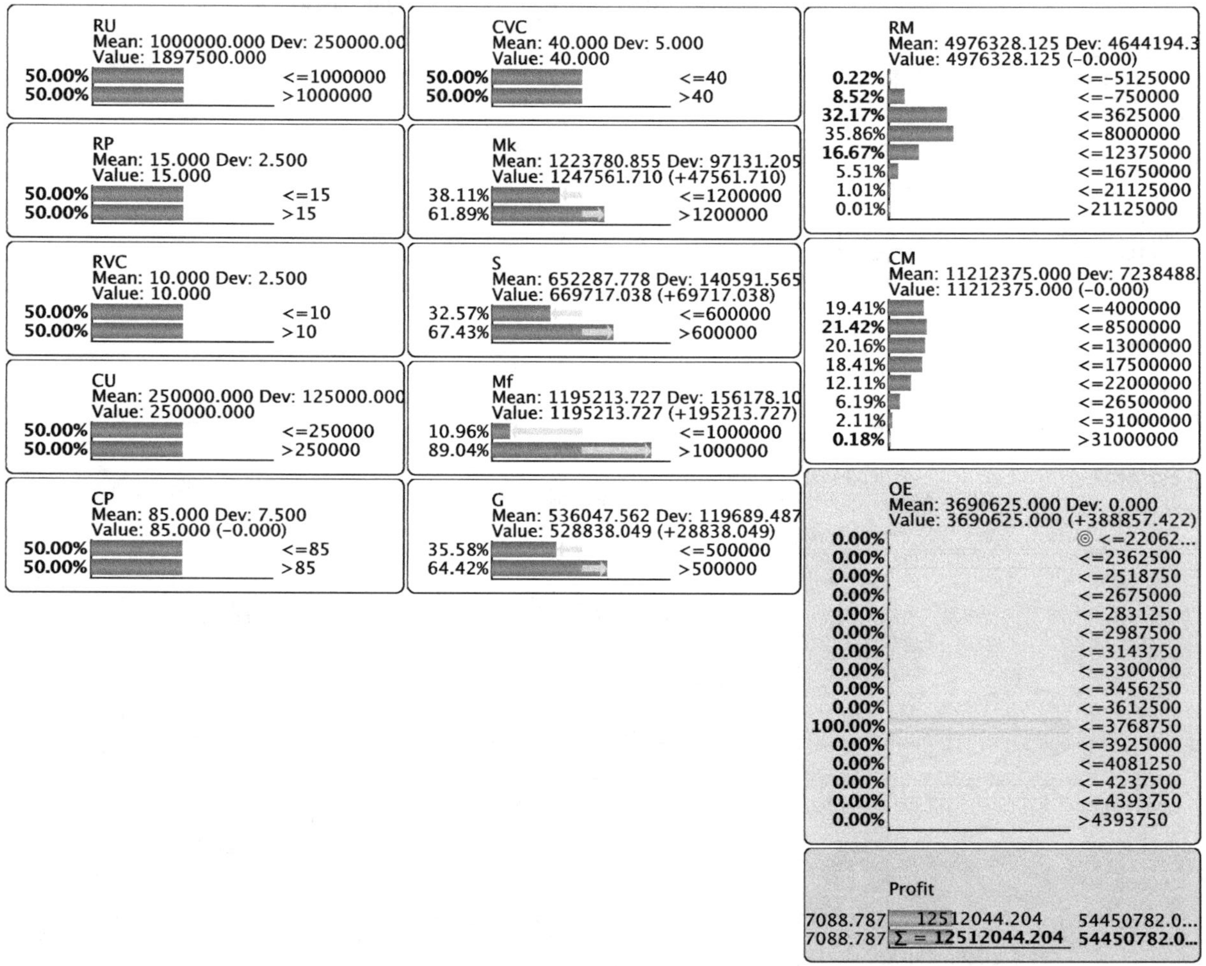

FIGURE 6.9 Drawing inferences from operating expenses of \$3.7m.

6.3 MARKETING USES THE "PROSECUTOR'S FALLACY"

Suppose the analyst begins by meeting with Sales and Marketing managers to gain a deeper understanding of the causes of retail sales volume. These managers explain that, for the most part, the firm's marketing efforts go toward maintaining the status quo. That is, all Slimtree's competitors advertise and offer promotions. The result is a fairly stable market share. Should marketing expenses be cut below the planned expected level of $1.2m, then Slimtree would certainly lose a significant portion of share. At the same time, managers do not believe raising expenditures above this point is profitable (i.e., marketing expenditures are at the critical point of negative marginal returns).

However, the managers go on to explain that, the agency always does a good job but every once in a long while a particular campaign hits the market just-so and goes "viral". Achieving viral status is rare and virtually impossible to induce by choice – historically, only 1 in 1,000 campaigns go viral. When it does, the impact on sales is huge and positive. To account for this effect, the analyst is advised to add a node for *Marketing Effectiveness* (*ME*) with two states: *normal* and *viral*. This node is a cause of *Unit Sales*. The expert managers also suggest breaking *Unit Sales* into three states, *low*, *medium*, and *high* corresponding to unit sales ranges of 0–1m, 1m–2m, and 2m–3m. In their judgment, the conditional probability table for *RU* given *ME* is as shown in the CPT at the bottom of Figure 6.10. In other words, when *ME* = *normal*, the probability is only 1% that *Unit Sales* will fall in the 2m–3m range. Alternatively, when a campaign is so effective that it goes viral, the probability is 100% that *Unit Sales* will be in that range!

Now, suppose that some time later, it becomes clear that sales are going to fall in the *high* (2m–3m) range. The Marketing manager walks into the CEO's office and – waving around the CPT in Figure 6.10 – claims that it's obvious that this year's marketing campaign was so effective that it went viral. "The odds are 100

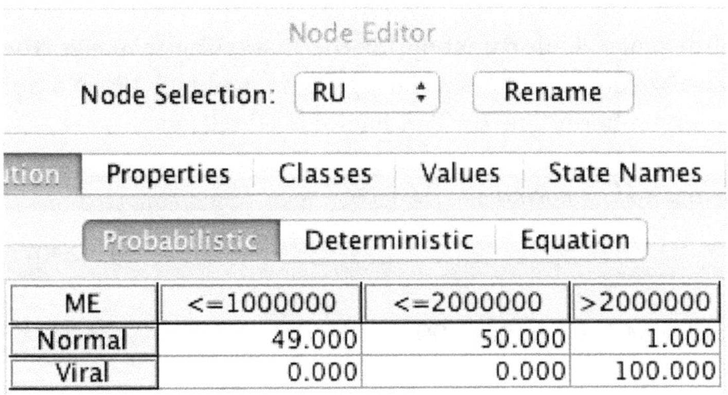

ME	<=1000000	<=2000000	>2000000
Normal	49.000	50.000	1.000
Viral	0.000	0.000	100.000

FIGURE 6.10 Influence of *Marketing Effectiveness* on *Retail Units*.

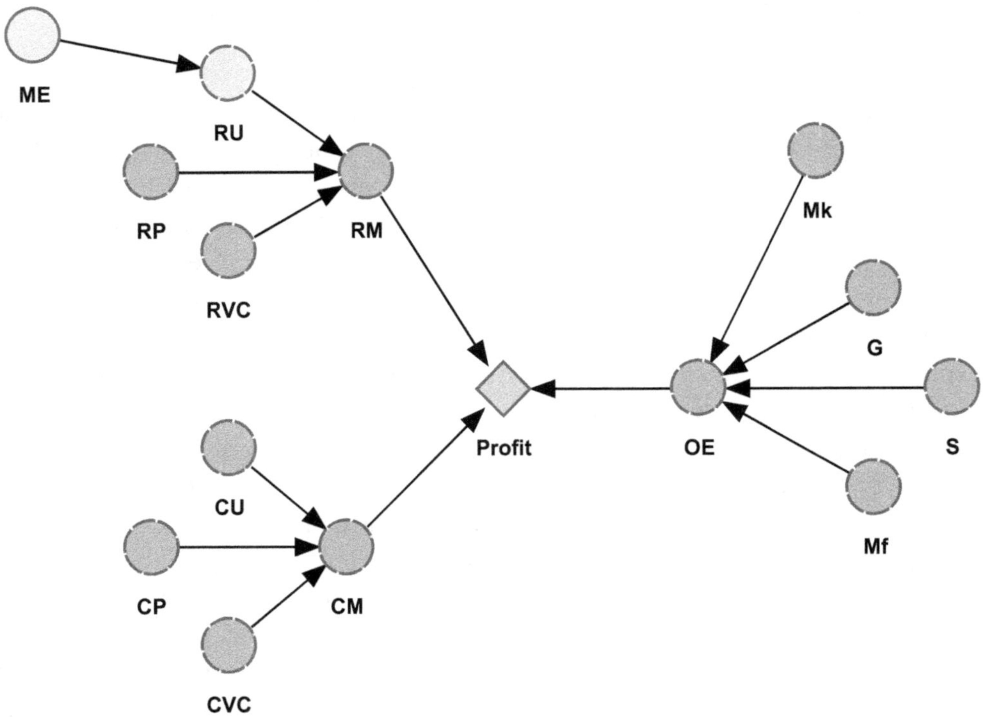

FIGURE 6.11 Model extended to include effectiveness of *Retail Marketing*.

to 1 that my program was super-effective. I want a bonus!" The CEO is happy to give the bonus, provided it is more likely than not that the marketing program was, indeed, effective (i.e., will give a bonus if the probability $OE = viral \geq 51\%$. Is the Marketing manager's claim correct? Should the CEO pay the bonus?

With the automated quantitative causal model in hand, it is an extremely simple matter to answer this question. The expanded graph is shown in Figure 6.11, along with the marginal probabilities for the two variables of interest, *ME* and *RU*. Figure 6.13 shows the updated prior for *ME* when the state of *RU* corresponds to the 2m–3m range. Unfortunately for the marketing manager, the probability that *ME* = *viral* given that *RU* is in its *high* state **is only 9.1%**. No bonus after all.

What is going on here? The CPT in Figure 6.10 tells us

$$P(RU = high|ME = normal) = 0.01$$

and

$$P(RU = high|ME = viral) = 1.00.$$

What that table does *not* tell us is $P(ME = \text{viral}|RU = high)$! True, it seems reasonable to infer that observing *Units Sold* in the 2m–3m range should cause

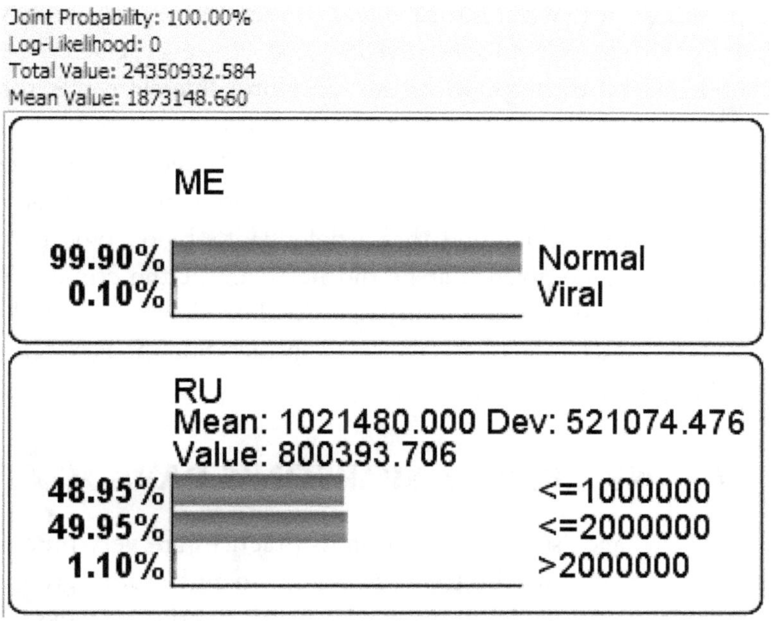

FIGURE 6.12 Marginal probabilities of *ME* and *RU* in extended model.

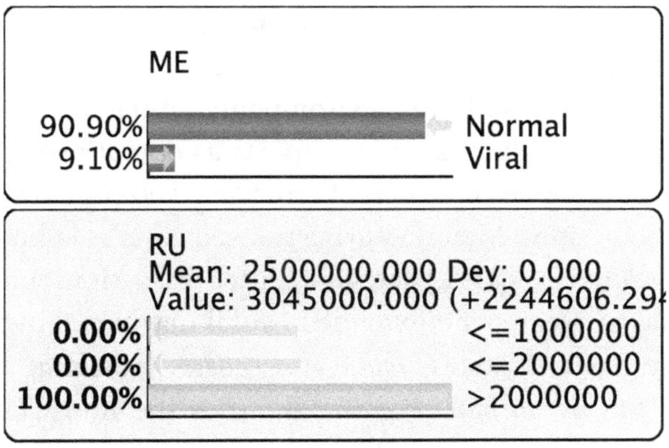

FIGURE 6.13 Updated priors on *ME* given $RU = high$.

us to increase our belief that $ME =$ viral. However, we know from Bayes' rule (recall equation 4.6) that to get the exact number, we need to compute:

$$\begin{aligned} P(ME = viral|RU = high) &= \frac{P(RU = high|ME = viral)P(ME = viral)}{P(RU = high)} \\ &= \frac{1 * 0.001}{1.1} \\ &= 9.1\% \end{aligned}$$

It is worth pointing out that our intuition was correct: the probability of viral went from 0.1% to 9.1%. That is a huge relative increase. However, given the extremely low likelihood of viral to begin with, the most probable state was that normal marketing effectiveness happened to trigger high sales.

The Marketing manager's claim is known as the **Prosecutor's Fallacy**. This is because expert witnesses in a criminal investigation often give testimony in which they state, e.g., that the likelihood that a defendant's blood type matches that found at the crime scene given that he did not commit the crime is incredibly small. However, to assess the probability that the defendant committed the crime given a blood type match, one must take into consideration the base rate of people in the general population with that blood type.

6.4 GREEN INK CREATES SIMPSON'S PARADOX

These days, "green" and "sustainable" business practices are very much *en vogue*. Suppose that, in keeping with this trend, Slimtree's manufacturing group discovers a supplier of environmentally friendly ink with which to print its books. During the course of the year, the Manufacturing department initiates a program in which the retail and college marketing managers can choose to publish a book title in "green" ink (not the color) if they so wish. Although the use of environmentally friendly ink is likely to be more costly due to additional setup and handling costs, Slimtree wants to be a good corporate citizen and, moreover, believes there may be positive externalities for marketing with the environmental initiative.

At the end of the year, Manufacturing reports its operating costs categorized by ink type (normal versus green). Assume there were 100 production runs, 80 retail titles and 20 textbook titles. Manufacturing tracked costs as falling into either a *low* ($5k–$10k) or *high* ($10k–$15k) category. After discussions with Manufacturing manager, the Analyst creates the extended causal diagram shown in Figure 6.14. The *Title* variable has two states, *retail* and *college*, indicating the type of title. The *Ink* variable can be *normal* or *green*. *Run*C is the *Run Cost*, which has the two states corresponding to $5k–$10k and $10k–$15k. Because the Marketing managers for retail and college books may have different preferences for green ink, *Ink* may be influenced by *Title* type. Since textbooks and retail books typically have different setup costs, *Title* also influences *Run*C. Finally, given these extensions, *Mf* is redefined to be *Run*C × 100 (the total number of runs).

Now suppose that, contrary to initial expectations, using the environmentally friendly ink seems to have reduced average run costs. Specifically, during a review of the program and its costs, Manufacturing presents Table 6.1.

Apparently, the use of environmentally friendly ink increases the likelihood that *Run Costs* are in the *low* range (from 45.7% to 66.7% of runs). Not surprisingly, the Manufacturing manager asks for a raise and suggests switching all runs to the *green Ink Type* – a win-win because the firm will be greener and more profitable!

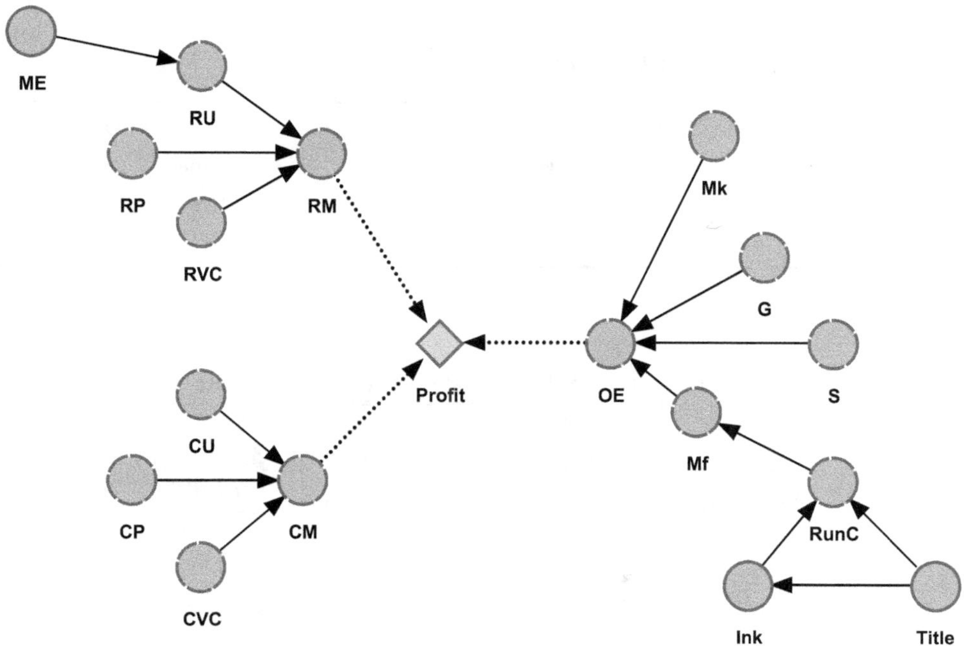

FIGURE 6.14 Model expanded to account for production runs.

Table 6.1
Average run costs.

Ink Type	Run Cost	
	low	high
normal	45.7%	54.3%
green	66.7%	33.3%

The CEO, who operates on the premise, "if something seems too good to be true, it is probably not true," asks the Analyst to dig deeper into the data.

With the causal model in hand, the Analyst knows exactly what data to request of Manufacturing – namely, that required to fill in the CPTs for each node. These are: (1) for the *Title* node, the percentage of *retail* and *college* runs; (2) for the *Ink Type* node, the percentage of *green* and *normal* runs by market type (*retail* or *college*); and (3) for the *Run*C node, the run cost categories by ink and market type. We already know that retail comprised 80% of the runs, so this can be entered into the model. The ink categories by market type are shown in Table 6.2.

Moving on, the Analyst constructs the CPT for the *Run Cost* node using raw data by run. The CPT is presented in Figure 6.15. This data is puzzling because it says

Table 6.2
Calculating CPT for *Ink Type* node.

	# of Runs			CPT	
Market	normal	green		normal	green
retail	8	72	⇒	10%	90%
college	16	4	⇒	80%	20%

Title	Ink	<=10000	>10000
Retail	Normal	75.000	25.000
	Green	69.000	31.000
College	Normal	31.000	69.000
	Green	25.000	75.000

FIGURE 6.15 CPT for *Run Cost* based upon manufacturing data.

that, within both market categories, the effect of using environmentally friendly ink on run costs was to *increase*, rather than lower, them. Among *retail* runs, using *green* ink lowered the likelihood of *low* costs from 75% to 69%. Similarly, the effect on *Run Costs* for *college* books was to lower the likelihood of *low* costs from 31% to 25%.

This seems to fly in the face of the information reported by Manufacturing, leaving the Analyst to wonder if the data presented in Table 6.1 was erroneous. The Analyst decides to double check the data by using the extended causal model to compute the relevant marginal probabilities. First, the data in Table 6.1 is checked by setting the *Ink Type* as "evidence" and then seeing the effect this has upon *Run Cost*. The results are shown in Figure 6.16. These calculations validate the Manufacturing department's original numbers: given *normal Ink*, the marginal probability of *low* cost is 45.7%; given *green Ink*, the marginal probability of *low* cost increases to 66,7%. Next, the Analyst checks the effects when both *Market Type* and *Ink Type* are known. The results for the retail market are presented in Figure 6.17. These computations also validate the finding that, within the *retail* market category, using *green* ink has a negative expected effect on costs. The same is true for the *college* category. Apparently, another manager will not be getting a bonus this year.

The Analyst has discovered an effect known as **Simpson's Paradox**. This is a problem that arises when the category of the variable receiving the treatment (the *Market Type* of the run) also has an effect on the likelihood of that treatment happening in the first place (the *Ink Type* used in the run); i.e., the true causal relationships induce misleading averages at the aggregate level. On average, *retail*

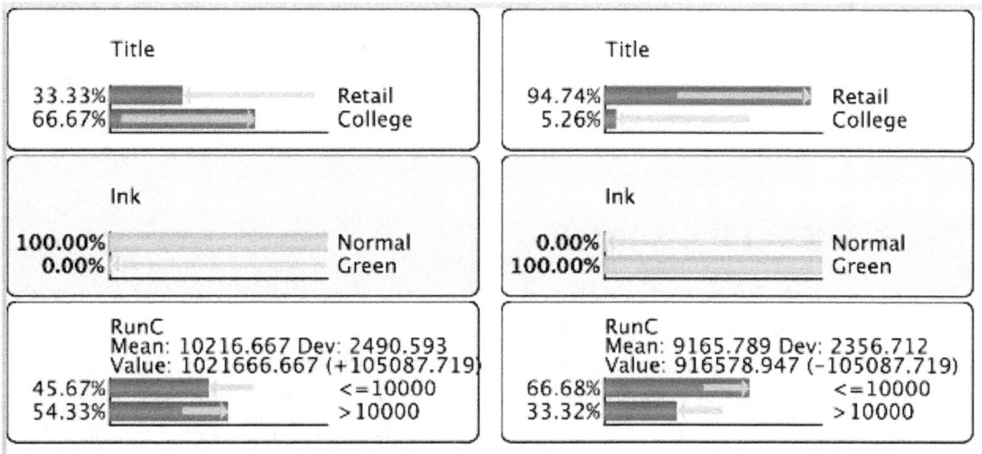

FIGURE 6.16 The effect of *Ink Type* on *Run Cost* when *Title* is unknown.

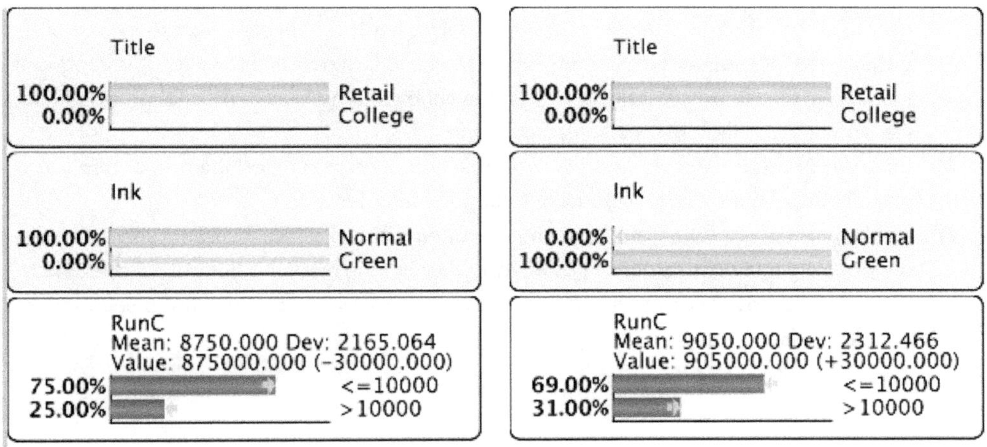

FIGURE 6.17 The effect of *Ink Type* on *Run Cost* when *Title* is known.

runs are much cheaper than runs for *college* textbooks. Simultaneously, *retail* runs are more likely to use environmentally friendly ink. At the aggregate level, this makes it appear as if *green* ink is causing the *low Run Costs*. But, this is not the case. Rather, the use of *green* ink is just correlated with the type of run that has lower costs.

Simpson's Paradox is not an esoteric statistical artifact, but rather an important difficulty that frequently turns up in data analysis. For example, pharmaceutical firms must be very careful how they interpret the results of drug trials in which patients self-select into the trial (for example, women may have inherently greater resistance to a disease and be more likely to try a new treatment for that disease, resulting in a misleading overall result of drug efficacy). The same is true for marketing focus groups, advertising, and customer loyalty programs. The problem also occurs generally; i.e. this is not a problem specific to Bayesian networks, though the method to check for these effects is facilitated by the structure of Bayesian networks.

KEY CONCEPTS: CHAPTER 6

- For causal models with more than a few variable and a few states per variables, analysis requires the use of specialized software, and there are many options available.
- The most common tool for financial modeling is using spreadsheets; however, they fail to sufficiently capture the influence relationships among the variables.
- The calculations embedded in a spreadsheet imply a functional relationship that can be converted into a causal model with appropriately specified conditional probabilities. These can readily be expanded to be more insightful for inference and the efficacy of different interventions.
- The Prosecutor's Fallacy denotes a situation in which the probability reported is the wrong conditional probability. When a particular outcome is extremely sensitive to one of its parents, the inclination is to attribute that outcome to that parent. However, it is important to incorporate the base probability of that event by finding the conditional probability of the parent given the observed outcome of the child.
- Simpson's Paradox occurs when the variable you are trying to test the role of also has a causal effect on the variable you are testing for sensitivity.
- Thinking about causation, and utilizing conditional probabilities appropriately, provides deeper insight into the relationships of variables and the meaning behind collected data. Some results may be counterintuitive, and it is in exactly those cases that a solid formal foundation is most important.

Chapter 7

Single-Agent Interventions

A managerial intervention problem has two essential features: (1) some element(s) in the environment can be controlled, chosen or "set" by a manager; and (2) this decision is intended to optimize some objective. Thus, causality is at the heart of the problem – implementing an optimal plan hinges critically upon understanding the consequences of one's actions. Thus, having a clear understanding of the relevant causal forces is essential to the success of such enterprises.

Outside of situational analyses, it is hard to think of a managerial decision that is not an intervention problem in some form or another. Managers create incentive plans (set compensation policy), make hiring decisions (set the characteristics of the employee population), innovate (set the active collection of R&D projects), enter new markets (set the markets served by the firm), and so on. All of these actions are interventions.

Typically, before deciding upon which state(s) to set, and set to what, for a particular intervention, managers must first decide *which* interventions to even attempt. This is due to constraints on managerial resources. That is, of the universe of interventions available to a given manager at a given moment, time, skill, and knowledge constraints place severe limits on the number of interventions that are actually available for consideration.

Causal models are useful in both situations (i.e., deciding which interventions to make and, having decided that, choosing optimally). The quantitative influence diagram (QID) is the appropriate decision tool either way.[1] The information required to analyze a decision using a QID is the same as that required to set up a decision tree. The differences are that: (1) knowledge with respect to the problem domain is presented in an intuitive, accessible form; and (2) the representation is compact and efficient (especially for complex, real-world problems).

1 A QID is a quantitative causal model that includes at least one strategic option node and one objective node per strategic agent modeled.

QIDs are quite flexible and capable of solving complex managerial intervention problems. The situation can be a single-agent or multi-agent decision problem. It can involve numerous, uncertain environmental factors, as well as those that behave according to a known functional form. Decisions can be multi-stage and information-dependent. We begin with the basic, single decision-maker case. In Chapter 9, we consider multi-agent, interactive decisions (i.e., the game theory version). As above, we start with an easy structure to introduce the concepts and calculations required, and then work through a more complicated model to illustrate more details and provide examples of more sophisticated situations.

7.1 ONE DECISION, NO INFORMATION

In order to elaborate the use of causal models to analyze managerial interventions, let us begin with the simplest sufficient case to be a QID: one decision, one environmental, and one objective node. The decision is represented by a strategic option node (rectangle), the states of which are the feasible options available to the manager. Uncertain elements affecting the outcome of the manager's choice are represented by an environmental node (circle or oval). The payoff to the manager (utility, profit, etc.) is represented by an objective node (hexagon) which, in all cases, will be a function of other nodes.

Using the methodology described in Chapter 5, each available strategic option is associated with an expected payoff by calculating through the probabilities of the environmental variables. The highest expected payoff is the optimal choice. For example, suppose Apple Inc. must decide whether or not to launch its new tablet product in Canada. The decision hinges on whether the current market is strong or weak. If the market is weak, the launch has a net present value (NPV) of –$100 (USD). If it is strong, then the NPV is $50. The state of the market is uncertain, though Apple management is 60% certain the market will be strong. Apple's options are either to launch in Canada or not. If the launch is not undertaken, the NPV is 0.

The QID setup is as shown in Figure 7.1. The *Launch* decision node has two states: *yes* and *no*. The *Market* also has two states, *strong* and *weak*. The objective node is a function that assigns a *Net Present Value* (*NPV*) to all combinations of the *Market* and *Launch* states. The *Market* is an environmental variable and, as such, is associated with a probability table. The table associated with the objective node has a row for each state combination of all nodes with an arrow to *NPV*, a column for the probability of that state, and a column for the payoff associated with that state.

Notice that, **for each value of the decision node, the probabilities over system states sum to 1**. This will always be true if the model is set up properly. By multiplying each row payoff by the row probability and summing for a given decision, we obtain the expected value of that decision. In this example, the

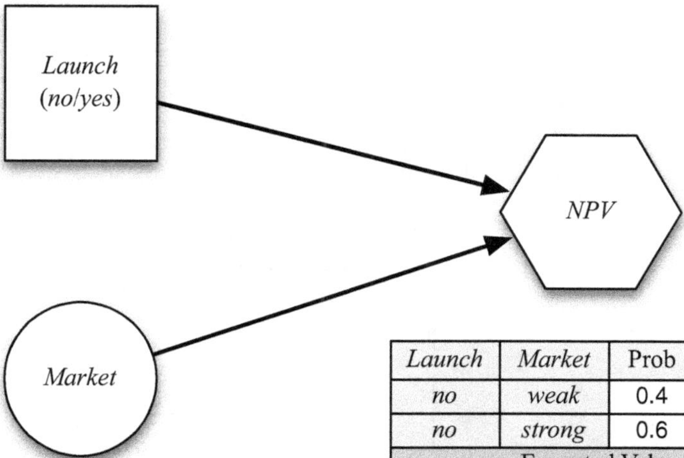

Market	
weak	*strong*
0.4	0.6

Launch	*Market*	Prob	*NPV*
no	*weak*	0.4	0
no	*strong*	0.6	0
	Expected Value:		0
yes	*weak*	0.4	-100
yes	*strong*	0.6	50
	Expected Value:		-10

FIGURE 7.1 Should Apple launch a new tablet in Canada?

expected *NPV* of not launching is: $E(no) = 0$ and the expected *NPV* of launching is $E(yes) = -10$. Therefore, the optimal decision is not to launch.

7.2 MULTIPLE DECISIONS WITH INFORMATION

Let us now extend the previous example to include another decision, and also add some information at the launch decision. The introduction of multiple decision nodes and informational arrows brings up important considerations touched upon briefly in Chapter 2. Recall that arrows pointing to environmental variables, either from decision nodes or other environmental variables, indicate a causal relationship, but arrows pointing into decision nodes indicate information known at the time of that decision. Decision nodes do not have conditional probability tables; people are free to decide what they want based on the information received. For situations in which one agent is making multiple decisions (like the ones in this chapter), it is assumed that all previous decisions made are known to the the decision-maker at that node. We will revisit each of these points as they come up in building the model and solving the optimization problems presented below.

The models presented in this chapter, actually in this book, are "pure" intervention problems. A decision problem is *pure* when the focal issue is deciding which variable (among a known set of variables – possibly just one) to set the state of, and what state to set that variable to.

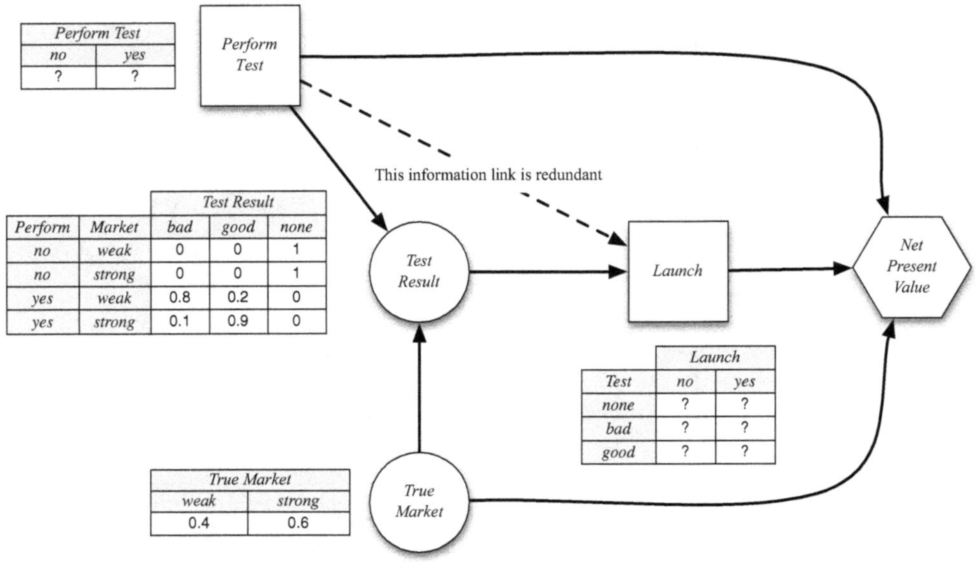

FIGURE 7.2 Opportunity for pre-launch market research.

7.2.1 The Extended Model

Suppose that, prior to launching, Apple can do some test marketing to assess whether the actual market will be strong or weak. The market *Test* results are either *good* or *bad*. Of course, the test is prone to some degree of error – 20% of the time research concludes *good* when, in fact, the *Market* is *weak*. 10% of the time the research concludes *bad* when the *Market* is actually *strong*. The test costs $1.

The new causal model is presented in Figure 7.2. We must add two nodes: one for the decision to *Perform* a market test, and one for the information provided by that *Test*. The states of the decision to *Perform* the market test are *yes* and *no*. Furthermore, because the market test requires a cash investment, which affects *NPV*, there is a link from the decision to *Perform* with the market Test to the *NPV* objective node.

The second node, for the market *Test* results, must be added to reflect the information generated for use in the *Launch* decision as a result of the *Perform* market test decision. Here, we must take care to ensure that the states at this node are MECE (mutually exclusive and collectively exhaustive). If the market *Test* is conducted, the result is either *good* or *bad*, but if the test is not conducted, however, no information is generated. This possibility must be included as an information state at the time of *Launch*. Thus, there are three states for the *Test* market observation variable: *bad*, *good*, and *none*.

Changing the *Perform* market test decision has a direct effect on the state of the market *Test* variable. The only other variable that co-determines the value of the *Test* variable is, of course, the actual *Market* strength. Because the market *Test* variable has two parents, its CPT requires a row for each possible combination of *Market* and *Perform*. The conditional probability table attached to the Market Test node must accurately reflect the influence of its parents. As shown in Figure 7.2, this is achieved through the use of the *none* state. That is, if the decision on whether to *Perform* a test is *no*, then the state of *Test* is *none* with complete certainty (probability = 1). If the test is done, then the test outcomes happen according to the error probabilities described above.

Apple knows the value of the market *Test* at the time of the *Launch* decision (which is the whole point of including it in the first place). Therefore, *Launch* is a child of *Test* (with an information arrow connecting them). The model is completed by ensuring that there is a link to the *NPV* objective node for every decision and variable that has an effect on that objective. We already pointed out that *Performing* the market test costs $1. We also still have, from the previous model we are extending, the cost of approving the *Launch*, and the result of the launch depending on the true *Market* conditions. All three of those must feed into the calculation for *NPV*, the results of the *Test* itself do not get directly factored into the objective.

Whenever an opportunity arises to simplify a model with no impact upon the analysis, it should be taken (and taken carefully). Technically, there should be an information arrow from the *Perform* test decision to the *Launch* decision (under the no-forgetting rule, Apple is assumed to remember the first decision at the time of the second decision). Note, however, that the state of the *Test* variable fully reveals the preceding state of the *Perform* test decision variable: the state of the *Test* variable is *none* only when the state of the *Perform* test variable is *no*; for any other state of the *Test* variable, the state of the *Perform* variable must be *yes*. Thus, having a link from the *Perform* variable to the *Launch* variable is redundant with the arrow from *Test* to *Launch*, and no harm is done by omitting it the informational link.

Finally, because there is an information link pointing to the *Launch* decision, we include a **decision table** at that node to keep track of the values of choices contingent upon what is known at the time of this decision. A decision table is built just like a conditional probability table for the decision node, except instead of probabilities we will fill it with expected values. Prior to any analysis, the cells of this table are blank. (The purpose of the analysis is to fill them in!) Formally, this decision table needs to have a row for every combination of input values – causal and informational – and *Perform* is also an input, but we can leave its states out of the decision table for the same reason we can remove the arrow from the diagram.

7.2.2 General Solution Procedure

We are interested in determining which *set of decisions* optimizes our objective. For the same reasons that we do not use joint probability tables to do situational assessment, we will not find the optimal strategy by finding the value of each possible combination and pick the highest one. The structure of the diagram simplifies the calculations and provides more natural intuitions about the effects of decisions along the chain.

We begin by listing the specific steps required to evaluate the optimal decision strategy in an influence diagram. There are actually several methods that can be used (see, e.g., Shachter, 1986). The one employed here is the most straightforward and intuitive, though not necessarily the fastest or most efficient. Efficiency considerations really only come into play when you are analyzing a large system, and in that case, you will likely be using software engineered by people who already use the most efficient algorithms available.

In what follows, it is important that the strategic option (decision) nodes be ordered in some specific sequence. If there is no explicit sequence to the decisions (e.g., because the order in which they are made does not matter or because you consider them simultaneous decisions), simply impose one because in those cases it won't matter. Keep in mind the "no forgetting" assumption: even though the information links may be omitted to keep the diagram simple, everything known at an upstream (ancestor) option node is also known at all downstream (descendant) option nodes.

1 Starting with the last strategic option (the one having the fewest steps away from the objective node), set the states of all strategic options and all information nodes to the ones that correspond to the first row, first column of the decision table at that node. Due to the "no forgetting" rule, this first step implies a row for every strategic option and every (relevant) information node value in the model.
2 The decisions corresponding to the current cell of the decision table can be considered "evidence" in the sense that they set the state of variables. And so you can use the network propagation methods described in Chapter 5 to update the marginal probabilities across the states of the parents of the objective node given this "evidence". It is worth noting here that the objective node is never itself observed (in the sense of evidence setting). Therefore, given its network position as a leaf node, and according to the definition of c-independence, the objective node blocks all paths through itself and never transmits messages to other nodes.
3 Use these updated marginal probabilities to calculate the expected value of the objective node for the current case. This is done by the standard expected value calculation: for each combination of system states matching the current

case, multiply the probability of that case by the monetary value assigned to that case by the objective function values, and then adding those probability-weighted values together. Enter this value in the corresponding cell of the decision table.

4 Holding all the other values in the system fixed (i.e., staying in the first row), repeat this calculation for each state of the last strategic option node. This will fill in the first row of the decision table for that node. When the row is complete, the strategic option (which is just the state of the decision node) corresponding to the column with the largest number is the optimal decision given the information corresponding to that row.
5 Move to the next row and repeat the process for each state of this decision node under the conditions corresponding to this row. Continue for each row in the decision table for that node until you've filled in the table completely. At this point you will have identified the best strategic option for each row; i.e., given you find yourself at this point of the decision problem, you have a complete contingency plan of what action to take in order to maximize your objective, given any possible information received.
6 Remove the objective node and replace the last decision node with a new objective node. Because you have identified the best action for any scenario, there isn't any decision left to take – just pick the action that optimizes your objective. The rows of this new objective node are the same as the decision table of the strategic option it replaces, and its values correspond to the best possible expected values noted in the decision table (that is, the expected values for the action you would take in each scenario/row).
7 Any barren nodes that result from removing the old objective node (except the new objective node) can now be removed. A **barren node** is a node without any successors. If the removal of a barren node creates new barren nodes, they can be removed as well.
8 Repeat the process with the modified causal model. Continue until an optimal, information-contingent decision is identified for all information states of all strategic options.

Essentially you are backtracking up the causal diagram, starting at the objective node, identifying an optimal contingency plan (or strategy) for what to do at that time given any possible information state you could be in at that time. You continue to replace your choices with optimal plans until you have a complete plan for the whole situation; i.e., an optimal action for every decision you will face. Let us illustrate this procedure with the extended Apple product launch problem.

> **Definition 7.1** *Barren node: a node that has no successors after the removal of an objective node from an influence diagram.*

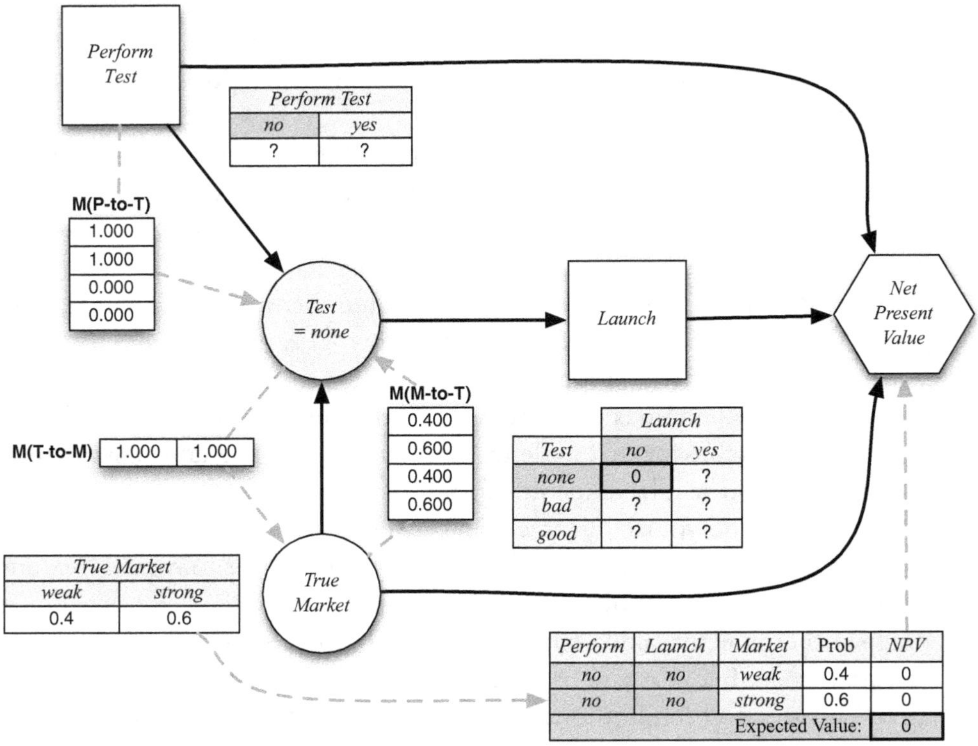

FIGURE 7.3 Expected *NPV* when *Perform* = *no* (and hence *Test* = *none*) and *Launch* = *no*.

Set States of Options and Information Nodes

In this model, there are two strategic option nodes and one information node. We must set each of these to one of their states and record the resulting expected value of the objective node. The starting states can be chosen arbitrarily. To be systematic, however, we recommend starting with the states corresponding to the first row, first column of the decision table at the last strategic option. In this example, this cell in the decision table is *Test* = *none* and *Launch* = *no*. As we noted earlier, *Test* = *none* implies that *Perform* = *no*. The initial setup is illustrated by the gray node in Figure 7.3.

Compute Conditional Probabilities for Parents of Objective Node

The parents of the objective node are: (1) the decision to *Perform* market test (because running a test incurs a cost), (2) the *Launch* decision node, and (3) the *Market* condition node. Both strategic option nodes have their states set to *no* (i.e., the probabilities of these states = 1). Thus, the only non-trivial conditional probabilities that need to be calculated are the ones for *Market* given the values of *Perform*, *Launch*, and *Test*.

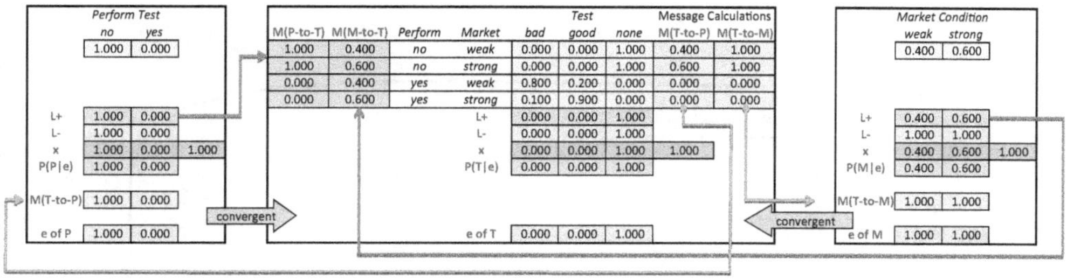

FIGURE 7.4 Updating the probability of *Market* given the initial configuration.

Since *Launch* is c-independent of *Market* given the evidence that *Test* = *none*, we need not consider *Launch* in evaluating *Market*. This leaves us with a convergent node pattern (*Perform* → *Test* ← *Market*) with evidence on both *Perform* and *Test*. The calculations are illustrated in Figure 7.4. The updated probabilities are, as shown in Figure 7.3, *weak* = 0.4 and *strong* = 0.6 (i.e., no change).

Compute Expected Value of Objective Node

Figure 7.3 shows the expected *NPV* calculation for the case *Perform* = *no* and *Launch* = *no* under the conditional probabilities calculated for *Market*. Of course, in this case the calculation is trivial because the *NPV* is 0 in all states (no launch and no cost for any market test). This value is entered (as shown in Figure 7.3) in the first row, first column of the decision table at the *Launch* node.

Repeat Step 3 for all Options

Holding all other node states fixed, cycle through the columns of the first row of the decision node closest to the objective. In this example, there is only one other option: *Launch* = *yes*. In this case, the *NPV*s do vary with the states of the *Market* node. The expected value calculation for this case, −10, is illustrated in Figure 7.5.

Repeat Steps 3 and 4 for all Rows of the Decision Table

You can save calculation time by noting decisions that result in deterministic outcomes and simply enter the corresponding values in the decision table. For example, if there is no launch and no market test, then the *NPV* is zero with certainty (as computed for the first row, first column). If there is a test but no launch, then the *NPV* is the cost of the test (i.e., −1 with certainty). Therefore, the first column of the *Launch* decision table can be filled in accordingly without further ado.

The remaining cells, corresponding to the expected *NPV*s when the product is launched in response to the market test results must be computed. These calculations are illustrated in Figures 7.6–7.7. For clarity, we take a closer look

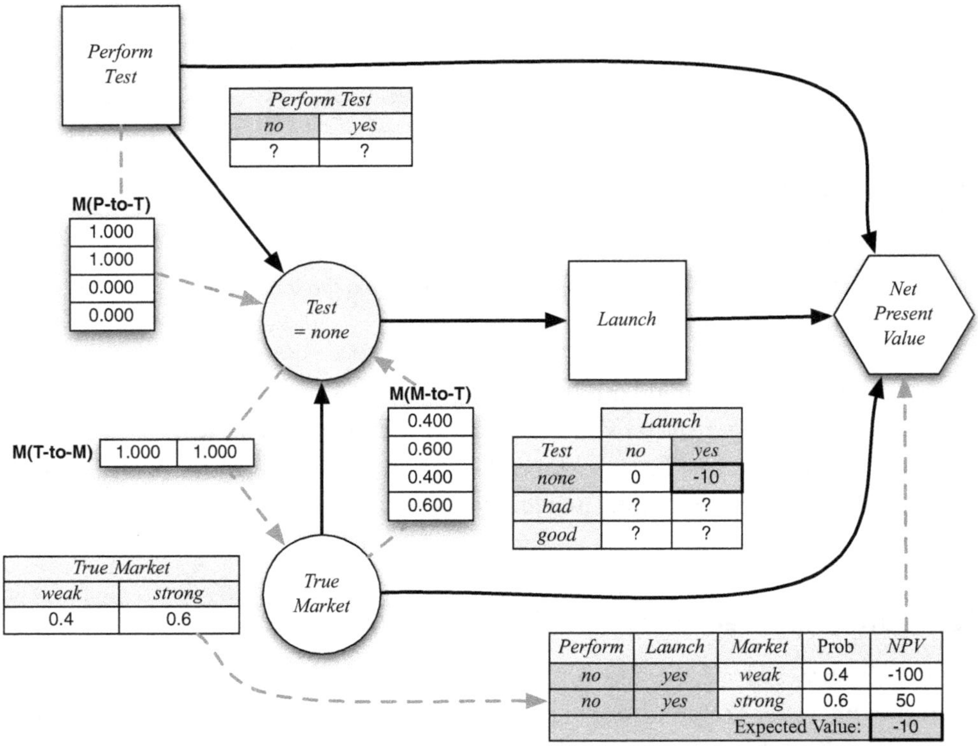

FIGURE 7.5 Expected *NPV* when *Perform* = *no* and *Launch* = *yes*.

at the calculation for when *Perform* = *yes*, *Test* = *bad*, and *Launch* = *yes* (second row, second column of the decision table). As before, the only variable of chance involved is *Market*. We know the initial probabilities for a *weak* and *strong Market* are 0.4 and 0.6, respectively, but for use in calculating expected values these need to be updated by the result of the market *Test*.

This kind of updating was explained in great detail in section 5.2.3, so here we will just skim over the details. Because we observed that the true state of *Test* is *bad*, we know the upstream message to *Market* is [0.8 0.1], which becomes *Market*'s L^- row. The L^+ row is the initial probabilities, and these are multiplied together column-by-column to get the x values. The normalized x values are the updated probabilities for *Market* given *Test* = *bad*: 0.842 for *weak* and 0.156 for *strong*.

If the market is in fact weak, then the *NPV* is $-\$100 + -\$1$ (for the test) = $-\$101$. If the market is strong, then the *NPV* is $\$50 + -\$1 = \$49$. From those evaluations we calculate expected value in the standard way: $(0.842 \times -101) + (0.158 \times 49) = -\$77.3$. Because $-\$1$ is larger than $-\$77.3$, if we were to find ourselves making the *Launch* decision with a market *Test* result of *bad*, then we would set the *Launch* decision to *no*.

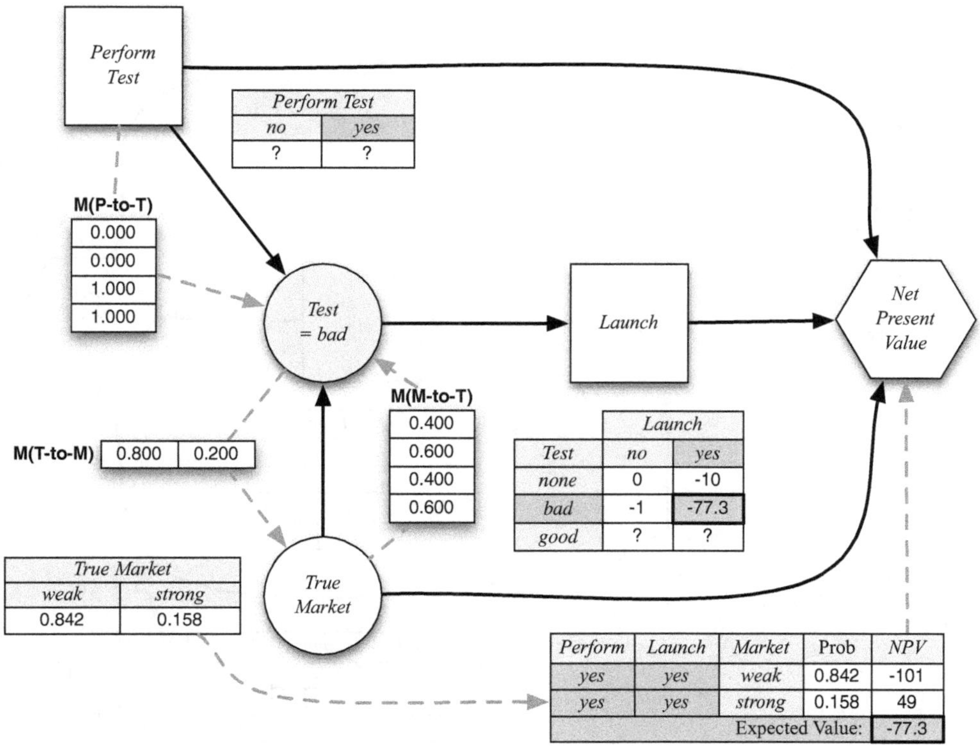

FIGURE 7.6 Expected *NPV* when *Perform* = *yes*, *Test* = *bad*, *Launch* = *yes*.

We will not go through the calculation for the final cell of the decision table, but they follow the same procedure just described (with slightly different numbers). This final calculation is presented in Figure 7.7, so we recommend that you perform this calculation and end up with the same result shown there.

Node Removal and Replacement

The decision table in Figure 7.7 details the optimal information-contingent decision strategy at the *Launch* node: given knowledge corresponding to a row, choose the option that maximizes the expected value. In this case, if there is no test (i.e., *Test* = *none*), do not launch (i.e., set *Launch* = *no*) because $0 > –$10. If the test is conducted and the result is *bad*, do not launch because –$1 > –$77.3. If the test result is *good*, however, go ahead and launch the product because –$1 < $29.7. The optimal information-contingent decision strategy indicates the expected payoffs for each state of the parent nodes; those three decisions together (one for each row) make up the **optimal strategy** for this node.

Because we always want to optimize our objective, and we know what action we would do for every contingency we might face at this juncture. Thus, the *NPV* objective node can be removed and the *Launch* decision changed into a new

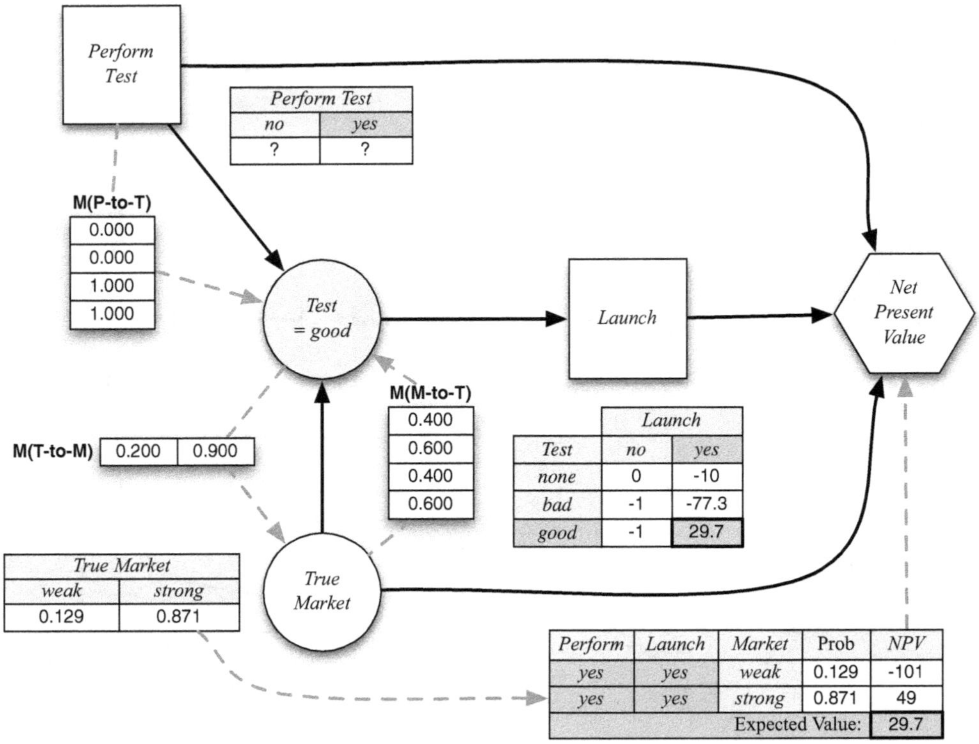

FIGURE 7.7 Expected *NPV* when *Perform* = *yes*, *Test* = *good*, and *Launch* = *yes*.

objective node that works under the assumption that an optimal decision will, in fact, be made at the *Launch* decision juncture. This results in the new, simplified causal model shown in Figure 7.8.

Notice the new table for the *Launch* node. As always there is a row for every possible input, but the columns have changed slightly. Now there is a column for what launch action *will be taken* for each input, and a column for the expected value of that input. There is also a column in Figure 7.8 which is used in the next step to optimize the *Perform* decision node.

> **Definition 7.2** *Decision strategy: a strategy is a complete contingency plan for a decision node; given any possible input there is an action preselected for that input.*

Repeat Steps 1–6 for each Decision Node

Once again, set the states of the strategic options and information nodes in the order indicated by the decision table of the next strategic option up from the one

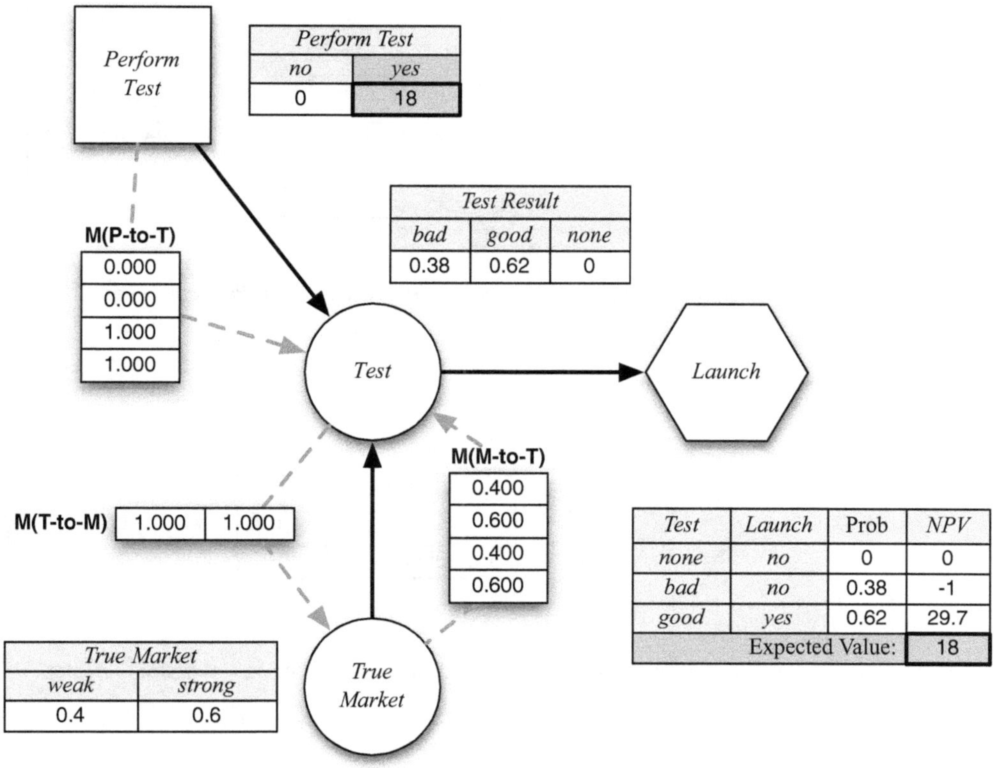

FIGURE 7.8 The model after changing the *Launch* decision node into a new objective node and the removal of the *NPV* objective node. Also shows expected *NPV* when *Perform* = *yes*.

just replaced. In our example, there is only one such node – the *Perform* market test decision node. This decision is either *yes* or *no* and is not contingent on any evidence in the model (i.e., no information links).

Important: note that the *Test* node in this new model is no longer information linked to any strategic option. It is now treated like any other environmental node. Also, it is the sole parent of the new objective node. (When you remove a node you also delete all the connections going into it.) Thus, to compute the expected value of the new *Launch* objective node, we must determine the marginal probabilities of the states of *Test* conditional upon the *Perform* market test decision.

It should be easy to see that, given a decision not to perform a market test, the marginal probability that *Test* = *none* is 1. Because if there is no test there is also no launch, and so no costs incurred and no money made; therefore the expected value of setting *Perform* = *no* is 0.

Calculating the updated marginal probabilities of the market *Test* given a decision to set *Perform* = *yes* is non-trivial, but it follows the procedures already outlined. The calculation is shown in Figure 7.9 (which, again, is a convergent node computation).

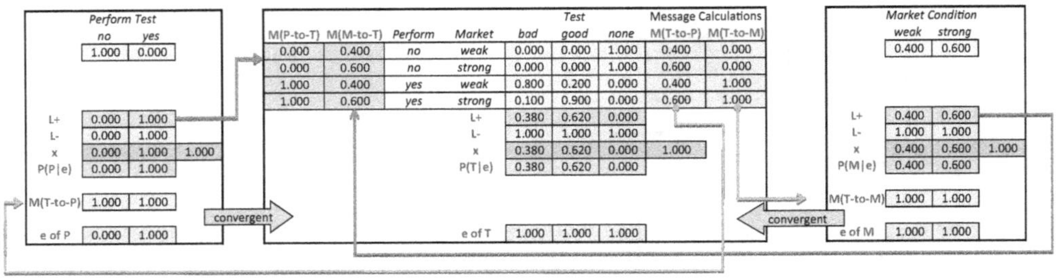

FIGURE 7.9 Updating the probabilities of the *Test* results given *Perform* = *yes*.

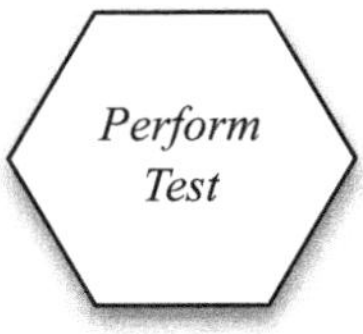

Perform	Prob	*NPV*
yes	1	18
Expected Value:		18

FIGURE 7.10 Transform *Perform* decision node into an objective node – a trivial final step.

Be careful to note the differences between Figures 7.4 and 7.9. The marginal probabilities for the *Test* node given *Perform* = *yes* are 0.38 for *bad* and 0.62 for *good*. Because we do not have new evidence regarding the state of *Test*, *Perform* and *Market* are still causally independent and so there is no update to make to *Market*.

Using these updated marginal probabilities for *Test* (the only input to the *Launch* objective node) and the dollar values for each case previously calculated, we can calculate the expected value of setting *Perform* = *yes*: $(0.38 \times -\$1) + (0.62 \times \$29.7) = \$18$. This calculation, and the way to input it into the simple decision table for *Perform* are also shown in Figure 7.8.

Final Step

At the *Perform* test option, then, the optimal decision is *yes*, yielding an expected *NPV* of \$18. This assumes, of course, optimal information-contingent decisions at the *Launch* option! Although we would not bother to do so in practice, for completeness, it might be helpful to note that, technically, the *Perform* node can, at this point, become a new objective node. It would also become the *only* remaining node. The result is the single-node model shown in Figure 7.10.

KEY CONCEPTS: CHAPTER 7

- An intervention problem requires that at least one element is decided or "set" in order to optimize some objective.
- Choosing the best intervention requires knowing the consequences of each possible action, which requires performing inferences across the influence relations among all variables involved.
- The general procedure for solving an intervention model is:

1. **Set states of options and information nodes:** Starting with the last strategic option, set the states of all strategic options and all information nodes to the ones that correspond to the first row, first column of the decision table at that node.
2. **Compute conditional probabilities:** Use belief propagation to update the conditional probabilities of the states of the parents of the objective node. Use c-independence to your advantage.
3. **Calculate expected value of objective:** Use the probabilities in Step 2 to calculate the expected value of the objective node.
4. **Fill in first row of decision table:** Holding everything else fixed, repeat Step 3 for each state of the last strategic option.
5. **Complete decision table:** Move to the next row and repeat Steps 1–4 until decision table is complete.
6. **Simplify the model:** Remove the original objective node and replace the last strategic option node with a new objective node that corresponds to the optimal information-contingent decision strategy for that node.
7. **Rinse, repeat:** Continue Steps 1–6 until all strategic options are eliminated. The final, one-node model indicates the expected value of the optimal decision strategy.

CHAPTER 7 REVIEW QUESTIONS

7.1 Part of the reason to get more education is in order to improve one's job prospects. However, it also requires taking time away from work which interferes with experienced-based promotions and results in foregone earnings. Furthermore, school isn't free – it often costs \$100 (dollar values in thousands) and keeps one from working for 2 years (for most Master degree programs). Further complicating the decision, the opportunity costs of going to school depend on the strength of the market during one's program.

If one decides not to enroll and the *Market Strength* is *weak* then a candidate can be assumed to make \$70 per year for the foreseeable future (a "glass ceiling"). If the *Market Strength* is *strong* then one's *Income* would instead be \$130 per year (which assumes that the effect of working during the *strong* market period "sticks"). If one does decide to enroll in the program, then the person misses any *strong* market effect and will make $120 per year regardless of the *Market Strength*. Keeping in mind that people change careers regularly, use non-discounted 10-year income totals to determine the value of each outcome, then use this information to determine the optimal decision using the expected value of each decision.

Market Strength	
weak	*strong*
0.8	0.2

Get Education	
no	*yes*
?	?

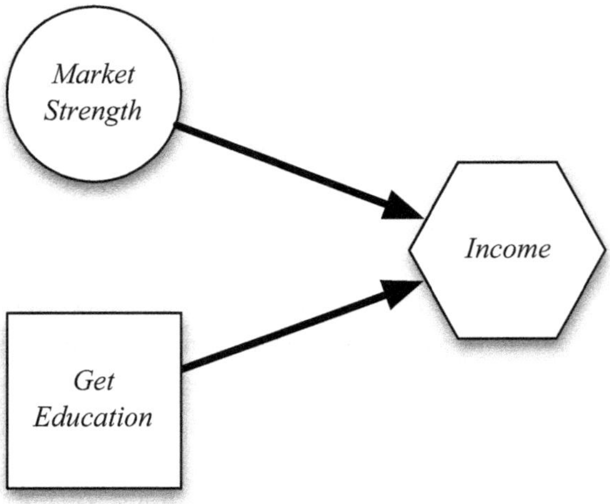

7.2 Refining the situation from the previous questions, note that not all of the benefits of education are derived from completing the degree: one may learn skills, make valuable contacts, or even drastically change career goals through school-related activities that have nothing to do with expected earnings from having the degree. With this in mind, we can separate the decision to *Start* an educational program from the decision to *Finish* it; and for simplicity, we can keep the objective in terms of *Income*.

Let's assume that some person currently makes $70 (in thousands) in a *weak Market* and will continue to earn at this level without more education. However, if the *Market Strength* becomes *strong* then the person makes $130. Once a person starts a program, they can complete the first year (paying only half the $100 tuition and passing up just one year's income) and observe the *Market Strength* before deciding to continue. If the *Market* is still *weak*, then rejoining half way will now provide $90 per year because you can leverage some contacts you've made. If the *Market Strength* is now *strong* then rejoining half way through the boom would provide a $115 annual salary. Finishing the program costs another year of tuition and foregoing another year of income, the payoff is a $120 income per year after graduating.

Use this information to determine the optimal decision strategies, and decide whether a person in such a position should start the program. Finally, compare the expected values to the previous question: does having the option to leave the program early convey any benefit?

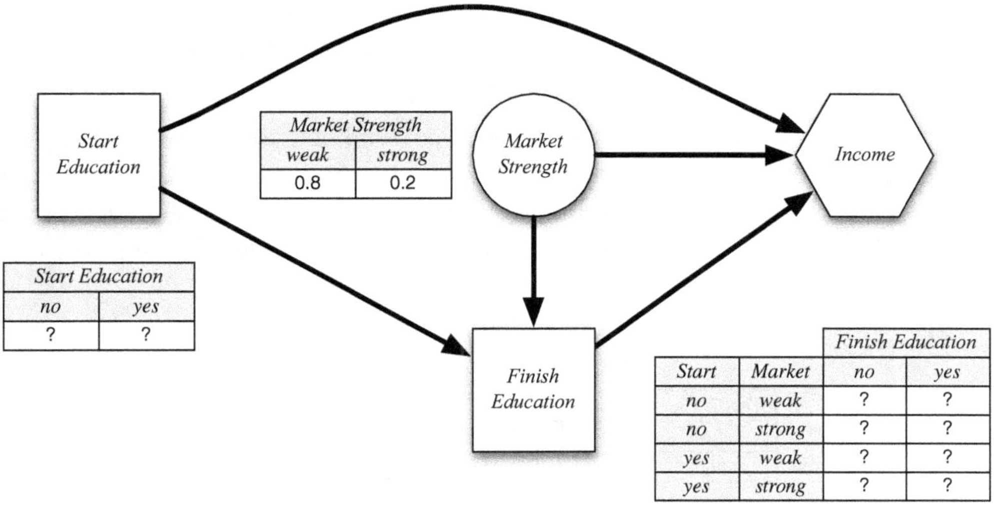

Market Strength	
weak	*strong*
0.8	0.2

Start Education	
no	*yes*
?	?

		Finish Education	
Start	*Market*	*no*	*yes*
no	*weak*	?	?
no	*strong*	?	?
yes	*weak*	?	?
yes	*strong*	?	?

7.3 In this scenario, you are responsible for marketing and advertisement decisions for a consumer electronics company. You are currently considering whether or not to launch a major roadside billboard advertisement campaign. The effectiveness of such a campaign depends on the traffic flow in the target area. The *Traffic Flow* can be either *low* or *high*, but you cannot observe this value before making your decision about the *Advertising Campaign* (*yes* or *no*). If you choose not to advertise, then there are no costs and no related revenues. If you choose to advertise and the *Traffic Flow* is *low*, then you'll make \$500 in extra revenue. If you advertise and the *Traffic Flow* is *high* then your revenue increases by \$2,000. Either way, designing the ads, having them put up, and renting the billboard space costs \$1,000.

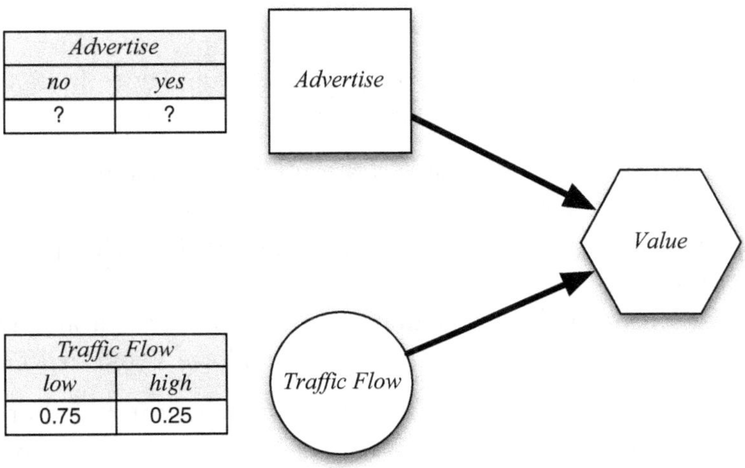

Advertise	
no	*yes*
?	?

Traffic Flow	
low	*high*
0.75	0.25

Use the value/cost data and quantitative influence diagram presented above to answer the following questions:

(a) What is the expected value of choosing *Advertising Campaign* = *yes*?
(b) What is the expected value of choosing *Advertising Campaign* = *no*?
(c) What, therefore, is the optimal decision?

7.4 After making a deal with the regional government, they agree to give you the data they collect on the number of cars going through the area using road sensors. The sensors are not perfectly accurate, and the conditional probabilities of your *Sensor* data given the actual *Traffic Flow* are provided in the diagram below.

	Sensor	
Traffic Flow	*low*	*high*
low	0.75	0.25
high	0.1	0.9

Advertise	
no	*yes*
?	?

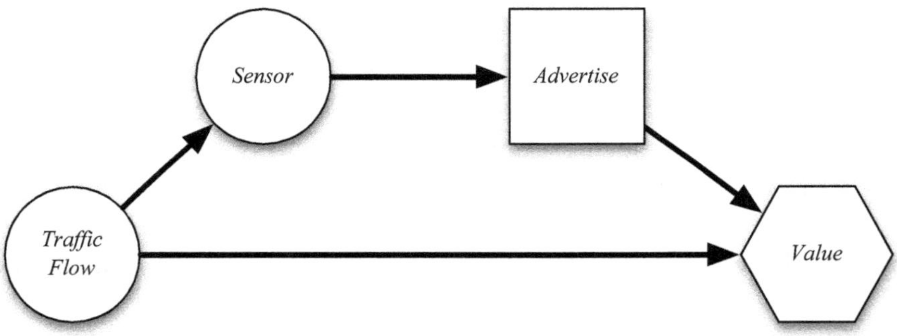

Traffic Flow	
low	*high*
0.75	0.25

Use this information and the cost and revenue values from the previous question to answer the following questions:

(a) What are the four possible strategies in this situation?
(b) What are the expected values of each strategy?
(c) What, therefore, is the best *Advertising* decision policy (i.e., the optimal strategy)?
(d) Compared with the situation without the sensor data, are there any cases for which having this extra information lowers your expected value?

7.5 Somebody at the regional government office gets smart and/or greedy and now you have to pay to get the road sensor data. The cost to observe the sensor data is $300. So now you have to decide whether or not to *Buy* the sensor data, and also whether or not to *Advertise* in each possible information state.

Buy Data	
no	*yes*
?	?

Advertise	
no	*yes*
?	?

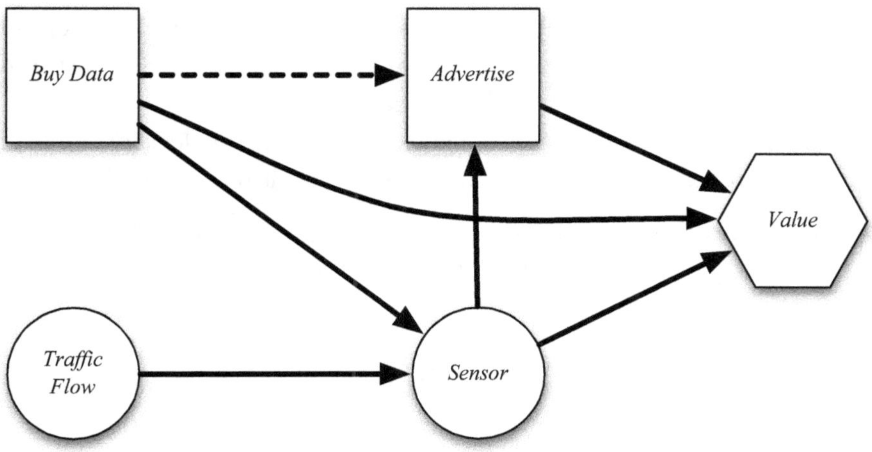

Traffic Flow	
low	*high*
0.75	0.25

		Sensor		
Buy Data	*Traffic*	*low*	*high*	*none*
no	*low*	0	0	1
no	*high*	0	0	1
yes	*low*	0.75	0.25	0
yes	*high*	0.1	0.9	0

Use this information and the values from the previous questions to answer the following questions:

(a) Which node do you need to eliminate first?
(b) What does the causal diagram look like after the first elimination (draw the remaining nodes and connections)?
(c) Calculate the optimal decision strategy for the *Advertise* decision node.
(d) Now calculate and report the expected value of the optimal *Buy Data* decision?
(e) Let's say that rather than spending the money on the road use data, you can instead reduce the cost of running the advertising campaign by \$300 (from \$1,000 to \$700). Is this a better investment of the money? Does this change your optimal action?

7.6 It turns out that the sensors are being used on another stretch of road and won't be available in time for you to make your decision. It is still the case that the *Traffic Flow* can be either *low* or *high*, but to estimate these states you need to procure another data source. *Traffic* is actually determined by three factors: the *Weather* (*fair* or *poor*), *Local Events* (*few* or *many*), and *Construction* (*no* or *yes*). You cannot observe the *Traffic Flow* directly, but you can choose to observe these variables. The costs to observe each variable is: *Weather* = \$0, *Local Events* = \$200, and *Construction* = \$400.

(a) Because of the time constraints in getting your advertising campaign going as quickly as possible, you can only observe at most, one of the three variables. Draw an influence diagram that represents this new situation (you don't have the probability tables, just draw all the appropriate nodes and connections).

(b) Given this situation, construct the (empty) decision node table you would fill in to determine the optimal *Advertise* decision.

7.7 In the process of optimizing the following decision problem, what will the influence diagram look like after the first elimination?

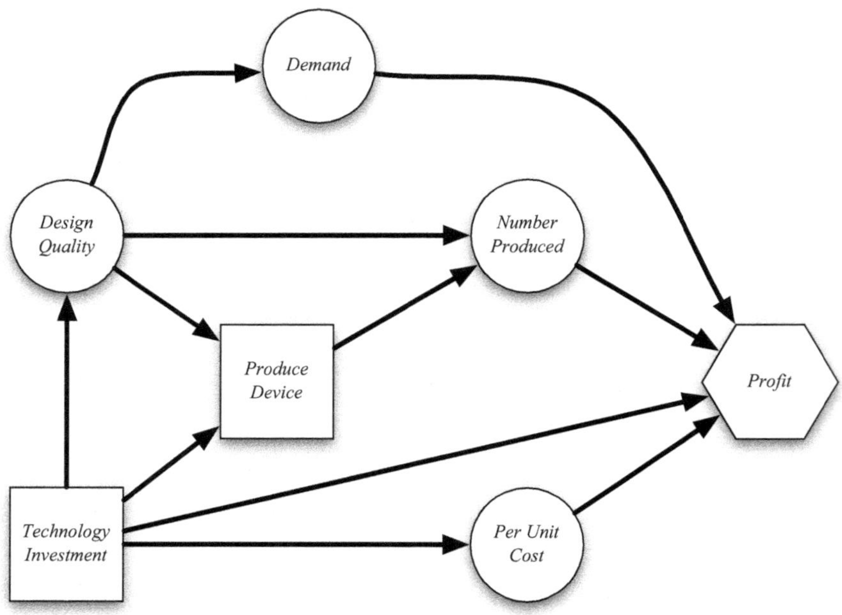

7.8 Penurious Insurance Inc. is interested in reducing the losses they face from fraudulent insurance claims (e.g., padding the loss amount, claiming pre-existing damages, and falsifying documents). Because insurance claims are frequently handled by adjusters (which are people), there is a chance that a fraudulent claim will be recognized by the adjuster. Insurance companies can also choose to buy filtering software that identifies fraudulent behavior by detecting patterns in claimant properties, and using the patterns to determine the probability that a claim is fraudulent.

If Penurious Insurance catches none of the fraudulent claims, then it makes \$80 per annum (figures in millions). Depending on the number and value of the fraudulent claims that are caught, the *Savings* can be either *small* = \$2 or *large* = \$10. If they choose to purchase the *Fraud Detection Technology*, there is a reoccurring outlay of expenses for server maintenance, profiling filter updates, training, etc. of \$1. What is the optimal decision regarding the purchase of *Fraud Detection Technology* in order to maximize company *Profit*? For a more advanced business problem (beyond what we'll ask you to do, but worth thinking about), what could the software company charge for their product and it still be worth purchasing for Penurious Insurance?

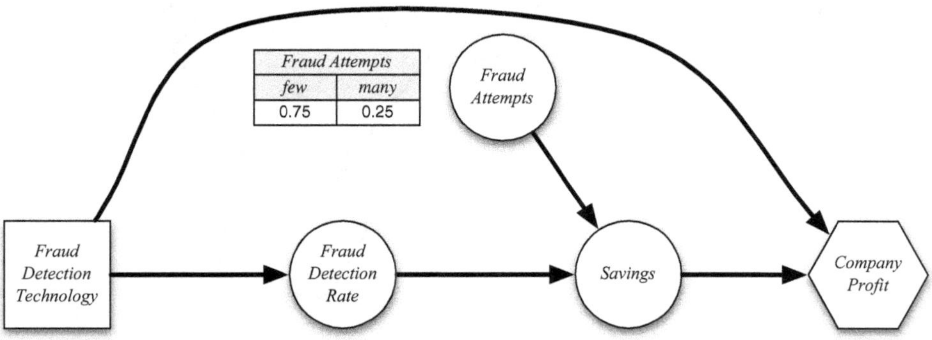

Fraud Attempts	
few	many
0.75	0.25

Buy Technology	
no	yes
?	?

	Detection Rate	
Technology	low	high
no	0.8	0.2
yes	0.1	0.9

		Savings	
Detection	Attempts	small	large
low	few	0.7	0.3
low	many	0.9	0.1
high	few	0.35	0.65
high	many	0.05	0.95

7.9 National defense contractors typically engage in large engineering projects to bring available technology to market, but they also may contain some portion of research and development work. The *Benefits* of the R&D work must (like all other aspects of business) outweigh the costs. In these situations, the *Benefits* are determined by the *Level of R&D* and the number of *Research Scientists* employed. It has been observed that companies with *high Levels of R&D* also have *many Research Scientists*.

If the *Level* is *low*, and *few* scientists are working on R&D, then the *Benefit* exceeds the labor costs by \$15 (in millions). However, if the company hires *many Research Scientists*, but only wins a few R&D grants (which makes the *R&D Level* = *low*), then the net *Benefit* is \$–10. When the *R&D Level* is *high* then either there are too *few Research Scientists* to complete them and hence the *Benefit* is a loss of \$25, or there are *many Research Scientists* and the *Benefit* is \$35. The natural *Level of R&D* fluctuates, but the company has the option to officially *Promote R&D Projects*. If they decide to intervene they can set the state of *R&D Level* to *high*. Examine the two distinct causal scenarios below and answer the following questions:

Scenario 1

	R&D Level	
Promote	*low*	*high*
no	0.5	0.5
yes	0	1

	Research Scientists	
R&D Level	*few*	*many*
low	0.45	0.55
high	0.2	0.8

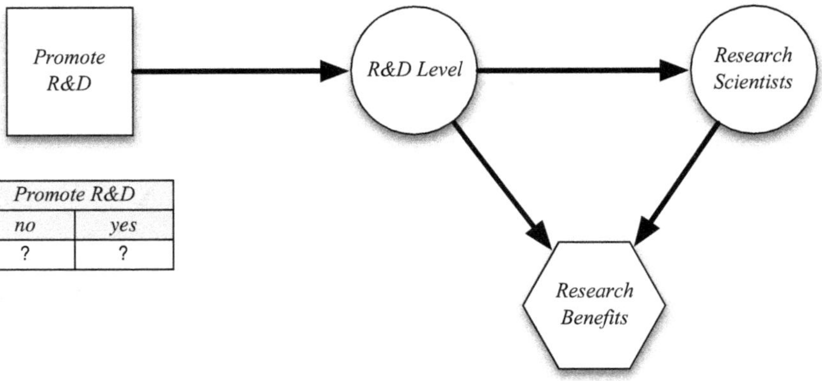

Promote R&D	
no	*yes*
?	?

Scenario 2

		R&D Level	
Promote	*Scientists*	*low*	*high*
no	*few*	0.8	0.2
no	*many*	0.1	0.9
yes	*few*	0	1
yes	*many*	0	1

Research Scientists	
few	*many*
0.57	0.43

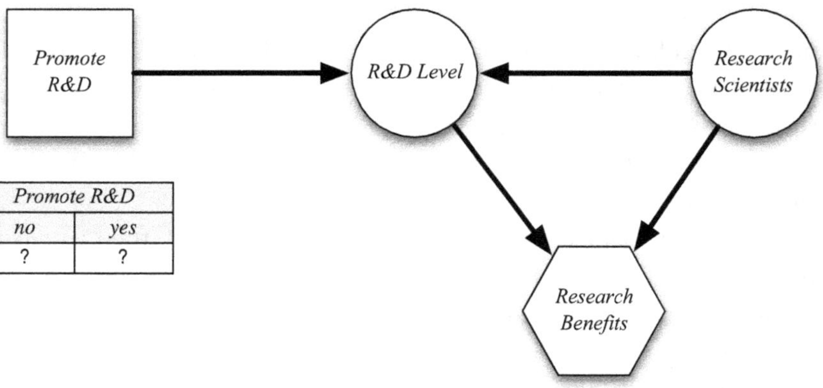

Promote R&D	
no	*yes*
?	?

(a) Is either one of the two scenarios obviously a better explanation or more intuitive than the other?

(b) When there is *no* intervention (i.e., R&D is not promoted), what are the marginal probabilities for the *R&D Level* in each scenario?

(c) Calculate the expected benefit of the intervention (i.e., setting *Promote R&D Projects?* = *yes*) compared to leaving the system to run on its own for both scenarios. What is the optimal policy in each scenario? What does this tell you about getting the causality correct when intervening in systems?

7.10 Returning to the production and warehouse problem, rather than rely on the individual departments to make production and facility decisions based on the *Expected Sales* figures from Accounting, you now have the option to set the number of *Units* to *many* and/or the *Warehouse Size* to *large* – or neither. Because the intervention takes up your time and has associated administrative and monitoring expenses, setting a variable costs \$50 (so setting both costs \$100). Also, the *Combined Unit Cost* variable now translates into a *Profit* level: *Cost* = *low* → \$500; *Cost* = *med* → \$300; and *Cost* = *high* → \$100. Use this information to determine the optimal decision of which variables (if any) to perform an intervention on.

		Units		
Set Units	Exp Sales	few	mod	many
no	low	0.35	0.60	0.05
no	high	0.05	0.45	0.50
yes	low	0.00	0.00	1.00
yes	high	0.00	0.00	1.00

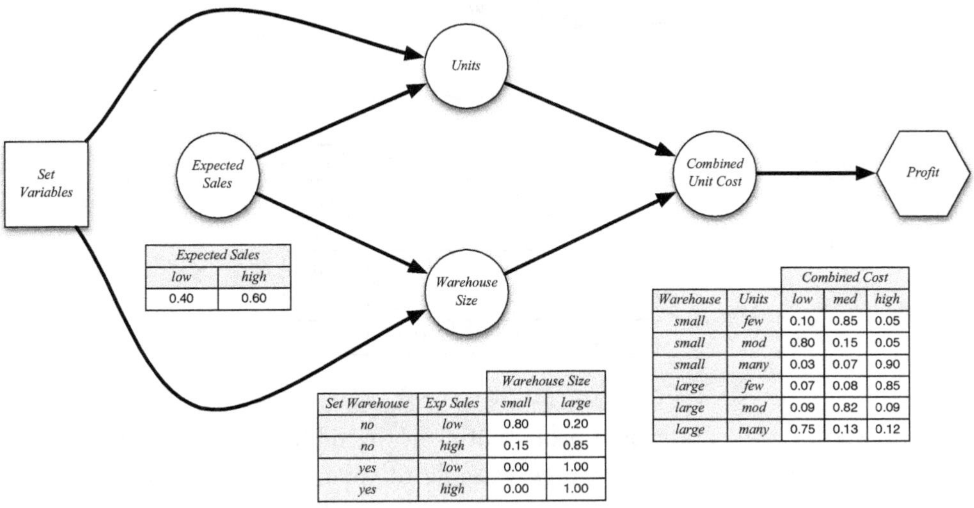

Expected Sales	
low	high
0.40	0.60

		Warehouse Size	
Set Warehouse	Exp Sales	small	large
no	low	0.80	0.20
no	high	0.15	0.85
yes	low	0.00	1.00
yes	high	0.00	1.00

		Combined Cost		
Warehouse	Units	low	med	high
small	few	0.10	0.85	0.05
small	mod	0.80	0.15	0.05
small	many	0.03	0.07	0.90
large	few	0.07	0.08	0.85
large	mod	0.09	0.82	0.09
large	many	0.75	0.13	0.12

7.11 In this problem, you are the manager of a "pre-owned motor-vehicle sales establishment". The causal model below depicts an intervention problem in which you must decide whether or not to send employees to spread flyers around town. Sometimes the employees decide to do this on their own, depending on how many people are working that day. The more people who are working, the more people are available to service customers; and the flyers do have an effect on sales as well. Obviously the more vehicles that are sold the greater the profits, but the employees work purely on commission (and so don't affect profits directly). Copies are cheap, but not free, so the *Flyer Campaign* affects profits as well.

Require Flyers	
no	*yes*
?	?

		Flyer Campaign	
Require	*Staff*	*no*	*yes*
no	*low*	0.85	0.15
no	*high*	0.75	0.25
yes	*low*	0	1
yes	*high*	0	1

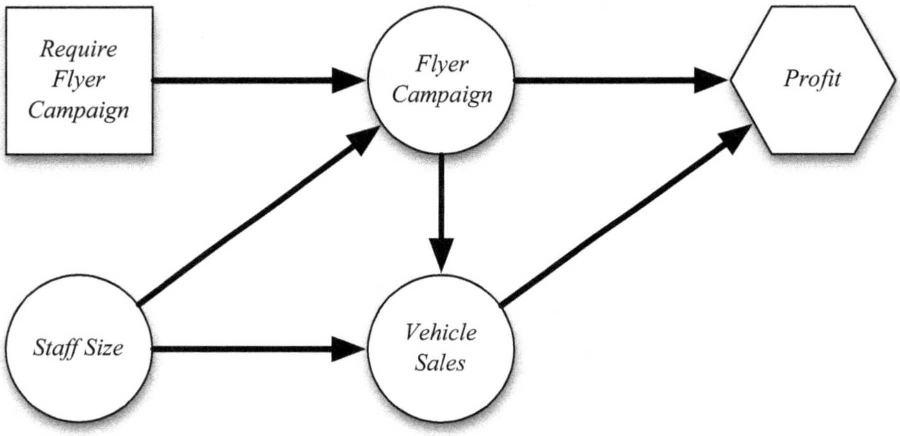

Staff Size	
small	*large*
0.75	0.25

		Vehicle Sales	
Staff	*Flyers*	*low*	*high*
small	*no*	0.9	0.1
small	*yes*	0.75	0.25
large	*no*	0.65	0.35
large	*yes*	0.3	0.7

Based on the conditional probabilities in the tables above, will intervening by setting the *Flyer Campaign* variable to *yes* tend to increase or decrease the marginal probability that *Vehicle Sales* is *high*?

7.12 Consumers for electronic goods have become increasingly design conscious; having a pleasant design strongly increases the chance that the *Market Demand* for the product is *strong*. Before releasing a new product, your company has the opportunity to invest in a new cooling technology that (if successful) will allow you to produce thinner and lighter laptop computers that also run more quietly, while using less power. It therefore also opens up options for improved visual design.

The *Technology Investment* costs \$100, and *Start Production* costs \$350 (both \$0 if *no*). If the *Market Demand* is *weak*, then the product will earn you \$250 in revenue. If the *Market Demand* is *strong*, then the product will earn you \$600. If you decide not to *Start Production* then you earn \$0 in revenue.

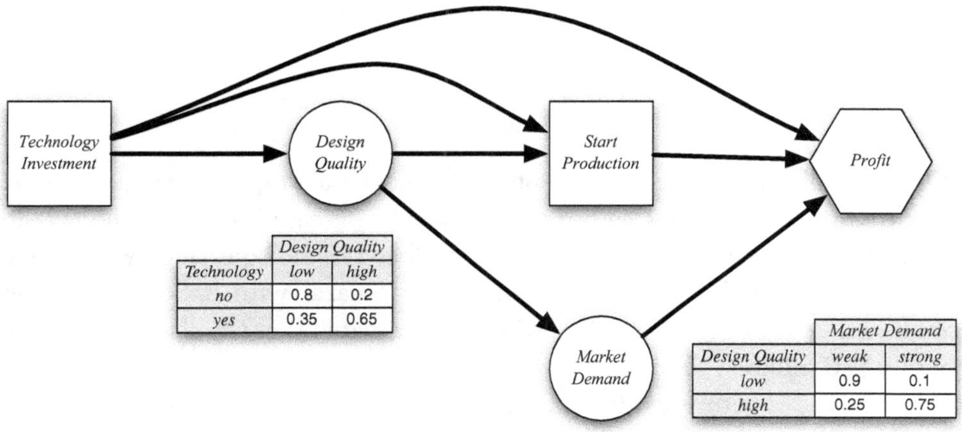

Technology	Design Quality: low	Design Quality: high
no	0.8	0.2
yes	0.35	0.65

Design Quality	Market Demand: weak	Market Demand: strong
low	0.9	0.1
high	0.25	0.75

(a) Write down the payoff table associated with the objective node.

(b) Now write down the decision table for *Start Production*, and fill it in with the expected values for each decision.

(c) What is the optimal decision **strategy** for *Start Production*?

(d) Draw the causal model (nodes, connections, and conditional probability tables) after the above calculation has been made and the objective node removed.

(e) Now solve the problem for the optimal decision at the *Technology Investment* node. What is the best decision and what is the expected profit of that decision?

Chapter 8

Application

Disrupting the Taxi Business

At the time of this writing, smartphone technologies are rapidly changing the way many traditional industries do business. A significant case-in-point is the taxi and limousine business. In 2009, UberCab (later, Uber Technologies, Inc.) was founded by Garrett Camp, Travis Kalanick, and Oscar Salazar with backing from well-established venture capitalists. By 2012, it is estimated that the startup had raised roughly $50 million in funding.[1]

The basic idea is to provide smartphone users with a "taxi-hailing" application such that a user can simply call it up, press a button, and a taxi is sent to the user's location (typically identified via the phone's GPS system). The fare (plus tip) is paid automatically via credit card through the app. At any given time, the user can call up a map of her or his location to see where the Uber cars are in relation to them and see an estimated time-of-arrival. Once the car arrives, the user is sent an e-mail or text message indicating the car is waiting. For the tech-savvy consumer's point-of-view, this service streamlines the taxi experience and eliminates the uncertainty of when the ride will arrive. Also, the automatic payment and receipt by e-mail is useful for business travelers.

Uber launched operations in San Francisco in 2010, and by 2012, had entered a number of other cities (including New York, Chicago, Toronto, and London). Over this period, a number of competitors offering similar services sprang up, including U.K.-based Hailo Network Holdings Ltd., Toronto-based Winston Inc., New York-based Get Taxi Inc., and others. These startups present varied approaches to implementing the same basic idea. Uber, for example, enrolls independent taxi and limo drivers. These drivers are free to go about their business normally but also have the option to sign into the Uber system as they see fit during the day, in order to increase their utilization using Uber clients. Uber, for its part, supplies the smartphone technology that coordinates rides and handles the automatic billing.

1 Wikipedia (2012)

Competitors run the gamut from partnering with existing taxi companies to supply them with Uber-style apps to purchasing their own taxi licenses and establishing taxi services along more traditional lines.

This new technology has not been without controversy. Traditional taxi companies have petitioned local governments and initiated lawsuits to ban Uber on the grounds that they are, in effect, running a taxi service without the government-required licenses and permits. This, argue the incumbents, creates unfair competition and raises public safety concerns. According to Joshua Gans, the Chaired Professor of technical innovation and entrepreneurship at the Rotman School of Management, such incumbent reactions are not unusual, "You can go back all the way to the introduction of the automobile and we had similar issues way back then. All manner of people tried to put restrictions on cars. This happens time and time again. It's very hard to change these sorts of industries." Dobby (2012).

8.1 AN ALLOCATION OF RESOURCES DECISION

With this background in mind, let us construct a software-based causal model to illustrate the analysis of single actor intervention decisions in a more complex setting. Consider a fictional, new entrant into the taxi-hailing market called Hack-a-Hack Corporation. HaHCo's business plan is based upon the Uber model: independent drivers will be enrolled; these drivers will be free to tap into the HaHCo system whenever they wish; HaHCo will run the scheduling and billing operations through their technology; customers will access the system via a smartphone app.

Assume that first-round venture capital investors have provided enough funding for HaHCo to establish sufficient infrastructure to launch its service in one or both of San Francisco and New York. HaHCo's venture investors require a proof-of-concept demonstration, involving the enrollment of 200 drivers. Therefore, the immediate decision facing HaHCo management is how to split the 200 driver positions between San Francisco and New York.

HaHCo believes that, initially, drivers who experience working within their system will choose to use the HaHCo system exclusively (i.e., full-time). Later, as the number of enrolled drivers increases substantially, the firm expects that a subset will opt into the system on a less than full-time basis. However, for the purposes of the 200 driver launch, managers are confident that all drivers will work with HaHCo full-time.

HaHCo market research indicates that, on average, a taxi in either of these markets make 18.8 trips per shift. The average number of miles per trip is 9.6. Therefore, the number of paid miles driven per shift is 180. A typical driver works 226 shifts per year (accounting for holidays, vacations, and sick days). Thus the typical driver will generate an expected 40,680 paid miles per year.

The experience of companies offering similar services suggests that the customers who use taxi hailing services are willing to pay a 50% to 75% premium over standard rates. Based upon this, management has decided to set fares at \$3.75/mile in San Francisco and \$3.00/mile in New York City. Based upon an expected 40,680 paid miles per driver per year, each driver enrolled in San Francisco is expected to generate fare revenues of $40,680 \times \$3.75 = \$152,500$/year. New York drivers should generate average revenues of \$122,040/year due to their lower rate of \$3.00/mile.

HaHCo will operate on a simple fare-sharing basis with drivers. Legally, all the enrolled drivers in a given city must be paid the same percentage – HaHCo cannot cut different deals with different drivers. Management's best estimate is that 10, full-time drivers can be enrolled for a 50% split (i.e., drivers receive 50% of the total fare income they generate). It is economical for the firm to enroll drivers for a given city in groups of 10. For every 10 drivers beyond the first 10, it is estimated that the sharing percentage must be increased by two percentage points. In other words, in order to induce 20 drivers to enroll in San Francisco and work within the HaHCo system full-time, all 20 drivers would have to be offered a 52% share of fare revenue. Alternatively, if the firm decided to enroll all 200 drivers in New York, HaHCo would have to offer them all an 88% share of revenues earned.

Based upon this information, how should the 200 drivers be allocated between San Francisco and New York? As we have set the problem up, this is clearly a single actor intervention problem, where the intervention is deciding how to set the allocation of enrolled drivers between two markets. At the moment, the problem is sufficiently simple to be solved either analytically, via a spreadsheet program like Excel, or using a causal model. Let us begin with the causal model and then check its results using more conventional approaches.

The causal model is illustrated in Figure 8.1. The *D-SF* variable is the number of drivers enrolled in San Francisco. It is setup as a decision node with values ranging from 0 to 200 in steps of 10. The *D-NY* variable is the number of drivers enrolled in New York. Since the venture capitalists require the firm to launch with 200 drivers, *D-NY* can be setup as a function node: $D\text{-}NY = 200 - D\text{-}SF$.

The next step is to compute the average cost per car, per year. At this point, the only cost we are considering is the share of revenue that must be paid to drivers on an annual basis. As described above, this varies in a direct way, depending upon the number of cars enrolled in the city. Therefore, we can also setup the average cost nodes as functions of the numbers of drivers in each city. Specifically,

$$AC\text{-}SF = \$152{,}550 \times (0.48 + 0.02 \times D\text{-}SF)$$

$$AC\text{-}NY = \$122{,}040 \times (0.48 + 0.02 \times D\text{-}NY).$$

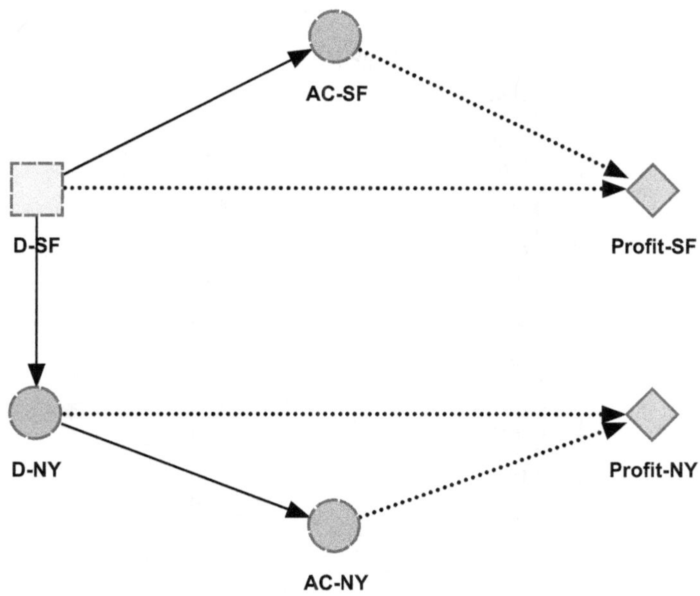

FIGURE 8.1 Model of HaHCo driver enrollment decision.

Thus, if 10 drivers are enrolled in San Francisco, the sharing percentage is 50%. Based upon their expected fare revenues of $152,550, the amount paid out to each driver would be $76,275.

Last, the payoff nodes are defined, one for each city. Again, these are simple functions of the variables for the numbers of drivers and the average costs:

$$\textit{Profit-SF} = (152{,}550 - \textit{AC-SF}) \times \textit{D-SF}$$

$$\textit{Profit-NY} = (122{,}040 - \textit{AC-NY}) \times \textit{D-NY}.$$

Again, it is worth noting that the annual fare revenue per driver is constant in each market. The assumption is that the numbers of drivers being enrolled are sufficiently small relative to the total market that fares are not affected (also keep in mind that these are already independent drivers, so they are simply substituting away from their regular taxi fares and into HaHCo-related fares, so the overall effect on market pricing is negligible).

Once the model is set-up, it is a simple enough matter to solve it using the software tool.[2] The results of the analysis are presented in Figure 8.2. It shows that the optimal number of enrollments for San Francisco is between 100 and 110. Recalling that the firm plans to enroll drivers in groups of 10, this indicates that

2 In the case of BayesiaLab, the model is put into "validation mode," and "Learn Static Policy" is selected from the "Learning" tab on the main menu. The optimal decision choice is then highlighted when the decision node is inspected in "modeling mode."

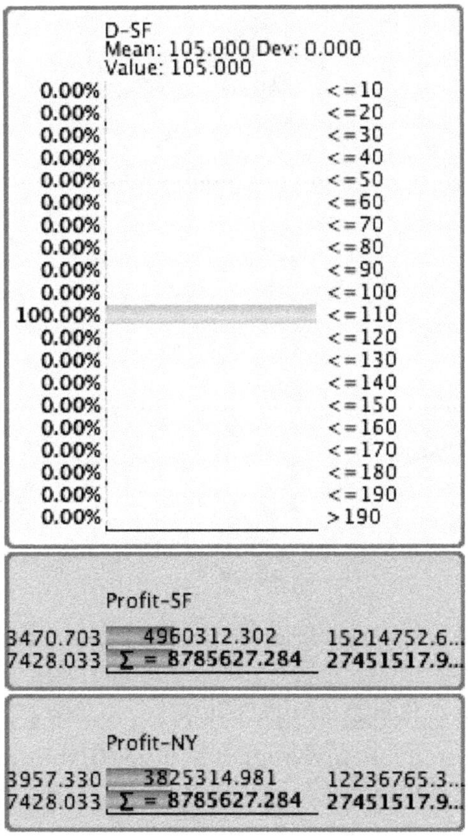

FIGURE 8.2 Optimal decision: enroll 110 drivers in SF, 90 in NY.

110 drivers should be enrolled in San Francisco and 90 in New York. The expected aggregate profit is, according to the analysis, \$8,785,627.

At this stage, the problem is sufficiently simple that we can check the causal analysis using more conventional means. One obvious way is to use a spreadsheet. The calculations for a spreadsheet analysis are presented in Figure 8.3, over the range of 100 to 110 drivers allocated to the San Francisco market. As we see, the exact optimal number of drivers to enroll in San Francisco is 103, which correctly falls within the 100–110 range indicated by the causal analysis.

This problem is also simple enough to solve analytically. The profits in each market expressed as a function of the number of drivers enrolled are:

$$\textit{Profit-SF} = (152{,}550 - 152{,}550 \times (0.48 + 0.002 \times \textit{D-SF})) \times \textit{D-SF}$$

$$\textit{Profit-NY} = (122{,}040 - 122{,}040 \times (0.48 + 0.002 \times \textit{D-NY})) \times \textit{D-NY}.$$

Aggregate profit is the sum of these which, using the fact that *D-NY* = 200 – *D-SF*, can be expressed as a function of one variable. Taking the first derivative of aggregate profit and setting it equal to zero yields:

$$113{,}497 - 1{,}098 \times D\text{-}SF = 0.$$

Solving this for *D-SF* yields a value of 103.3.

Notice that, by choosing to enroll drivers in groups of 10 rather than one at a time, HaHCo loses roughly $24,000 in aggregate annual profit due to the difference between enrolling 103 optimally in SF versus 110. This demonstrates that in any analysis, choosing an appropriate level of precision is an important consideration with real consequences. In this case, the difference between the optimal choice of 103 versus the coarser choice of 110 is only 0.3% – presumably a sufficiently small difference to warrant simplifying computations by breaking enrollments into 10-unit blocks.

It is also worth noting that the expected value according to the causal model ($8,785,627) is off from the actual value of enrolling 110 drivers in San Francisco ($8,768,574) by roughly $17,000, or 0.2%. This difference is due to the fact that the software is estimating the average value over the 100–110 driver interval. Loosely, the expected value reported by the causal model is around the midpoint of this interval (105 or 106, though the exact point will depend upon the solution algorithm of the software being used).

In summary, all the usual methods result in suitable answers. Indeed, if there were nothing more to the problem than the facts considered thus far, one would be hard-pressed to argue in favor of using the less familiar causal modeling approach over the straightforward – and easily understood – spreadsheet model. However, we will now extend the model to permit us to consider a greater array of real-world concerns. Chief among these will be uncertainty with respect to important new variables. These added levels of complexity quickly render spreadsheet and calculus approaches much less appealing.

8.2 PRICE UNCERTAINTY AND MARKET RESEARCH

One of the interesting features of the app-based taxi-hailing business is the ability of firms like Uber to set prices in real-time in order to equilibrate supply and demand. Economists call this "peak-load" pricing. Higher prices during busy periods induce more drivers to opt into the system and, at the same time, skim off those customers with the highest willingness-to-pay. For example, for New Year's Eve 2011, Uber charged up to seven times its normal rates in New York City.

While this is good news for our fictional HaHCo, it does imply that average prices over the course of a year are subject to a fair amount of uncertainty. For example, the number of bad-weather days and when those days occur during the year (even the time of day the inclement weather hits) have a substantial impact on taxi demand. Another key uncertainty is not only the extent to which smartphone

San Francisco											
Drivers enrolled	100	101	102	103	104	105	106	107	108	109	110
Revenue/Driver	152,550	152,550	152,550	152,550	152,550	152,550	152,550	152,550	152,550	152,550	152,550
Total Revenue	15,255,000	15,407,550	15,560,100	15,712,650	15,865,200	16,017,750	16,170,300	16,322,850	16,475,400	16,627,950	16,780,500
Sharing %	68.0%	68.2%	68.4%	68.6%	68.8%	69.0%	69.2%	69.4%	69.6%	69.8%	70.0%
Cost/Driver	103,734	104,039	104,344	104,649	104,954	105,260	105,565	105,870	106,175	106,480	106,785
Total Cost	10,373,400	10,507,949	10,643,108	10,778,878	10,915,258	11,052,248	11,189,848	11,328,058	11,466,878	11,606,309	11,746,350
Profit	4,881,600	4,899,601	4,916,992	4,933,772	4,949,942	4,965,503	4,980,452	4,994,792	5,008,522	5,021,641	5,034,150
New York											
Drivers enrolled	100	99	98	97	96	95	94	93	92	91	90
Revenue/Driver	122,040	122,040	122,040	122,040	122,040	122,040	122,040	122,040	122,040	122,040	122,040
Total Revenue	12,204,000	12,081,960	11,959,920	11,837,880	11,715,840	11,593,800	11,471,760	11,349,720	11,227,680	11,105,640	10,983,600
Sharing %	68.0%	67.8%	67.6%	67.4%	67.2%	67.0%	66.8%	66.6%	66.4%	66.2%	66.0%
Cost/Driver	82,987	82,743	82,499	82,255	82,011	81,767	81,523	81,279	81,035	80,790	80,546
Total Cost	8,298,720	8,191,569	8,084,906	7,978,731	7,873,044	7,767,846	7,663,136	7,558,914	7,455,180	7,351,934	7,249,176
Profit	3,905,280	3,890,391	3,875,014	3,859,149	3,842,796	3,825,954	3,808,624	3,790,806	3,772,500	3,753,706	3,734,424
Aggregate Profit	**8,786,880**	**8,789,992**	**8,792,006**	**8,792,921**	**8,792,738**	**8,791,457**	**8,789,077**	**8,785,599**	**8,781,022**	**8,775,347**	**8,768,574**

FIGURE 8.3 Spreadsheet analysis of driver allocation.

users are willing to adopt the technology, but also whether those who do adopt it tend to be at the high or low end of the demand curve.

Let us illustrate how such considerations can be handled in a causal model by expanding our example. Suppose each city can be categorized as to whether it is tech-savvy. A tech-savvy market is one that is more likely to contain upscale HaHCo customers; i.e., users whose willingness-to-pay is high, resulting in higher average prices during peak-load periods. Because HaHCo has not yet operated in either market, it does not yet know into which category San Francisco and New York fall.

Suppose HaHCo management estimates that San Francisco has a 90% probability of being tech-savvy and that New York has only a 50% probability. Thus, they expect San Francisco to be much more likely to be tech-savvy. If it does turn out to be tech-savvy, it is estimated that the average fares in San Francisco, accounting for peak-load pricing, will be $4.25/mile. If not, average fares will be $3.25/mile. Assume the corresponding fares for New York are $3.50 and $3.25/mile.

Prior to launching in a market, HaHCo can conduct market research to determine whether it is tech-savvy or not. These market tests are reasonably accurate: the likelihood of correctly identifying the market type is 90%. For example, if the true market type is *savvy*, the market research will indicate this 90% of the time. The cost to conduct this type of pre-entry market research is $50,000.

In this situation, there are three interventions that must be considered by HaHCo management: (1) conduct a *Market Test* for SF; (2) conduct a *Market Test* for NY; and (3) based upon the results of 1 and 2, allocate the 200 drivers between SF and NY. The expanded causal model is shown in Figure 8.4. The *Savvy* variables take one of two states (*savvy* or *not savvy*) with the probabilities mentioned above. The state of the markets affects their average prices which, in turn, affect profit.

The firm can choose to *Test Market* or not. The decision to test in combination with the true state of the market determines the test *Result*. The conditional probability table for San Francisco test results (*Result-SF*) is shown in Figure 8.5. If the test is run and the true market type is *savvy*, then the test result is *good market* with 90% probability. Notice that the *Result-SF* variable requires a *no result* state that occurs with certainty (100% likelihood) if the test is not run. Of course, the decision on whether or not to test has a direct impact on profit: if the decision is to test, then $50,000 must be subtracted from profit to reflect the cost of the market research. Note also that the results of the market tests are available at the time of the driver allocation decision (hence, the arrow pointing from the *Result-SF* node to the *D-SF* node and likewise for NY).

With the model completely setup, the solution can be computed, and here are the conclusions. First, the optimal testing plan is to test in NY but not in SF. This is not surprising. SF already has a very high likelihood of being a tech-savvy market.

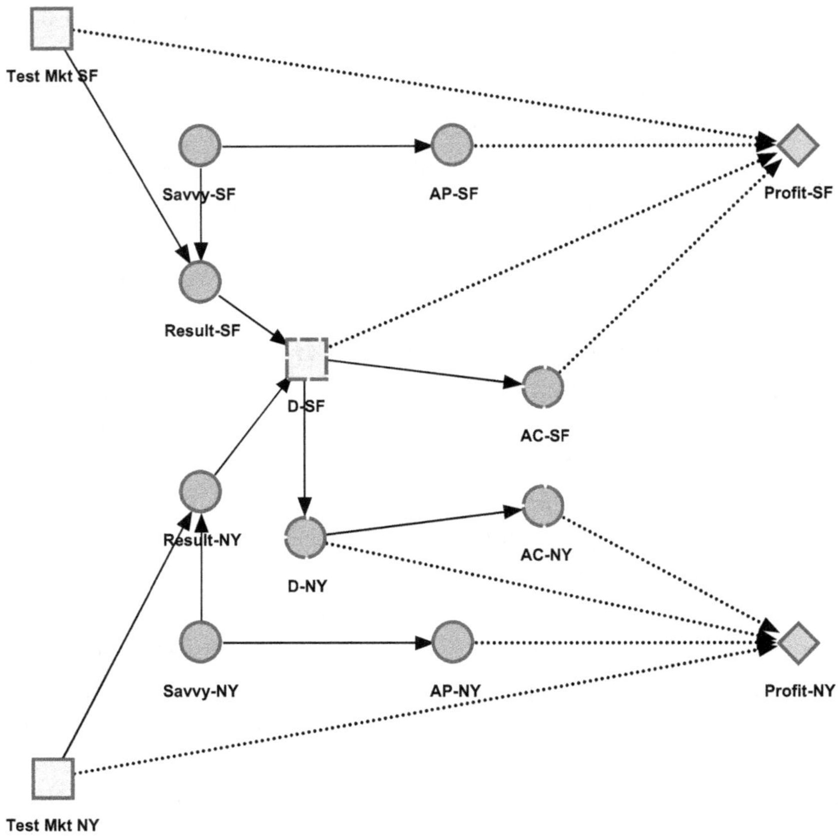

FIGURE 8.4 Model with market uncertainty and research options added.

Test Mkt...	Savvy-SF	No Result	Good Mkt	Bad Mkt
Test SF	Savvy	0.000	90.000	10.000
	Not Savvy	0.000	10.000	90.000
No Test SF	Savvy	100.000	0.000	0.000
	Not Savvy	100.000	0.000	0.000

FIGURE 8.5 Conditional probability table for *Result-SF*.

Therefore, it is not worth the $50,000 to conduct a test for the purpose of verifying this. NY, on the other hand, only has a 50% chance of being tech-savvy. Here, the pre-launch market research is valuable. The optimal allocation of drivers depends upon the results of the NY test.

The driver allocation results are highlighted in Figure 8.6. The software indicates the optimal decision for every possible state of information that could be available at the time of the driver allocation decision. We already know that it is optimal to conduct the NY test but not the SF test. Given these decisions, the potential information states at the time of the allocation decision are: (1) *No-Result-SF* &

Node Editor

Node Selection: D-SF Rename

States | Expected Values | Properties | Classes | Values | State Names | Filtered State | Comment

Decision | Equation

Result-SF	Result-NY	<=90	<=100	<=110	<=120	<=130	<=140
No Result	No Result	9977167.531	10295150.539	10504673.598	10588530.048	10556726.625	10454491.699
	Good Mkt	11792645.983	11960668.746	12010545.696	11917333.716	11732385.130	11462980.510
	Bad Mkt	8052694.480	8534604.509	8910562.894	9152864.240	9291308.954	9343027.220
Good Mkt	No Result	10248042.349	10579999.638	10813887.520	10950039.449	10953876.554	10881079.770
	Good Mkt	12063520.801	12245517.845	12319759.618	12278843.117	12129535.058	11889568.582
	Bad Mkt	8323569.298	8819453.608	9219776.816	9514373.642	9688458.883	9769615.292
Bad Mkt	No Result	8556993.196	8693448.350	8733215.081	8669878.383	8475570.831	8203885.794
	Good Mkt	10372471.648	10358966.557	10239087.179	9998682.050	9651229.335	9212374.606
	Bad Mkt	6632520.145	6932902.320	7139104.377	7234212.575	7210153.160	7092421.316

FIGURE 8.6 Optimal allocations based upon test results.

Good-Mkt-NY and (2) *No-Result-SF* & *Bad-Mkt-NY*. In the former case, the optimal decision is to enroll 110 drivers in SF, as before. However, in the latter case, the optimal decision is to increase SF enrollment to 140.

The numbers in the table of Figure 8.6 present the expected values associated with the optimal decisions. Prior to taking any action, but conditional upon making the optimal decisions throughout, the software reports that the expected aggregate profit of this project is $10,676,786. If the NY market tests good, the expected aggregate annual profit is $12,010,545. If, on the other hand, NY tests bad, the expected profit drops to $9,343,027. None of the actions affect the value of the market, but rather they change your expectations of earns by updating beliefs based on new information, and allocating resources accordingly.

8.3 COMPETITOR LEGAL RESPONSE

As mentioned in the introduction, the incumbents in markets threatened by the new technology have frequently responded by filing lawsuits and lobbying local governments in attempts to shut the taxi-hailing entrants out of the market. Getting shut out of a market is a catastrophic event for one of these startups. Therefore, any realistic model of this business would have to include consideration of such a potential eventuality. Let us extend our model to illustrate the inclusion of a potential lawsuit and analyze its effect on the HaHCo's decisions.

To keep things as simple as possible, assume that the management of HaHCo is relatively certain they are safe from lawsuits in New York. San Francisco, on the other hand, could be a problem. In that market, the use of drivers without taxi permits could result in a class action lawsuit brought by existing taxi companies charging HaHCo with unfair competition. One could imagine the headline: "HaHCo charged with taking food from the mouths of law-abiding taxi drivers!" In addition, there is a risk that HaHCo's app-based GPS metering technology could

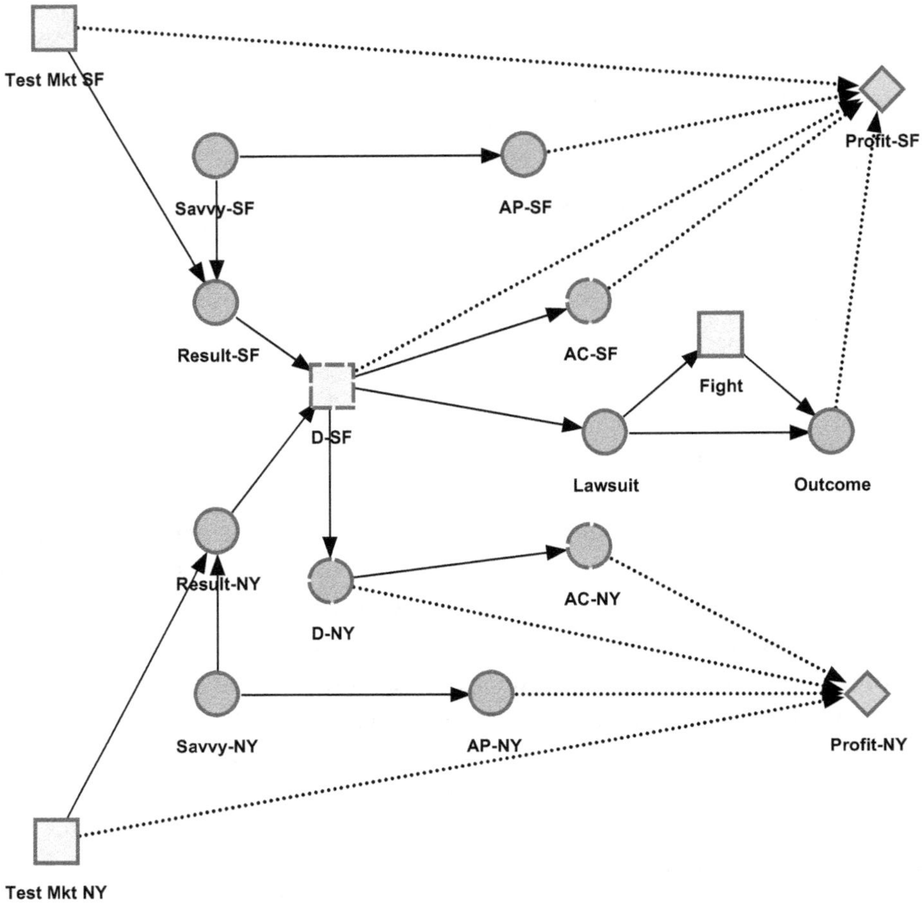

FIGURE 8.7 Model extended to include potential lawsuits.

be found unlawful since the local government regulates and monitors metering devices.

Suppose the likelihood of a *Lawsuit* in San Francisco increases as HaHCo's presence there increases. Specifically, below 100 enrolled drivers, there is no chance of a lawsuit. However, the probability of a lawsuit increases by 5% for every 10 drivers enrolled beyond that. For example, at 140 enrolled drivers, the likelihood of a *Lawsuit* is 20%. If HaHCo is sued, it can choose to *fight* or *settle*. The lawsuit *Outcome*, a *win* or a *loss* for HaHCo, depends upon whether they are sued and whether they choose to fight or settle.

Figure 8.7 presents the extended causal diagram, including that the outcome of the lawsuit affects profit.[3]

3 Technically, an influence arrow should also point from the *Fight* node to the iProfit-SF node in order to add the legal costs incurred should HaHCo decide to fight a lawsuit. This has been omitted to reduce clutter on the diagram.

Fight	Lawsuit	Win	Lose
Settle	Yes	0.000	100.000
	No	100.000	0.000
Fight	Yes	80.000	20.000
	No	100.000	0.000

FIGURE 8.8 Conditional probability table for *Outcome*.

The conditional probability table for *Outcome* is shown in Figure 8.8. If there is no lawsuit, then HaHCo *wins* automatically. If there is a lawsuit, then the outcome depends upon whether the firm *fights* or *settles*. If it *settles*, it *loses* with certainty. If it *fights*, then the likelihood that it *wins* is 80%. The *Profit-SF* node is computed as follows:

$$\pi_{SF} = (AP\text{-}SF \times 40{,}680 \times \delta_O - AC\text{-}SF) \times D\text{-}SF - 50{,}000 \times \delta_{T_{SF}}$$

where *AP-SF* is the average price per mile in San Fransisco, *AC-SF* is the average SF cost, *D-SF* is the number of SF drivers enrolled, and δ_O and $\delta_{T_{SF}}$ are variables whose values are 1 or 0 depending upon the states of *Outcome* and *Test Mkt SF*, respectively. Specifically, $\delta_O = 1$ if the state of *Outcome* is *win* and $\delta_{T_{SF}} = 1$ if the state of *Test Mkt SF* is *test*. Thus, if HaHCo loses a lawsuit, its revenues go to zero (note: costs are still incurred). Similarly, if HaHCo decides to test in SF, it incurs the $50,000 market research cost.

Using the software to solve for the optimal decisions, we come to the following policy:

- No market test in SF;
- Perform a market test in NY;
- Choose the following for driver enrollment:
    - If *Test-NY* = *good*: enroll 100 drivers in SF & 100 in NY;
    - If *Test-NY* = *bad*: enroll 120 drivers in SF & 80 in NY;
- If sued, fight.

Without the threat of a lawsuit, the optimal policy was to enroll 110 drivers in SF in response to a positive market test in NY and 140 drivers in response to a negative test. The threat of a lawsuit reduces both of these numbers. The reason is that the probability of a lawsuit increases as the number of enrollees in SF increases, which therefore increases the opportunity cost of increasing this number. As a result, the marginal value of an additional enrollee in NY increases as HaHCo enrolls SF above the 100 driver level. Because the chance of winning a lawsuit by fighting is high

(80%) and, in addition, the financial consequences of losing one are so profound (losing all SF revenue), it makes sense to fight a lawsuit if sued (but keep in mind, we have not included legal costs in this example – a key omission in this calculation).

KEY CONCEPTS: CHAPTER 8

- Building a causal model using software is best done in steps of increasing complication, starting with nodes having a known functional relationship.
- Because the states of variables in causal models bin continuous values into categories, the results often vary slightly from exact analytical solutions; however, the slight reduction in accuracy affords more usefulness in flexibility and insight into the results of interventions.
- Many values in real-world problems are uncertain, and changing circumstances can shift the distributions of values. Causal modeling is especially useful for capturing this type of system behavior.
- As new information and relevant considerations come into play, a causal model can be extended by altering just the portions affected. Because calculations are performed locally (at the node level) the whole model doesn't have to be changed/rethought with every expansion.
- Software tools easily handle multiple decision nodes and provide optimal strategies across the whole set of decisions.

Chapter 9

Multi-Agent Interventions

In Chapters 7 and 8 we illustrated how to solve quantitative causal models in which one or more of the model's variables are under the control of a single decision maker (hence, the label "single-agent intervention"). An implicit assumption in such models is that the states of those variables outside the control of the decision maker are safely assumed to occur according to some fixed stochastic process. This assumption is consistent with a wide range of managerial problems. In a large class of decision problems, however, managers find themselves in situations where the states of some variables are set independently by other agents – agents who are likely to set the variables they control according to their own objectives.

For example, a seller typically makes an offer so as to attain the highest price possible while still making the sale. Her buyer seeks to consummate the transaction at the lowest price possible and will certainly refuse an offer that is above his willingness-to-pay. Alternatively, a monopolist may threaten aggressive pricing in the hopes of deterring an entrant. Yet, an entrepreneur may enter anyway in the hope that the monopolist was only bluffing.

Notice that in each of these examples, the agents' payoffs depend upon each others' actions. The seller makes an offer, the buyer decides whether to accept or decline. The economic values ultimately enjoyed by each party jointly depend upon the choices they both make. The incumbent takes actions suggesting severe competition in the event of entry; the entrepreneur decides whether or not to enter; if entry occurs, the incumbent decides whether or not to retaliate. The payoffs of both the incumbent and the entrepreneur depend upon this entire sequence of decisions.

A situation in which a number of agents make decisions, and the consequences of their decisions affect each others' payoffs, is known as an *interactive decision problem*. In such situations, each agent's own objectives will drive his or her decisions. Therefore, when faced with an interactive decision problem, we would like to adopt an analytical approach that helps us predict what the other agents are likely to do in order that we may implement the best decisions possible. This takes us straight into the realm of *game theory*, a vibrant subfield of economics.

In this chapter, we show how to represent and solve two-player games using causal models. Prior knowledge of game theory is not required here, though some familiarity with it is certainly complementary to understanding our material (and visa versa). Game theorists have developed their own analytical formalisms, known as *normal* and *extensive* form games. These provide elegant, general frameworks for representing and solving interactive decision problems. Our purpose here is not to present a substitute approach *per se*. Rather, by illustrating the use quantitative causal models to solve games, we aim to provide managers with a practical tool with which to analyze interactive decision problems in the real world (in particular, managers who have access to causal modeling software).

9.1 ELEMENTS OF A GAME

A standard interactive decision problem, or game, is composed of several basic elements. These elements include: (1) the agents whose decisions affect each others' payoffs, known as *players*; (2) the sequence of player decisions, called **moves**; (3) an elaboration of the *actions* available to each agent at each decision; (4) an elaboration of payoff-relevant exogenous events and the probabilities with which they occur, referred to as *Nature's moves*; (5) what each agent *knows* at the time of his or her decision; and (6) the payoffs accruing to each agent for every possible combination of actions (including those by Nature). To be analytically useful, a representation scheme must provide formal definitions for each of these elements.

The extensive form is a representation that, like causal models, uses a graph (*aka* network) to depict the essential elements of a game.[1] In the extensive form, each node corresponds to a decision point for one player. The arrows emanating out from the node correspond to the actions available to the player at that decision. The nodes and arrows of an extensive form game form a tree, so that each branch of the tree represents one combination of decisions. At the very end of each branch, the payoffs corresponding to that branch's actions are listed for each player. The knowledge of a player at the time of her decision is indicated by grouping nodes into "information sets" that correspond to her ability to distinguish her true branch from other branches at which she might be located.

To illustrate the standard approach, consider a simple Entry Game. There are two players: an entrant and an incumbent. In this situation, there is a market for some product that is initially served by the incumbent. The entrant must decide whether

1 As mentioned above, the other common representation is the "normal" form. This form is typically used for games in which the players make their decisions simultaneously, though it is not restricted to such situations.

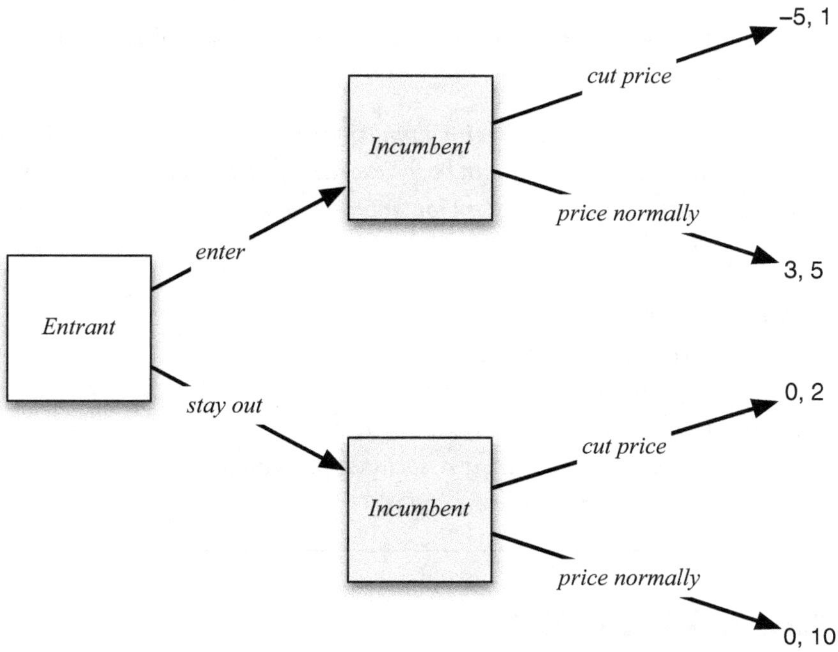

FIGURE 9.1 Extensive form version of an Entry Game.

or not to enter that market and compete with the incumbent. If entry occurs, the incumbent need not passively stand by to see its market share get eaten up by the newcomer. If it so chooses, the incumbent can employ a price retaliation tactic that ruins the entrant's profitability. Unfortunately, by cutting prices, the incumbent also experiences low profit. Alternatively, the incumbent can accommodate the entrant by pricing normally. If it accommodates, it loses market share but makes out better profit-wise than under the retaliation option. Of course, the entrant also does better under this option. If the entrant does not enter, then the incumbent still has the option of cutting prices or pricing normally.

These details are encoded in the extensive form game in Figure 9.1. The first node represents the *Entrant*'s decision. If it does not enter, it receives a payoff of 0 no matter what the *Incumbent* does. If entry does occur, and the *Incumbent* accommodates (*price normally*), the *Entrant* receives payoff of 3 and the *Incumbent* a payoff of 5. If it retaliate (*cut price*), the *Entrant* receives −5 and the *Incumbent* receives 1. If the *Entrant* stays out and the *Incumbent* prices normally it receives a profit 10. If it *cut price*, its profit is 2. Notice that, in this game, the *Incumbent* observes whether or not *entry* occurs. Thus, its decision nodes lie on the branches corresponding to the *Entrant*'s actions. The *Incumbent* would like to threaten price retaliation, deter entry and, then, *price normally* to receive the best payoff of 10. The key question from the *Entrant*'s point-of-view is whether the retaliation threat is, indeed, credible.

Definition 9.1 *Move: a decision point at which a player selects from a set of feasible actions.*

Definition 9.2 *Game: a situation involving two or more agents making decisions, and in which each agent's payoff can be written as a function of the decisions made by all agents. "Game" is another term for "interactive decision problem." Formally, a game is composed of:*

1. *A set of relevant players;*
2. *The sequence of players' moves;*
3. *The feasible actions at each move;*
4. *The information available at each move;*
5. *Any exogenous states and their associated probabilities;*
6. *Payoffs as a function of player actions.*

9.2 NASH SOLUTIONS

Up to this point, all we have done is enumerate the basic elements of a game. Recall that the whole motivation for thinking in terms of games is trying to predict the actions of other agents when you find yourself faced with an interactive decision problem. To do this, you must *solve* the game – that is, derive a prediction with respect to other players' actions and craft an optimal response.

There are many solution concepts for games, with different concepts more appropriate for some situations than others. For example, some solution concepts correspond to decisions by highly rational players, others to boundedly rational players, and still others to irrational players. Elaborating all of these is beyond the scope of this book. Instead, we will focus upon the most basic, well known, and widely used concept known as **Nash equilibrium**. Nash equilibrium is a good solution concept when players are rational.

There are two important concepts required to define a Nash equilibrium. The first is that of **strategy**. A player's strategy is a plan that describes what the player will do for every possible state of information at every move. For example, if a player only has one move available in the game and must make it with no information, then the strategy is simply the choice of action. However, if information is available to a player at the time of her move (e.g., observing the move of another player), then a proper strategy for that player provides an algorithm that specifies an action choice for every possible information state (e.g., what to do in response to each action chosen by the other player).

For example, consider a market with two PC processor manufacturers. Suppose *Firm A* is a technology leader and, as such, has the opportunity to make a public showing of a major commitment of resources to developing the "next

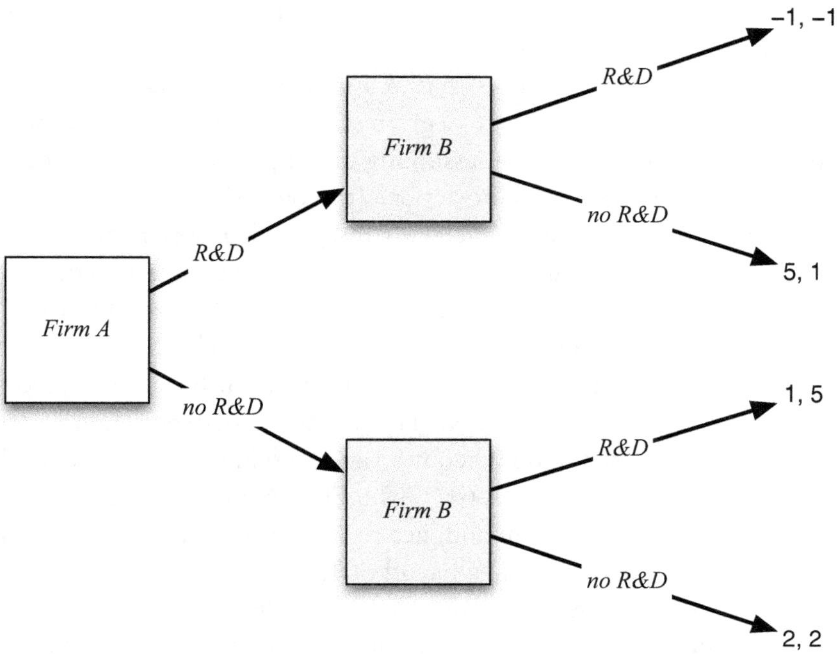

FIGURE 9.2 R&D Game.

generation" chip. Alternatively, it can choose to maintain the status quo and make no investment. *Firm B* observes *Firm A*'s decision and then decides whether it will make its own R&D investment in next generation technology. Figure 9.2 illustrates the game in the extensive form. The payoffs, shown in the order of *Firm A's payoff, Firm B's payoff*, reflect the idea that the R&D investment generates a positive return to a firm if it is the only one with advanced R&D; if both firms invest, buyers view the firms as indistinguishable and, as a result, neither earns a positive return on its investment.

In this example, a strategy for *Firm A* is simply a choice of whether to invest in *R&D* or not. However, a strategy for *Firm B* must specify an action for every information state. In this case, there are two such states: either it observes *R&D* investment by *Firm A* or it does not. In either state, *Firm B* must choose whether or not to make its own *R&D* investment. Therefore, there are $2 \times 2 = 4$ strategies:

1 Regardless of *Firm A*'s decision, invest in *R&D*;
2 Regardless of *Firm A*'s decision, make *no R&D* investment;
3 Only invest in *R&D* if *Firm A* invests;
4 Only invest in *R&D* if *Firm A* does not invest;

Again, it is extremely important to distinguish between moves (decision points) and strategies (complete contingent decision plans). The items above are strategies: each one provides a complete description of what Firm B will do at every information state (and, hence, at every move).

The second important concept is that of a **best reply**. A strategy for a player is said to be a best reply with respect to a given strategy for her opponent, if it results in the optimal payoff in the game, assuming the opponent plays according to the given strategy. In a game with finite actions (the only kind we consider), a player always has at least one best reply to every strategy of her opponent.

In our example, suppose Firm B plays strategy #1 above (invest no matter what). Then, it should be easy to see that Firm A's best reply with respect to that strategy is not to invest. Alternatively, suppose Firm A invests. Then, Firm B's strategies #2 and #4 are both best replies. In both strategies, Firm B does not invest, given Firm A's investment, which is optimal. The difference between the two strategies is what Firm B is supposed to do when it observes no investment by Firm A: in #2 it does not invest; in #4 it does invest. Given Firm A's strategy of investing, Firm B does not observe no-investment and, hence, it does not matter which of strategy #2 or #4 it chooses – both provide a payoff of 1.

We can now state the definition of a *Nash equilibrium*. In a two-player game, a Nash equilibrium is a set of strategies, one for each player, such that each is a best reply to the other. Notice that, in a Nash equilibrium, no player wants to change their strategy (hence, the term "equilibrium"). If players are rational, then the strategies that constitute a Nash equilibrium may provide a good prediction of how the agents will play. This prediction is stronger if the Nash equilibrium is unique. If the equilibrium is not unique, then the analyst must assess which is most likely.

In our first example of the Entry Game, the Nash equilibrium is for the *Entrant* to *enter* and the *Incumbent* to choose the strategy: *price normally* no matter what. The *price retaliation* threat is not credible because a rational *Incumbent*, faced with an actual *Entrant*, would not choose to *cut prices* because doing so would result in lower payoffs than accommodating by *pricing normally* (and clearly *pricing normally* is optimal when there is no entry). In the second, R&D example, the Nash equilibrium is for the technology leader, *Firm A*, to choose R&D and the follower, *Firm B*, to choose either strategy #2 or #4. By investing in R&D, *Firm A* pre-empts *Firm B*'s own investment.

Definition 9.3 *Strategy: a description of what actions a player will take for every possible state of information at every one of his moves.*

Definition 9.4 *Best reply strategy: a strategy that delivers the optimal expected payoff given the opponent's strategy.*

Definition 9.5 *Nash equilibrium: a set of strategies, one for each player, in which each strategy is a best reply to the others.*

9.3 CAUSAL FORM GAMES

The preceding material provides a brief review of several basic game theoretic ideas. Now, we wish to demonstrate how to represent and solve games using causal models. We will begin by converting extensive form Entry Game in Figure 9.1 into a multi-agent influence diagram, and showing how to solve it.[2]

Let us begin with the players' moves. In this game, are two players: the entrant and the incumbent. The entrant has just one move: the decision of whether or not to *enter* the market. **In a causal model, a player's move is represented by a decision node**. In this case, we create a decision node for the entrant, call it *Entrant*, and associate two possible actions to this decision: *enter* or *stay out*. The incumbent likewise has just one decision to make, and only two action options. Thus, we create a decision node for the incumbent called *Incumbent* and associate the actions *cut price* and *price normal* with that node.

As in a single-agent influence diagram, *information* available to a player about the states of other variables in the system are indicated with an arrow. That is, if a player is assumed to know the state of a variable at the time of her move, then an arrow is drawn from the observed variable to the agent's decision node. In multi-agent models we do the same things for other players' moves as well as moves by Nature, which we may refer to as environmental variables in the context of a causal model; i.e., variables with distributions that are exogenously determined. There are no environmental variables in this example.

Information-wise, the *Incumbent* observes whether or not the *Entrant* actually *enters*. Therefore, we would create an arrow from the *Entrant*'s move to that of the *Incumbent*. Notice that, in order to observe an opponent's move, that opponent must have chosen the move temporally prior to the move of the observing player. Thus, there is never a cycle: if the *Incumbent* observes the Entrant's move, the *Entrant* cannot observe the *Incumbent*'s move.

Naturally, we **use objective nodes to represent the players' payoffs**. In what follows, we adopt the convention of representing both players' payoffs with a single objective node. This facilitates manual analysis; the methods presented here may have to be adapted if one is using a particular causal software program for the analysis (we present an example using the BayesiaLabs program at the end of the chapter). Since the players' payoffs are functions of the *Entrant*'s move and the *Incumbent*'s move, arrows are drawn from both players' decision nodes to the *Payoff* node.

With the qualitative move and payoff structure from the extensive form game encoded in the causal graph, we must now add the tables for the nodes to complete the model. In general, this requires: (1) for environmental variables, a probability

2 For more advanced perspectives on the merging of game theory and causal models see, for example, Vickrey and Koller (2002), Koller and Milch (2003), and Penalva and Ryall (2008).

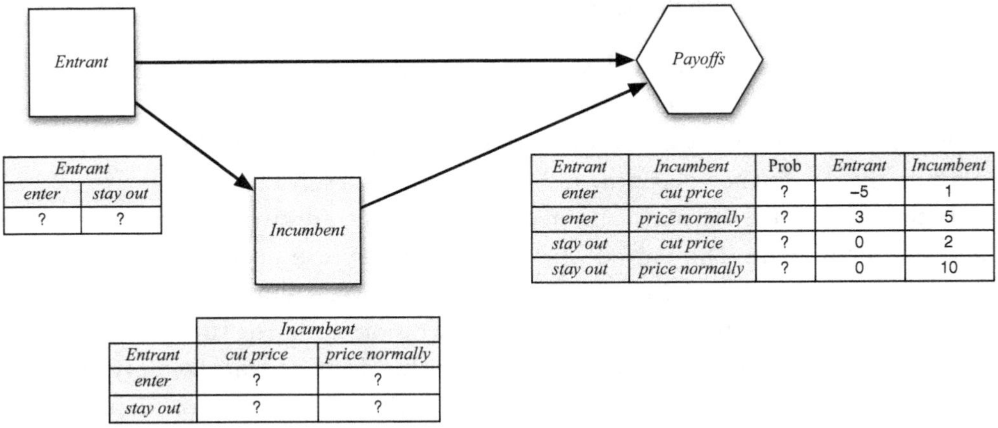

Entrant	
enter	stay out
?	?

Entrant	Incumbent	Prob	Entrant	Incumbent
enter	cut price	?	–5	1
enter	price normally	?	3	5
stay out	cut price	?	0	2
stay out	price normally	?	0	10

	Incumbent	
Entrant	cut price	price normally
enter	?	?
stay out	?	?

FIGURE 9.3 Entry Game in causal form.

table if the variable is a root node and a conditional probability table if it is an intermediate node; (2) for each decision node, a decision table; and (3) for the objective node, an expected value table.

Beginning with the *Entrant*'s decision node, we attach a decision table with the options *stay out* and *no*. Note that this is a root node: the *Entrant* has no information at the time of her move. We also add a decision table for the *Incumbent*'s move, with the actions *price normal* and *cut price*. Because the *Incumbent* observes the *Entrant*'s move, both of the actions are conditional on the information received from the *Entrant* regarding her action. Remember that even though there is nothing for the *Incumbent* to do if the entrant stays out (neither action affects the outcome in that case) the row corresponding to *stay out* must be included in the decision table for completeness.[3]

The *Payoff* node is influenced by both decision nodes. Therefore, it must include a row for each possible combination of states of each of them. There must be a column for the probabilities of each row (similar to the procedure in Chapter 7). In addition, there must also be **two value columns – one for each player**. In this example there are no random variables. This implies that all states happen with probability 1 or 0, depending upon the players' strategies. For this reason it is quite simple to fill in the value columns for both players for each action combination directly from the payoffs presented in the extensive form game in Figure 9.1.

3 As mentioned already, there are cases when redundant or unnecessary information can be left out of the diagrams to simplify things, once you have mastered the techniques then you are free to simplify as you please. For now, we include everything for completeness and clarity of exposition.

Figure 9.3 shows the complete setup, awaiting analysis. Before proceeding, let us check to see that all of the elements of the game are represented in the causal model:

1 The set of relevant players are indicated by their decision nodes;
2 The sequence of players' moves is indicated by the arrows from the *Entrant*'s node to the *Incumbent*'s node;
3 The feasible actions at each move are indicated by the columns in each of the decision tables;
4 The information available at the *Incumbent*'s move is indicated by the arrow from the Entrant's to the Incumbent's node, with the specific information states represented in the Incumbent's decision table;
5 There are no environmental variables in this example;
6 Payoffs as a function of player actions are summarized in the table for the objective node.

There are a couple of points worth highlighting. First, comparing the extensive form in Figure 9.1 with the causal form in Figure 9.3, we see that each *node* in the former corresponds to a *row* in a decision table of the latter. Specifically, in Figure 9.1 the *Incumbent* has two moves (nodes), one for each of the actions available to the *Entrant*. In Figure 9.3, the *Incumbent* has one decision node with a table attached. Each of the two rows in that table are "moves" in the sense that they correspond one-to-one with the two nodes in the extensive form of this game.

Second, it is worth pointing out that **partial information about a variable can be represented by adding an intermediating environmental variable** to the model. For example, suppose the game begins with three flips of a coin. There are eight possible sequences: *HHH, HHT, HTH, THH, HTT, THT, TTH, TTT*. Suppose a player must make a decision upon being told only the number of heads after three flips. This could be represented in a causal model by three variables, one for each coin flip. All of these would then point to a node that would serve as an information filter. The filter variable's conditional probability table would have rows that correspond to the eight states implied by the parent outcomes, and the columns would correspond to the four possible counts of *heads* (0, 1, 2, or 3). This filter variable would then point to the player's decision node to represent information regarding the number of heads.

9.4 SOLVING GAMES WITH CAUSAL MODELS

The first step in solving for the optimal decisions in a multi-agent causal model is the same as for the single-agent models in Chapter 7: (1) convert the closest decision node to the current objective node into a new objective node, (2) determine the optimal action for each input for this decision, and (3) remove the old

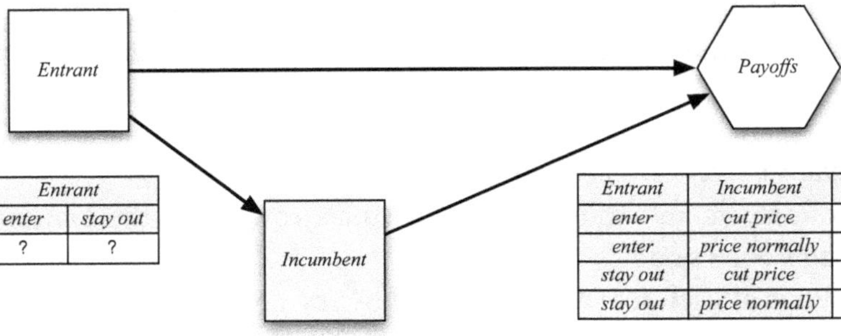

Entrant	
enter	stay out
?	?

Entrant	Incumbent	Prob	Entrant	Incumbent
enter	cut price	?	−5	1
enter	price normally	?	3	5
stay out	cut price	?	0	2
stay out	price normally	?	0	10

	Incumbent	
Entrant	cut price	price normally
enter	−5, 1	**3, 5**
stay out	0, 2	**0, 10**

FIGURE 9.4 *Incumbent*'s decision table completed.

objective node. In this simple model, every cell of the *Incumbent* node's decision table corresponds to a row of the *Payoff* table each one having probability 1 in that scenario. For reasons that will become clear later, we wish to include the payoffs to all players in each player's decision table, although a player will only optimize his or her own payoff. Following these construction rules produces the causal model in Figure 9.4 with the *Incumbent*'s decision table filled in.

Given the payoffs for the incumbent in the *Incumbent* decision node, it is easy to see that if the state of *Entrant* is *stay out*, then the *Incumbent*'s optimal action is *price normally*. If the state of *Entrant* is *enter* then the *Incumbent*'s optimal action is also to *price normally* (i.e., it is best to accommodate the entrant because 5>1). This results in payoffs of (3, 5) for the *Entrant* and *Incumbent*, respectively. Thus, the optimal **strategy** for the *Incumbent* is: **always choose to *price normally***.

The next step is to simplify the model by changing the *Incumbent* node into the new objective node with payoffs equaling the values of the determined optimal strategy. In this case, the objective node only needs to contain payoff information for the *Entrant* because only the *Entrant* has a decision left to make, although recording both may be useful for reference in larger models. The decision table for the *Entrant* node is completed in Figure 9.5. This tells us to expect that the entrant will choose *enter* and the incumbent choose *price normally*. The optimal policies identified in this analysis constitute a Nash equilibrium of the game.

9.4.1 Vicious Incumbent Entry Game

The previous games did not include any environmental variables, but seeing how they play into the games will be helpful and insightful. So now we extend the Entry Model to include the possibility that the incumbent CEO may simply gain personal satisfaction from putting competitors out of business; i.e., we add a chance

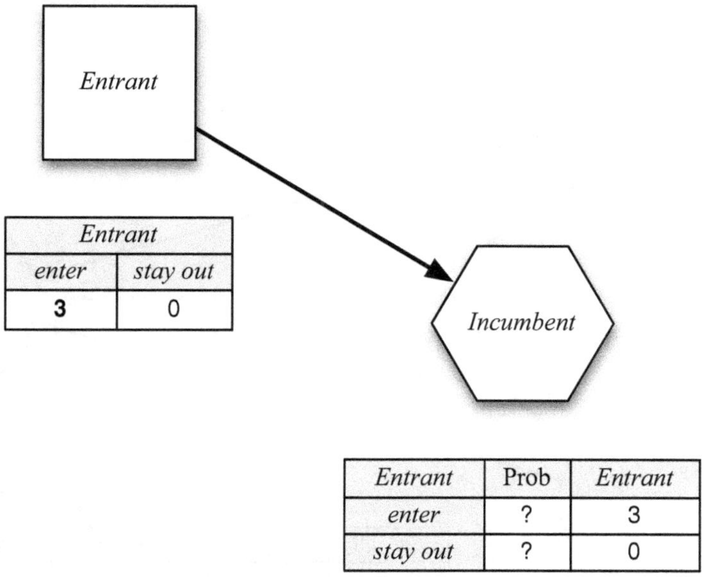

FIGURE 9.5 Simplified model with *Incumbent* converted into an objective node.

that the CEO is simply vicious. The idea is to create a model in which there may be some uncertainty in the entrant's mind as to the incumbent's willingness to accommodate. We build in this uncertainty by including an environmental node for incumbent's *Type* with two states: *vicious* and *rational*. By adding a probability table for which type is the true type of the incumbent, we create the desired uncertainty. Naturally, we assume that the incumbent knows his own type, but the entrant does not.

Figure 9.6 illustrates the model with *rational* and *vicious Incumbent Types*. The payoffs are shown in Figure 9.6. Notice that the *Incumbent*'s *Type* affects its payoffs. A *vicious Type* hates accommodating by choosing *price normal* when the *Entrant enters* and likes retaliating by *cutting prices* – flipping the preferences under entry versus a *rational Type*. Because payoffs are affected by *Type*, there is an influence arrow from the *Type* node to the *Payoff* node. Since the *Incumbent* knows its *Type*, an arrow must also be drawn from *Type* to the *Incumbent* node. According to the figure, at the time of making its decision, the *Entrant* does not know the true type of the *Incumbent*. However, the *Entrant* **does know the likelihoods** with which the *Incumbent* is one or the other *Type*. These probabilities may reflect: (1) the *Entrant*'s subjective beliefs, (2) actual experience across a number of new markets, or (3) a data-driven review of historic incumbent behavior.

To an outsider – say, one who only observes the direct financial implications of the *Incumbent*'s moves – *price cutting* might appear irrational. Technically, however, agents are "rational" in the game theoretic sense as long as they act to maximize their utility, whatever it may be that provides utility. In the real-world, a manager's

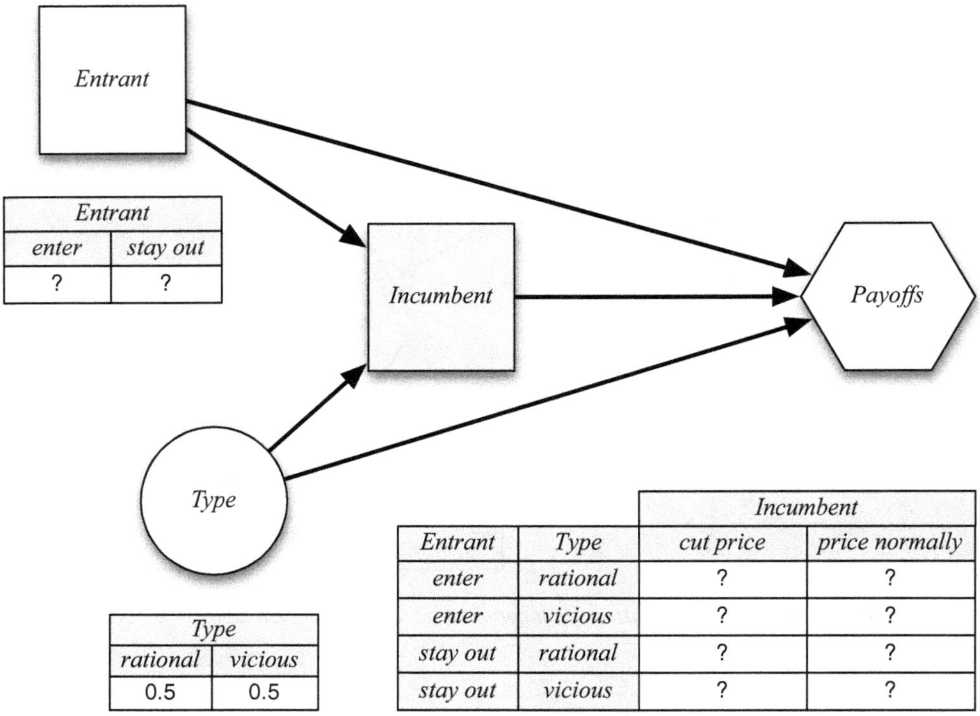

FIGURE 9.6 Modeling a chance that the *Incumbent* is *vicious*.

Entrant	*Type*	*Incumbent*	Prob	*Entrant*	*Incumbent*
enter	*rational*	*cut price*	?	–5	1
enter	*rational*	*price normally*	?	3	5
enter	*vicious*	*cut price*	?	–5	7
enter	*vicious*	*price normally*	?	3	–1
stay out	*rational*	*cut price*	?	0	2
stay out	*rational*	*price normally*	?	0	10
stay out	*vicious*	*cut price*	?	0	2
stay out	*vicious*	*price normally*	?	0	10

FIGURE 9.7 Payoffs in the Vicious Incumbent Entry Game.

behavior can be driven by a wide range of motivations, from maximizing short-term profit, net present value, market share, or share price to more personal benefits such as gaining prestige or harming a hated rival. From a modeling point-of-view it is crucial that the objectives for other agents accurately reflect their true motivations, whatever they may actually be.

The solution of the model proceeds using the same backward stepwise simplification procedure outlined for the single-agent case and the simple Entry Game. Players determine optimal strategies for each decision based upon their own expected payoffs, and these are worked up the causal model until an optimal solution to the whole model is found. If the *Incumbent* is of the *rational Type*, then the payoffs are the same as before. If the *Incumbent* is of the *vicious Type* then, if the *Entrant* decides to *enter*, he receives a payoff of 7 from retaliating (*cutting price*) and a payoff of −1 from accommodating (*normal pricing*). The key difference between this case and the previous model is that the *Entrant* must calculate **expected payoffs** for herself if she enters because she does not know for sure the *Type* of the *Incumbent*.

As always, the first step is to fill in the expected values of the decision table for the decision node closest to the current objective node – in this case it is still the *Incumbent* decision node. These cells are easy to fill in because each one corresponds to a pair of cells in the *Payoff* node. The *Incumbent* knows its *Type* as well as the decision of the *Entrant*. Once the *Incumbent* makes his own decision, the payoffs for both players are fully determined. The result of this stage of the analysis is presented in Figure 9.8 with the optimal action for each row is highlighted.

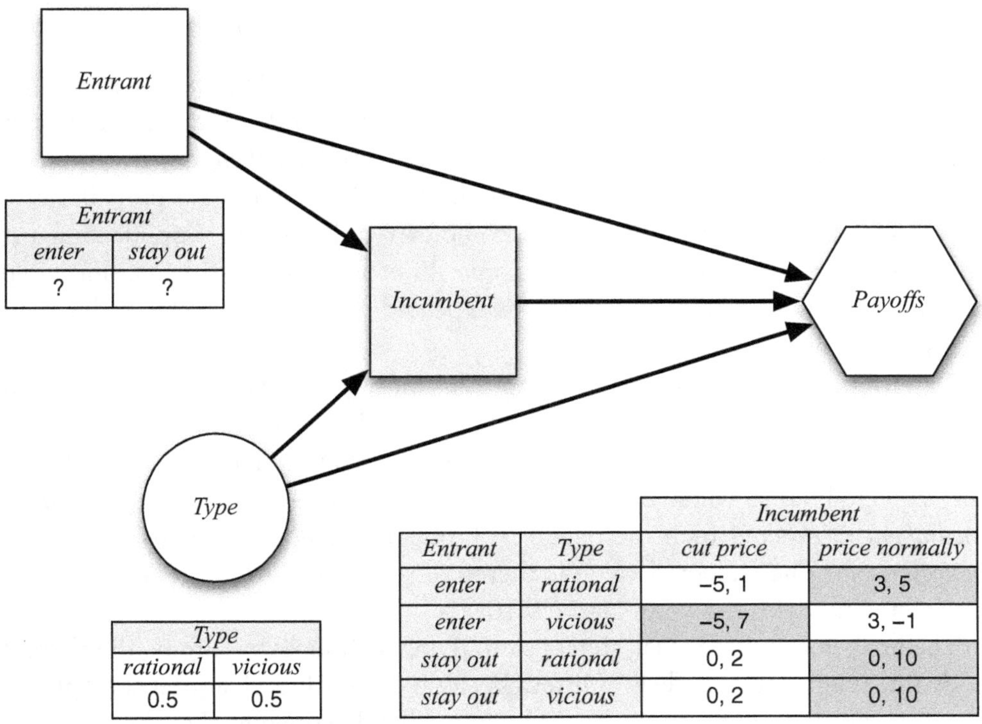

Entrant	
enter	stay out
?	?

Type	
rational	vicious
0.5	0.5

		Incumbent	
Entrant	Type	cut price	price normally
enter	rational	−5, 1	3, 5
enter	vicious	−5, 7	3, −1
stay out	rational	0, 2	0, 10
stay out	vicious	0, 2	0, 10

FIGURE 9.8 *Incumbent*'s decision table complete.

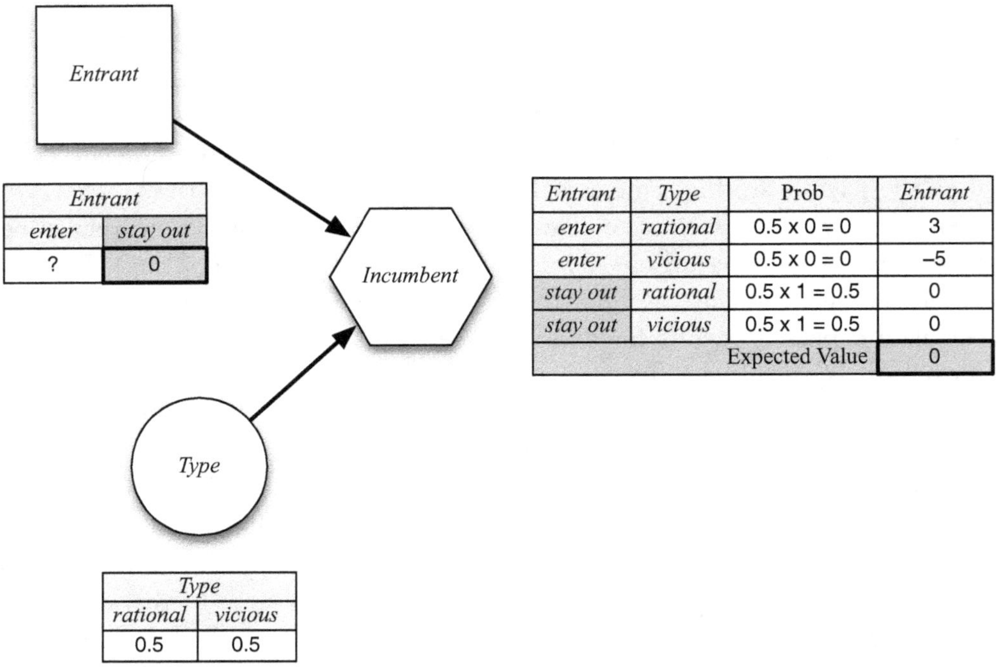

Entrant	
enter	stay out
?	0

Entrant	Type	Prob	Entrant
enter	rational	0.5 x 0 = 0	3
enter	vicious	0.5 x 0 = 0	–5
stay out	rational	0.5 x 1 = 0.5	0
stay out	vicious	0.5 x 1 = 0.5	0
		Expected Value	0

Type	
rational	vicious
0.5	0.5

FIGURE 9.9 Expected value when *Entrant* = *stay out*.

The next step is again the simplification step; transforming the decision node closest to the old objective node into the new objective node. In the previous step we determined the optimal action for each input at the *Incumbent* node, so now we simplify the model into a decision problem that involves the *Entrant* only. Because *Type* points to *Incumbent* it does not become barren during the simplification – it is still a factor in determining the outcome. In order to fill in the decision table for the *Entrant* we need to compute the expected value produced by each action across all states of *Type*.

Determining the expected value for the *Entrant* when she chooses *stay out* is still trivial because she gets 0 no matter the state of *Type*. However, for completeness we still show the full calculation for coming to this number in Figure 9.9. Calculating the expected value when *Entrant* = *enter* is just slightly more complicated. If *Incumbent* = *rational* then the payoff is 3, but if *Incumbent* = *vicious* then the payoff is −5. The expected value is therefore $(0.5 \times 3) + (0.5 \times -5) = -1$. The completed model is shown in Figure 9.10.

The final, in this case trivial, step of simplifying the model by turning the *Entrant* node into an objective node is omitted. Because $0 > -1$ it is clear that the optimal state for *Entrant* is *stay out*. It should be immediately apparent that the strategies identified constitute a Nash equilibrium of the expanded game.

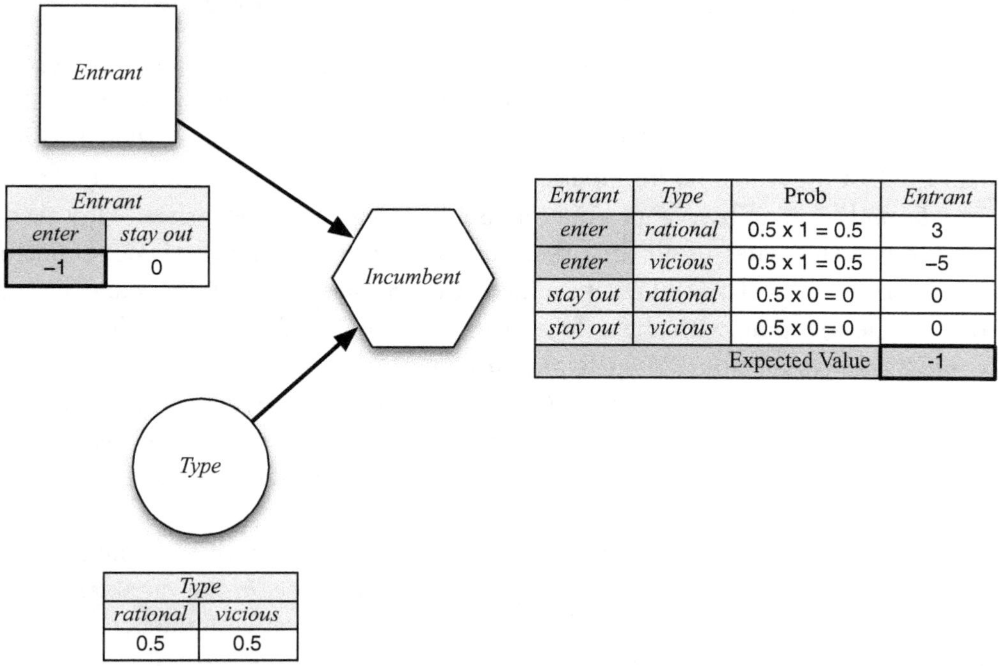

Entrant	
enter	*stay out*
−1	0

Entrant	*Type*	Prob	*Entrant*
enter	*rational*	0.5 x 1 = 0.5	3
enter	*vicious*	0.5 x 1 = 0.5	−5
stay out	*rational*	0.5 x 0 = 0	0
stay out	*vicious*	0.5 x 0 = 0	0
		Expected Value	-1

Type	
rational	*vicious*
0.5	0.5

FIGURE 9.10 Complete decision table for *Entrant*.

Given the 50 : 50 possibility that the incumbent is vicious instead of rational, the entrant's expected payoff is worse relative to the basic Entry Game, in which the incumbent was known to be rational. Thus, our model illustrates a case in which the threat of "irrational" behavior discourages market entry. The outcome is sensitive to the probabilities and payoffs assumed in the setup, specifically: (1) the actual likelihood of vicious incumbents, (2) the entrant's profit under accommodation versus retaliation, (3) the incumbent's intrinsic utility from crushing entrants. While we do not demonstrate it here, one valuable use for a causal model such as this one is to identify the value ranges for the key parameters over which, say, entry is deterred.

The preceding analysis raises the interesting possibility that the incumbent, knowing that being vicious can deter entry, might try to find ways to indicate to the entrant that he is vicious – even when his true type is rational. Such an indication, called a *signal* in game theory literature, will be the result of some action by the incumbent that is observed by the entrant. This would require adding a second decision node for the incumbent, representing an action that occurs prior to the entrant's decision. As it turns out, solving such a model requires some additional skills. In the next section, we develop those skills in a new setting (adding some variety to keep things interesting).

9.5 GAMES WITH NON-TRIVIAL STRATEGIES

The previous Entry Game model, even with the addition of the environmental variable, had a particular structure that allowed it to be solved using essentially the same methodology as that used for single-agent interventions. The *Entrant* had the first move and no information. Its strategy was simply a choice of action. The *Incumbent* did observe information. So, its strategy was not trivial. However, it observed the states of all the other variables prior to making its move. Since its action was the last state determined, the *Incumbent*'s decision table had a one-to-one mapping with the values in the table of the *Payoff* node. Thus, one could easily transfer the values to the *Incumbent,* determine the optimal actions for each row, make the relevant simplification, and be left with a simple decision problem for the *Entrant*. Everything was very straightforward – the distinction between a strategy and a move did not inject any substantive issues into the analysis.

Now, we present a model in which *both* players must formulate non-trivial strategies and operate on partial information about the states of the variables in the system. In such situations, we must augment the quantitative influence diagram to reflect the information and choice structure properly. We must also employ a different type of decision table to simplify the calculations. All of the steps necessary to solve such a problem are presented through a research company buyout problem described below.

Research scientists working at universities often spin off new inventions they discover to independent companies. In our scenario the basic research company is called *BuenoBrains Incorporated*. Their goal is to develop an idea sufficiently well to attract the attention of a large technology firm, *GrandeGadgets Corporation*. This kind of large firm sometimes purchases small basic research companies to gain the rights to their proprietary technology and bring it to market. When this happens, the start-up scientists typically cash in their interest at a reasonable profit.

It is the nature of basic research that not all ideas are created equal. Some ideas produce innovative "game changing" technologies. Still, even an incremental idea can have market value. Though GrandeGadgets employs scientists and engineers to evaluate the technologies it buys, they are not basic research scientists and may not even be specialists in the necessary fields. In some cases, there are clear benchmarks to evaluate the technology (such as the results of clinical trials in medicine or the performance of prototypes in hardware). In other cases, the product quality is more difficult to test (such as a test for whether a person will develop diabetes in five years, or a device to predict when the next volcano will erupt).

Because GrandeGadgets' engineers are not capable of evaluating the new capability itself, they instead scrutinize the development efforts of BuenoBrains and base their decision on the degree to which the idea has been completed. The reasoning behind this strategy is that good ideas are easier to develop than bad ideas, implying that the latter are not worth the effort based solely on the hope of being bought out. We will test this hypothesis by solving the game.

Of course, sometimes good ideas come from companies that cannot afford to develop them. And sometimes the developers themselves may not appreciate the full potential of the idea and stop short. For the scientists, selling their company without fully developing the idea brings in a lower price, but it saves them the additional development costs. A less-than-stellar idea may also be developed if it is expected to bring in some money for the scientists. If the idea is mediocre, the scientists must decide whether to try to "dress it up" and try to sell it for more than it would normally be worth.

All of these considerations can be modeled in a relatively simple influence diagram. We now provide all of the information pertinent to the game. We then identify the best-reply strategies for both BuenoBrains and GrandeGadgets. Once the best replies are identified, it is possible to solve for the Nash equilibrium.[4]

9.5.1 Setup for the model

The process starts with the new technology *Idea*, which has an 80% chance of being *incremental* and only a 20% chance of being *innovative*. Assume that the scientists can correctly assess the potential of the idea before they decide to *Develop* it *partially* or *fully*. Developing the idea *partially* costs nothing, regardless of how good the idea is, and it can be sold to GrandeGadgets or some other firm for $55 million (all payoffs are presented in millions in this problem). *Fully* developing an *innovative* idea costs only $10 because it can win external research grants and attract more talented researchers. However, *fully* developing an *incremental* idea costs $175 due to the likelihood of false starts and the difficulties involved in precisely measuring and validating small changes in results.

GrandeGadgets cannot observe the quality of the *Idea* directly, but learns the degree to which the research company *Develops* it. The large firm must decide whether to *Acquire* the technology or not (i.e., *no* or *yes*). If it decides not to purchase BuenoBrains, then nothing is gained or lost. If GrandeGadgets purchases BuenoBrains with *partially* developed technology, it pays $55 for it. If GrandeGadgets purchases a *fully* developed technology with the company, it pays $100. An *incremental* idea is never brought to market and earns the firm nothing. Ideas that are *innovative* reap significant rewards, including status, media attention and direct revenue – all of which are valued at $200.

These details describe all the elements of a game. The corresponding causal model is shown in Figure 9.11, with the payoffs table as shown in Figure 9.12. Notice that because *Idea*, *Develop*, and *Acquire* all (potentially) affect the rewards

4 For readers familiar with game theory, what follows is a signaling game and the solution presented will be a Bayesian Nash equilibrium.

Idea	
incremental	*innovative*
0.8	0.2

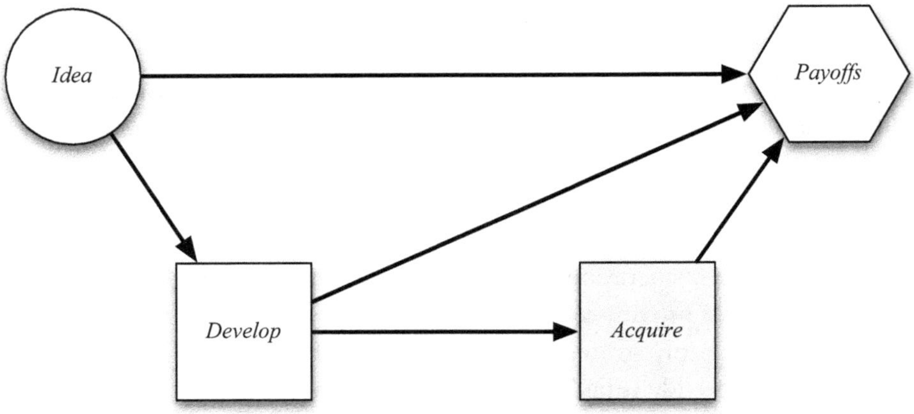

	Develop	
Idea	*partially*	*fully*
incremental	?	?
innovative	?	?

	Acquire	
Develop	*no*	*yes*
partially	?	?
fully	?	?

FIGURE 9.11 Causal model of technology development and acquisition problem.

Idea	*Develop*	*Acquire*	Prob	*BuenoBrains*	*GrandeGadgets*
incremental	*partially*	*no*	?	55 – 0 = 55	0
incremental	*partially*	*yes*	?	55 – 0 = 55	0 – 55 = –55
incremental	*fully*	*no*	?	55 – 175 = –120	0
incremental	*fully*	*yes*	?	100 – 175 = –75	0 – 100 = –100
innovative	*partially*	*no*	?	55 – 0 = 55	0
innovative	*partially*	*yes*	?	55 – 0 = 55	200 – 55 = 145
innovative	*fully*	*no*	?	55 – 10 = –45	0
innovative	*fully*	*yes*	?	100 – 10 = 90	200 – 100 = 100

FIGURE 9.12 Payoff table for technology development and acquisition problem.

to one or both players, they each have an arrow pointing to the *Payoff* objective node.

Notice that both players have non-trivial strategies (i.e., their action decisions are conditioned upon information states). Moreover, neither player has full knowledge of the states of the other variables in the system at the time of their move. BuenoBrains knows the quality of their idea, but GrandeGadgets does not. However, BuenoBrains does not know what GrandeGadgets will do. The backward optimize-and-simplify solution method no longer works. For example, in order to compute an optimal strategy now, GrandeGadgets either needs to know how good the idea is in addition to observing BuenoBrains' action, or it needs to know BuenoBrains' strategy (which amounts to the same thing).

To proceed, **we must transform the model into a multi-agent, pure intervention problem** – that is, a problem in which **each player has a single, root node decision** (i.e., one made without information about the values of other variables in the game). How is this possible? The answer is: by **adding in nodes that represent player strategies**. Remember, a strategy is essentially an algorithm that describes what action a player will take at every decision point for every possible information state that the player might observe. By adding in strategy nodes for the players, we effectively convert the model into one in which each player is faced with a single decision: pick your strategy. Since this decision is made at the start of play, no (in-game) information is available at the time the strategy is selected, and so the strategy nodes will be root nodes as required.

The first step in the transformation is to change the original strategic decision nodes into environmental variables (of the same name). The next step is to add new decision nodes, one for each player, representing the strategies available to them. Last, the strategy nodes are connected to the environmental variables that replaced the players' original decision nodes. The result of such a transformation on our current model is illustrated in Figure 9.13. Think of the strategy nodes as controlling what happens at the converted environmental nodes.

For each player we need to construct a **strategy table** that has columns corresponding to strategies and rows corresponding to the information states that they may observe throughout the game. These are not the decision tables for the nodes (those come soon), but rather a reference that assigns every possible combination of information-contingent plans of action to a unique strategy. By doing this we can use *the strategies* as the players' decision variable; i.e. this models players as choosing whole strategies.

First we build the strategy table for BuenoBrains. The only information BuenoBrains receives at the time of its move is the quality level of the research *Idea* which is either *incremental* or *innovative*. There are two actions that can be taken in response to these two information states, so there are $2 \times 2 = 4$ possible strategies. Each combination corresponds to a unique strategy: a specification of an action for each possible information state. Let us simply number these *strategy1*, *strategy2*, etc. For example, according to the assignment of actions to strategies in Figure 9.14, if BuenoBrains utilizes *strategy2* then it will *partially* develop an *incremental Idea*, and *fully* develop an *innovative Idea*.

We also need a strategy table for GrandeGadgets. Its action is contingent on information about the level of *Develop* (which has two possible states). Since it has two feasible actions at that move, it also has four possible strategies. Every possible combination of setting the value of *Acquire* to *no* or *yes* based on the level

Idea	
incremental	*innovative*
0.8	0.2

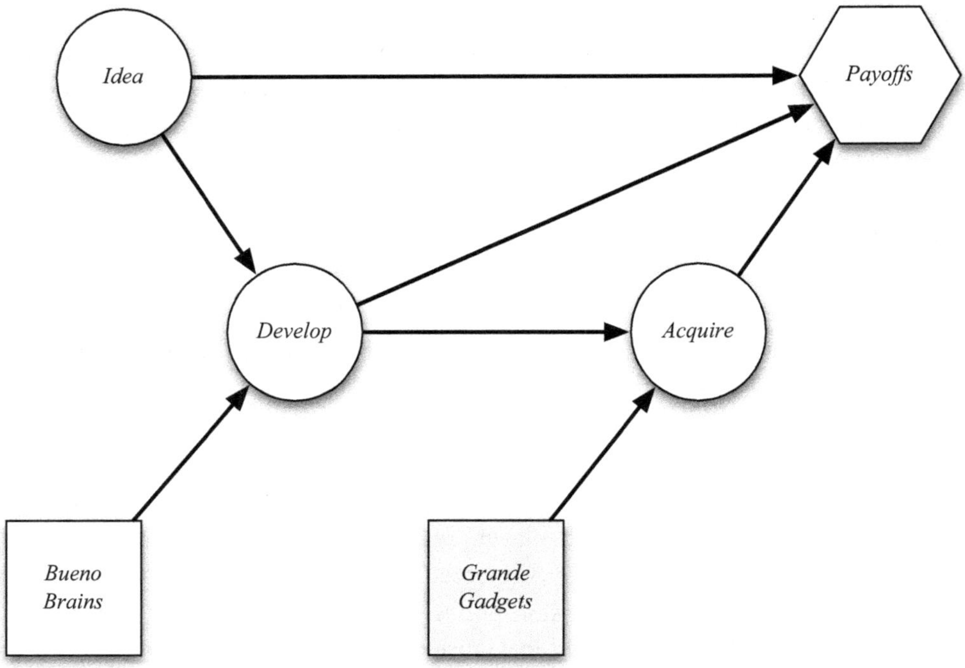

FIGURE 9.13 Transformed model – now a pure intervention problem.

	BuenoBrains			
Do if ...	*strat1*	*strat2*	*strat3*	*strat4*
incremental	*partially*	*partially*	*fully*	*fully*
innovative	*partially*	*fully*	*partially*	*fully*

	GrandeGadgets			
Do if ...	*strat1*	*strat2*	*strat3*	*strat4*
partially	*no*	*no*	*yes*	*yes*
fully	*no*	*yes*	*no*	*yes*

FIGURE 9.14 Strategy tables for both players.

of development is assigned to a strategy. GrandeGadget's strategies can also be seen in Figure 9.14. We reiterate that these are not the decision tables, the strategy tables simply define the actions players will take according to each strategy.

At this point, we know that each firm will choose one strategy of its four possible strategies. Therefore, we attach a decision table to each strategy decision node. The states of the decision table for each player are that player's strategies; i.e., *strategy1*, *strategy2*, etc. as defined by the strategy table for each player.

Now we can turn our attention to the environmental variables that replaced the original decision nodes and construct their condition probability tables. The formats for these conditional probability tables are the same as they always are for an environmental variable – the rows represent all the possible state combinations implied by the parent nodes. However, the probabilities of these tables differ from what we have seen so far. Remember that one of the input (parent) variables will always be the strategy decision node, and the strategy value is used to specify an action at this environmental node. Here's how that works.

To fill in these conditional probability tables, we examine each row and enter "1" in the column corresponding to the action required by that row's strategy (given its information state). That is, each row will have a particular strategy input, and that strategy specifies a particular action given all the other inputs. So the action (the column state) appropriate for this row's strategy happens for sure. Naturally we enter "0"s in the remaining cells for that row.[5]

For example, the conditional probability table for the *Develop* node has inputs from both *Idea* (two states) and *BuenoBrains' Strategy* decision node (four states). Therefore, there must be eight rows to handle all input combinations in the table attached to *Develop*. Each row tells which action is taken according to the row strategy and the value of *Idea*. The same procedure is taken to construct the conditional probability table for the *Acquire* environmental variable. Both are shown in Figure 9.15.

As it turns out, because the payoff table lists both player's payoffs, it will be simpler and still sufficient to work with a single, joint decision table for the players' strategy choices. This table is constructed by setting the first player's strategies as the rows, and the second player's strategies as the columns. Now, in this combined table, each cell will contain two numbers: the expected payoff for the row player given the strategies corresponding to that cell, followed by the expected payoff for the column player under those strategies. The completed model, with the joint decision table, is shown in Figure 9.16.

9.5.2 Solving the Reformulated Model

With the setup complete, we are now ready to fill in the decision table with the expected values of each strategy combination for both players, and then determine the best action(s) for each player. To solve this model, we proceed by applying the now familiar "rolling the marginals forward" technique – as was done to set up the inference problems in Chapter 5 (e.g., section 5.1).

To calculate the expected value for each player for the first cell (the one corresponding to *strategy1*, *strategy1*), we first find the marginal probabilities of the states of each environmental node given the "evidence" of the strategies chosen. For the *Develop* node, the probability that *BuenoBrains* = *strategy1* and the state of

5 Our method identifies *pure strategies*, those in which actions are chosen with probability 1 or 0. While we do not cover them here, it is also possible to incorporate *mixed strategy* solutions, in which actions are chosen with intermediate probabilities.

Idea	
incremental	*innovative*
0.8	0.2

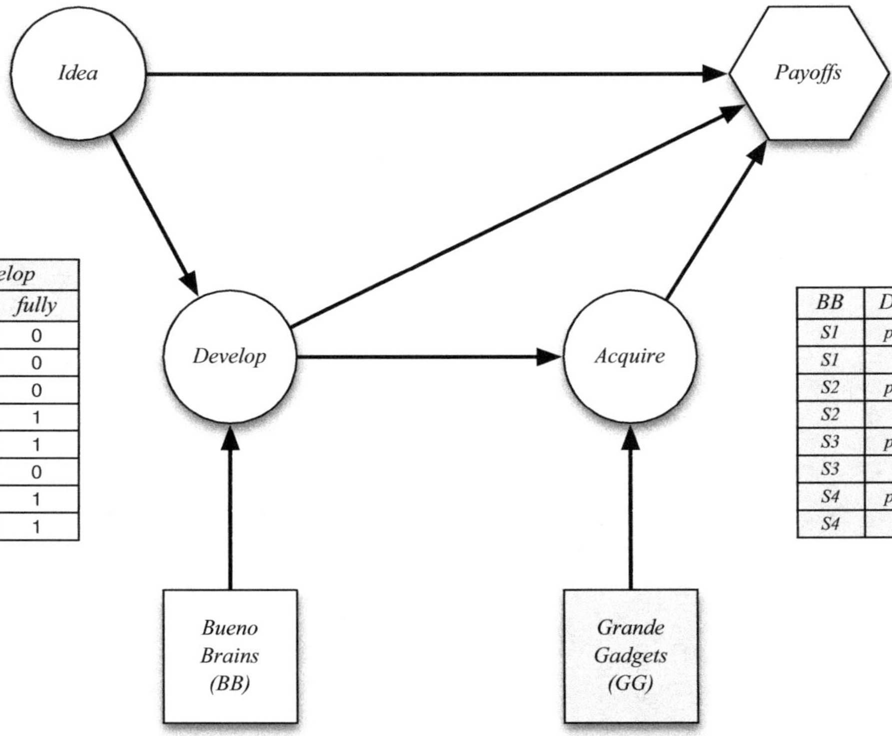

		Develop	
GG	*Idea*	*partly*	*fully*
S1	*incremental*	1	0
S1	*innovative*	1	0
S2	*incremental*	1	0
S2	*innovative*	0	1
S3	*incremental*	0	1
S3	*innovative*	1	0
S4	*incremental*	0	1
S4	*innovative*	0	1

		Acquire	
BB	*Develop*	*no*	*yes*
S1	*partially*	1	0
S1	*fully*	1	0
S2	*partially*	1	0
S2	*fully*	0	1
S3	*partially*	0	1
S3	*fully*	1	0
S4	*partially*	0	1
S4	*fully*	0	1

GrandeGadgets			
strat1	*strat2*	*strat3*	*strat4*
?	?	?	?

GrandeGadgets			
strat1	*strat2*	*strat3*	*strat4*
?	?	?	?

FIGURE 9.15 Model with decision tables and conditional probability tables added.

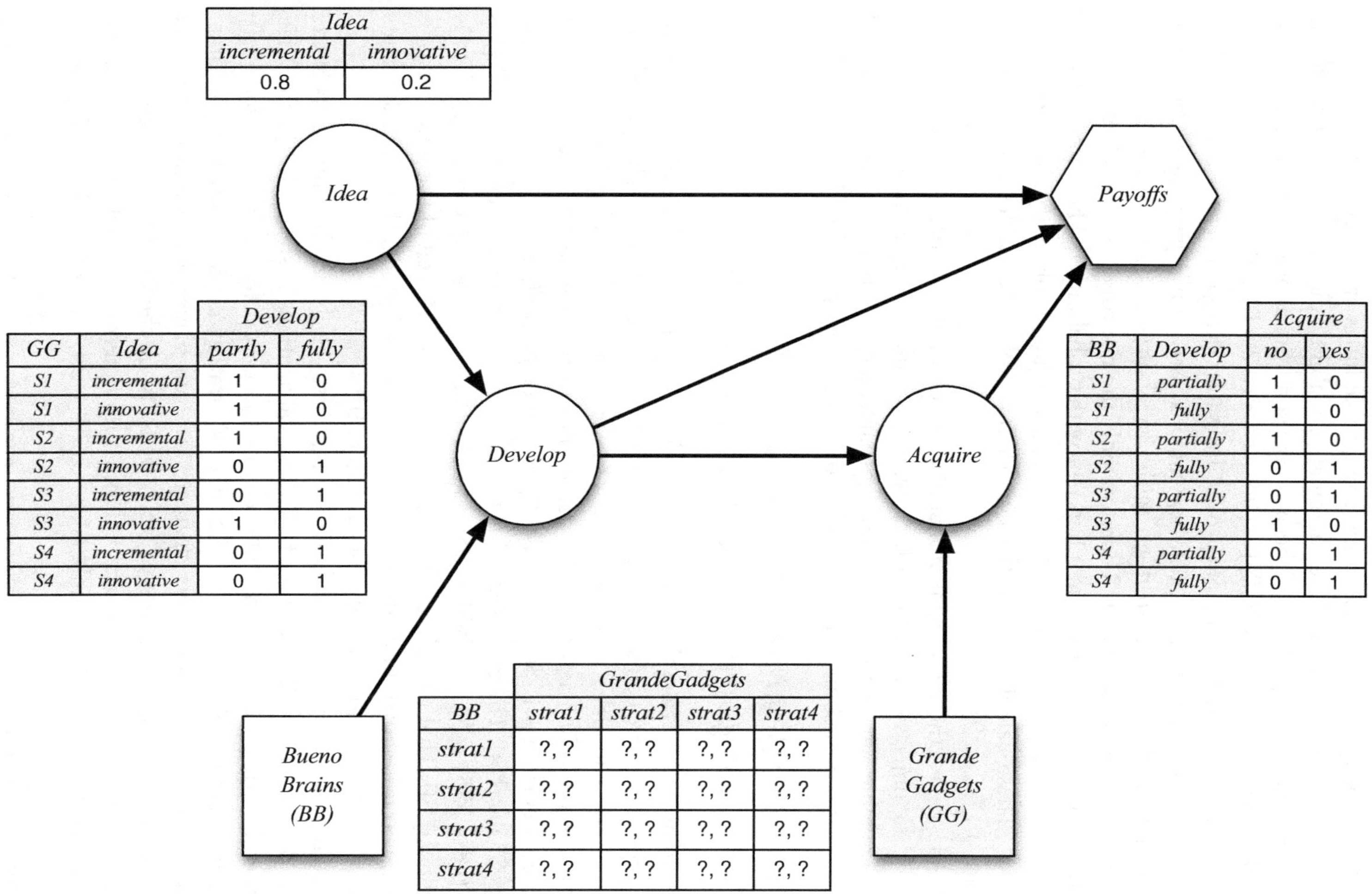

Idea	
incremental	*innovative*
0.8	0.2

		Develop	
GG	Idea	partly	fully
S1	incremental	1	0
S1	innovative	1	0
S2	incremental	1	0
S2	innovative	0	1
S3	incremental	0	1
S3	innovative	1	0
S4	incremental	0	1
S4	innovative	0	1

		Acquire	
BB	Develop	no	yes
S1	partially	1	0
S1	fully	1	0
S2	partially	1	0
S2	fully	0	1
S3	partially	0	1
S3	fully	1	0
S4	partially	0	1
S4	fully	0	1

	GrandeGadgets			
BB	strat1	strat2	strat3	strat4
strat1	?, ?	?, ?	?, ?	?, ?
strat2	?, ?	?, ?	?, ?	?, ?
strat3	?, ?	?, ?	?, ?	?, ?
strat4	?, ?	?, ?	?, ?	?, ?

FIGURE 9.16 Model can be solved with a single, joint decision table.

Idea = *incremental* is 0.8. The probability that *BuenoBrains* = *strategy1* and the state of *Idea* = *innovative* is 0.2. All other row probabilities are 0 because the strategy cannot be any other value. The marginal probability that *Develop* = *partially* is therefore $(0.8 \times 1) + (0.2 \times 1) = 1$ because the conditional probability of *partially* = 1 in those rows. Likewise $P(fully) = (0.8 \times 0) + (0.2 \times 0) = 0$.[6]

Now that we have the marginal probabilities for *Develop*, we can use them to find the marginal probabilities for *Acquire*, which is also a converging connection. The value of *GrandeGadgets* is *strategy1* with probability 1, and the value of *Develop* = *partially* with probability 1. Therefore only the one row corresponding to those particular values is possible, and it happens with probability 1. The state of *Acquire* in that row is *no* with probability 1, so the marginal probability that *Acquire* = *no* is 1. These calculations on the conditional probability tables can be seen in Figure 9.17.

Next, we use these marginal probabilities to compute the probabilities for each row of the objective node's payoff table; one calculation for each cell in the joint decision table. Though the tables have gotten larger, the approach here is identical to the previous section. The probability of a row in the payoff table is:

$$\begin{aligned} P(row) &= P(Idea, Develop, Acquire) \\ &= P(Idea)P(Develop|Idea)P(Acquire|Develop) \end{aligned}$$

We know for certain the value of *Develop* = *partially* and *Acquire* = *no*, so we only have to calculate for the two possible values of *Idea*. The probability of the row with *Idea* = *incremental*, *Develop* = *partially*, and *Acquire* = *no* is $0.8 \times 1 \times 1 = 0.8$. The payoff to BuenoBrains in that row is \$55 and the payoff to GrandeGadgets is \$0. Similarly, $P(innovative, partially, no) = 0.2$, which also pays \$55 and \$0. The expected value of this strategy combination for BuenoBrains is $(0.8 \times \$55) + (0.2 \times \$55) = \$55$. The expected value for GrandeGagdets is obviously \$0. That completes the calculation for the first cell of the joint decision table, and it is shown in Figure 9.17. The entire causal model showing the results of playing *strat1* and *strat1* appears as Figure 9.18.

We repeat the process for calculating that first cell for all the other cells in the joint decision table. The result of all these calculations is shown in Figure 9.19. Remember that this decision table tells us the payoff for each player for every possible strategy. By combining the players' strategy choices into a single table, it is easy to identify best reply strategies of each player. For example, if we fix *BuenoBrains*' decision to *strategy3*, then we focus on the third row of Figure 9.19, look at the second entry in each column, and determine which column has the highest payoff value. The numbers are 0, −80, 29, and −51. Clearly \$29 is the

6 This is a convergent connection, so the marginal probabilities are updated with a message combining of the L^+ values from both parents, multiplied and summed down each column's probabilities (as explained in section 5.2.3). The calculation is just much easier here because so many values are zero.

Row Prob	BB	Idea	Develop partly	Develop fully
1 x 0.8 = 0.8	*S1*	*incremental*	1	0
1 x 0.2 = 0.2	*S1*	*innovative*	1	0
0 x 0.8 = 0	*S2*	*incremental*	1	0
0 x 0.2 = 0	*S2*	*innovative*	0	1
0 x 0.8 = 0	*S3*	*incremental*	0	1
0 x 0.2 = 0	*S3*	*innovative*	1	0
0 x 0.8 = 0	*S4*	*incremental*	0	1
0 x 0.2 = 0	*S4*	*innovative*	0	1
		marginal	1	0

Row Prob	GG	Develop	Acquire no	Acquire yes
1 x 1 = 1	*S1*	*partially*	1	0
1 x 0 = 0	*S1*	*fully*	1	0
0 x 1 = 0	*S2*	*partially*	1	0
0 x 0 = 0	*S2*	*fully*	0	1
0 x 1 = 0	*S3*	*partially*	0	1
0 x 0 = 0	*S3*	*fully*	1	0
0 x 1 = 0	*S4*	*partially*	0	1
0 x 0 = 0	*S4*	*fully*	0	1
		marginal	1	0

Idea	*Develop*	*Acquire*	Probability	*BuenoBrains*	*GrandeGadgets*
incremental	*partially*	*no*	0.8 x 1 x 1 = 0.8	55	0
incremental	*partially*	*yes*	0	55	–55
incremental	*fully*	*no*	0	–120	0
incremental	*fully*	*yes*	0	–75	–100
innovative	*partially*	*no*	0.2 x 1 x 1 = 0.2	55	0
innovative	*partially*	*yes*	0	55	145
innovative	*fully*	*no*	0	45	0
innovative	*fully*	*yes*	0	90	100
			Exp Payoffs	**55**	**0**

FIGURE 9.17 Calculation for first cell of decision table – *strat1* and *strat1*.

best, and that matches *strategy3* for GrandeGadgets. What this says is that *strategy3 for GrandeGadgets is a best reply to BuenoBrains' play of its own strategy3.*

Suppose GrandeGadgets does play its *strategy3*? What is BuenoBrains' best response to that? Using the same reasoning, we see that *strategy1* for BuenoBrains pays better than all the others in that column, so that is its best response. There can be ties, in which case both actions count as best responses. We have gone through and identified the best responses for both players in each row and column, highlighted in **bold**, in Figure 9.19.

Thus, the completed decision table also indicates all the Nash equilibria.[7] There are two equilibria here: (1) BuenoBrains never *fully* develops any technology, and GrandeGadgets never acquires BuenoBrains; and (2) BuenoBrains *fully* develops all and only *innovative* ideas, and GrandeGadgets only acquires BuenoBrains if the technology is *fully* developed. These match *strategy1/strategy1* and *strategy2/strategy2*, respectively.

Both of these strategy combinations count as solutions, but as you can see from the decision table, these solutions are not equally desirable. In this case the *strategy2/strategy2* equilibrium yields a higher payoff for both players. One might

7 More accurately, it indicates all the pure strategy, Bayesian Nash equilibria.

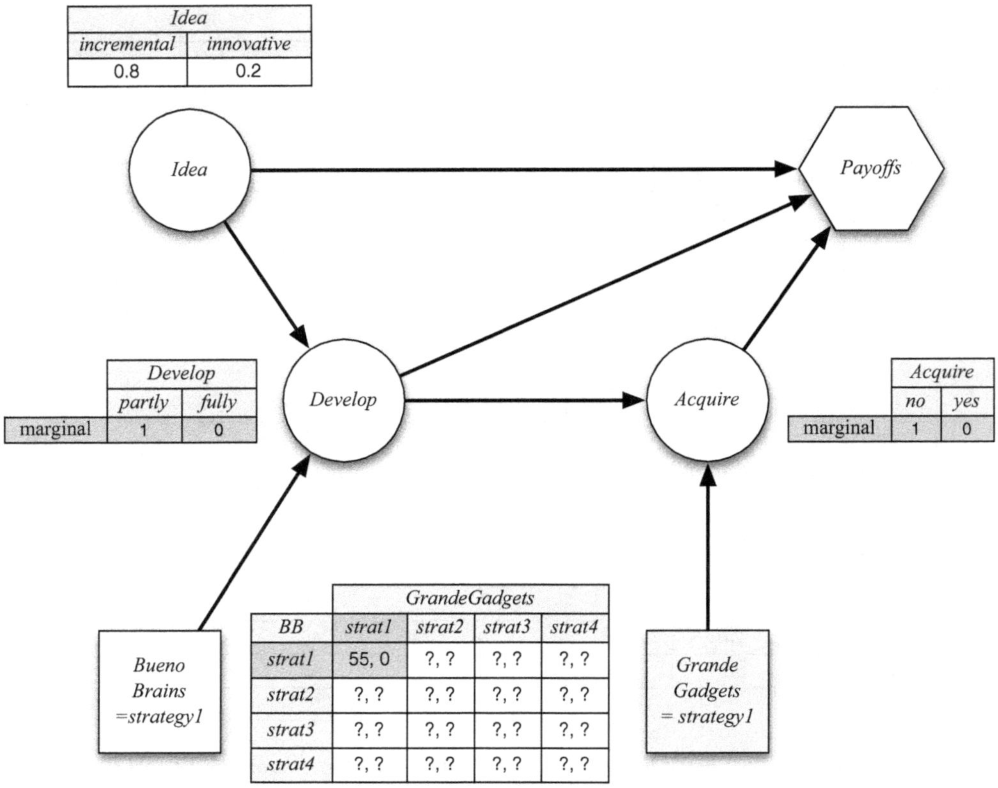

FIGURE 9.18 Causal model showing the results of playing *strat1* and *strat1*.

	GrandeGadgets			
BuenoBrains	*strat1*	*strat2*	*strat3*	*strat4*
strat1	**55, 0**	55, **0**	**55**, -15	55, -15
strat2	53, 0	**62, 20**	53, -44	**62**, -24
strat3	-85, 0	-49, -80	-85, **29**	-49, -51
strat4	-87, **0**	-42, -60	-87, **0**	-42, -60

First entry is Row Player Payoff

Second entry is Column Player Payoff

FIGURE 9.19 Complete joint decision table – best replies in bold, equilibria are shaded.

reasonably expect this to be the more likely outcome of the game. Of course, this is a judgment call on the part of the analyst – the model only identifies which strategy pairs are equilibria and stops there. Equilibrium refinement and equilibrium selection remain active areas of research in game theory.

Definition 9.6 *Best response: the strategy with the highest expected payoff given the strategy of the other player.*

Definition 9.7 *Bayesian Nash equilibrium: a strategy and set of beliefs for each player that maximizes the expected payoffs for each player given their beliefs about the strategy and type of the other player.*

9.5.3 Insights from Technology Development Problem

The game just presented is usually considered a challenging one to solve for newcomers to game theory. It includes stochastic/uncertain elements (the quality of the idea) and asymmetric information (GrandeGadgets does not know the true quality of the idea). Converting the game into a multi-agent influence diagram produces several benefits. It allows us to include those features, but also to reduce the game into an easy-to-solve two-player simultaneous game.

Because causal modeling is designed to deal with information flow, inference, and belief revision, the technique we have just shown to solve games using a causal model is an excellent approach for problems in the real world. With supporting software, many real-world complexities can be added to the analysis of a particular business problem. The use of player strategy nodes permits the analyst to identify best replies and Nash equilibria, both of which are helpful in assessing likely competitor behaviors.

Even in our simple example, useful insights about technology transfer markets arise. The first equilibrium suggests a situation in which research companies always develop their ideas as little as possible, and technology companies are never interested in buying them. Of course, our model also assumed that the scientists could sell their company/idea to *somebody*. This would be satisfied if potential buyers existed outside the scope of the model that did not fit the features of GrandeGadgets: other scientists, investment companies, overseas research companies, foreign technology companies, and so on. If this "bad" equilibrium prevails, then, the system of carrying ideas from the universities, through research companies, and into the market by large technology firms has collapsed – *complete market failure*.

The second equilibrium in our model captures a situation in which developing the technology is a clear signal that the idea is innovative, and these are the only ideas that the technology company is interested in. The market is efficient in the sense of weeding out the lower quality ideas, but the cost and effort to fully develop the idea, and thus signal its high quality, may be prohibitively (or at least punishingly) expensive.

This second equilibrium brings up an interesting and important point: the value of fully developing the research idea is completely due to the *signal* it sends to the technology firm; i.e., that the idea is of high quality. Even if the technology company cannot reuse any of the work done in developing the idea, and could just as easily proceed from a partially developed idea, the research firm must still fully develop it if it ever hopes to be purchased. Thus costly full development might occur, even in cases where it *adds no intrinsic value to the idea*.

Furthermore, given the setup of the game, if the difference in selling price between partially and fully developed ideas becomes too great, then "faking" good ideas by developing them fully begins to pay off. Once that happens, GrandeGagdets can no longer use full development as an unambiguous signal of idea quality. At this point, it may no longer be willing to acquire such research companies (depending on the expected value of the gamble). This devalues full development, and eventually leads to another market breakdown. This phenomenon is known as the "market for lemons" for which Akerlof, Spence, and Stiglitz shared the Nobel Prize in 2001.

The technique used for the technology example can be applied to a wide array of problems in information economics. For example, it is possible to analyze "principal-agent" problems with causal models. These are settings in which an agent working for a principle knows his or her own effort, but the principal only observes the resulting profit. This induces the principal to intervene with an incentive scheme, such as paying for quantity of output. We can also analyze "screening" games, such as when a health insurance company makes different offers to different types of customers. Individuals who purchase insurance know their own level of health risk, but the insurance company does not. Based on certain tests, the insurance company can attempt to screen individuals to assess which insurance packages to offer. If the company prices the packages correctly, customers purchase the insurance that matches their actual risk levels.

9.6 SOFTWARE SOLUTIONS

This chapter barely touched upon the power of using causal models to analyze game theoretic situations. The methodology can be generalized to accommodate more players, longer chains of causal relationships, more sophisticated payoff functions, and many other complications one might find in the real world. As we indicated in the introduction, the standard approaches used in game theory are both elegant and complete. The material in this chapter is useful for business managers who wish to incorporate game theoretic considerations (i.e., account for the strategic behavior of competitors) in their causal models. With the support of causal modeling software, it is possible to assess fairly complex, real-world situations. Still, we must point out that the causal modeling industry is still in its "early days" in terms of pulling together tools that seamlessly integrate game theory and causal models. As of the time of this writing, tricks, work-arounds, repurposing, and methodological "band-aids" are still the norm to get influence diagram solvers to work on game-theoretic causal models.

Let us demonstrate an example using the previous problem. At the time of this writing, the commercial-grade causal modeling software tools are only designed to analyze single-agent interventions. Therefore, without making things too complicated, one way of proceeding is to create mirror-image models, one for each player's decision problem, run them to identify best-response strategies and compare them to see which strategy pairs, if any, constitute Nash equilibria.

Figures 9.20 and 9.21 show the causal diagrams corresponding to our technology leader example for player 1 (BuenoBrains) and player 2 (GrandeGadgets), respectively. Let us focus upon the model for Player 1. We have added a node called

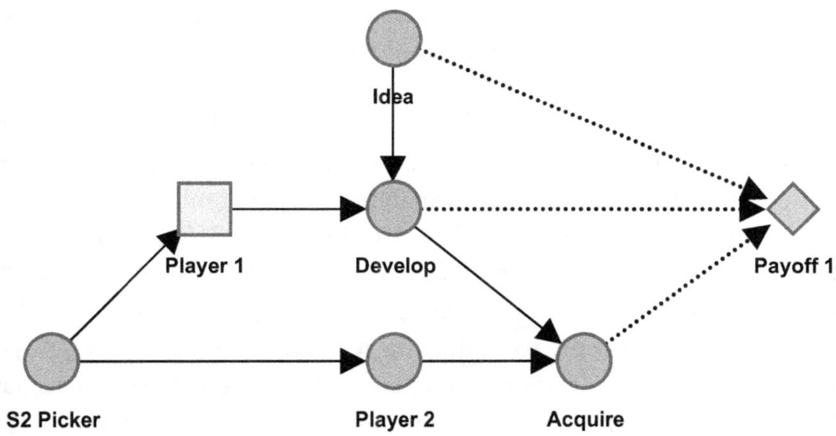

FIGURE 9.20 Model to compute best responses for Player 1.

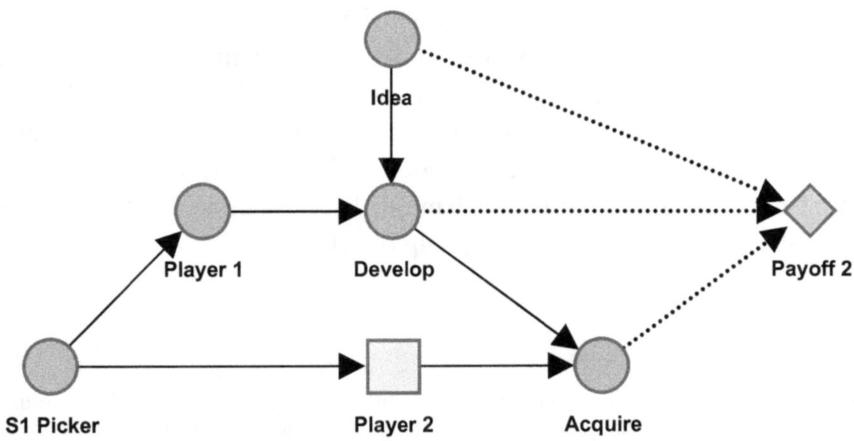

FIGURE 9.21 Model to compute best responses for Player 2.

S2 Picker	P2-S1	P2-S2	P2-S3	P2-S4
1	100.000	0.000	0.000	0.000
2	0.000	100.000	0.000	0.000
3	0.000	0.000	100.000	0.000
4	0.000	0.000	0.000	100.000

FIGURE 9.22 *S2-Picker* selects strategies to be played at *Player 2*.

S2-Picker. This node selects strategies for Player 2 with uniform probabilities (25% for each of the four strategies). The idea is to play all of Player 2's strategies with positive probabilities in order to identify Player 1's best replies to all of them in one run.

The *S2-Picker* node points to Player 2's strategy node (*Player 2*) which, in this case, is treated as an environmental node. When *S2-Picker* picks a random strategy, then *Player 2* plays it with certainty. The conditional probability table for *Player 2* is shown in Figure 9.22. The *S2-Picker* node also points to Player 1's strategy decision node (*Player 1*). What this setup does is treat Player 1 *as if* she knew Player 2's strategy at the time of her move. The software then finds the optimal choice of Player 1's strategy in response to Player 2's strategy. Since all of Player 2's strategies are setup to be played with positive probability, all best replies will be found.

The results of a solution run are shown in Figure 9.23. The columns indicate optimal strategy choices for Player 1 in response to strategy choices by Player 2 (rows). We see that *strategy1* is best reply to both *strategy1* and *strategy3* by Player 2, and *strategy2* is the best reply to both *strategy2* and *strategy4*. The corresponding table for Player 2's best replies is shown in Figure 9.24. Here, we see that Player 2's *strategy1* and *strategy2* are both best replies to Player 1's *strategy1*; *strategy2* is a best reply to Player 1's *strategy2*; *strategy3* is a best reply to Player 1's *strategy3*; and *strategy1* and *strategy3* are both best replies to Player 1's *strategy4*. Also note that the software has computed all the expected payoffs (compare the entries in Figures 9.23 and 9.24 with those in Figure 9.19).

By intersecting the best replies of each player, we identify the Nash equilibria. While it would be nice to have software capable of identifying Nash equilibria directly (and, perhaps, following the publication of this book the major developers will give it some thought), this method does work.[8] Moreover, it can be applied to models of much greater complexity than our little example – models that would be quite difficult to solve without the help of software.

8 Keep in mind, not all games have pure strategy equilibria. Sometimes, the only Nash equilibria are those in which players randomize over their actions. If the methodology presented here does not identify any equilibria, then the analyst knows the only equilibria are those that involve randomizing over actions.

S2 Picker	P1-S1	P1-S2	P1-S3	P1-S4
1	55.000	35.000	-85.000	-105.000
2	55.000	62.000	-49.000	-42.000
3	55.000	35.000	-85.000	-105.000
4	55.000	62.000	-49.000	-42.000

FIGURE 9.23 Best replies for Player 1.

S1 Picker	P2-S1	P2-S2	P2-S3	P2-S4
1	0.000	0.000	-15.000	-15.000
2	0.000	20.000	-44.000	-24.000
3	0.000	-80.000	29.000	-51.000
4	0.000	-60.000	0.000	-60.000

FIGURE 9.24 Best replies for Player 2.

KEY CONCEPTS: CHAPTER 9

- Game theoretic problems can be captured as multi-agent decision problems using the causal modeling approach.
- Games include: (1) players, (2) players' moves, (3) actions at each move, (4) environmental variables, (5) information available to players at each move, and (6) payoff for each player for each outcome.
- Moves in the game correspond to decision nodes, but because causal models represent information differently they typically allow for greater economy in model components.
- Causal-form games use an objective node to represent the payoffs received for each player for each outcome.
- Multi-agent decision problems, like any game, can be solved to find the Nash equilibria: sets of strategies that are all best responses to each other.
- Games with a particular information structure can be solved using the same stepwise process used for single-agent decision problems.
- Games in which both players operate on partial information require a conversion into a pure decision problem and non-trivial strategies for both players. This is done by adding strategy decision nodes and converting the decision into a variable that implements the chosen strategy.
- Using a joint strategy table, with payoffs for both players listed, makes identifying Nash equilibria easy and intuitive.
- The techniques presented here can be adapted to work with analysis software, and generalize to any number of players and decisions.
- The combination of game-theoretic and Bayesian network techniques allows you to incorporate both sophisticated evidence-based inferences of new beliefs as well as the actions of other agents in determining the optimal intervention in a system.

CHAPTER 9 REVIEW QUESTIONS

9.1 Which of the following is true regarding the relationship between extensive form games and causal form games.

(a) Each node in the game tree corresponds to a decision node in the causal diagram.
(b) The payoffs at the end of every branch are incorporated into a single objective node.
(c) Extensive form games cannot include random variables such as the environmental variables in causal models.
(d) The Nash equilibria of the extensive form game will often differ from the equilibria found in the causal model.
(e) Information sets are used to capture the arrows pointing into decision nodes in a causal model.

9.2 To solve a single-agent decision problem you identify the best solution by calculating expected values of optimal strategies at each decision node. For multi-agent decision problems, how is this technique adapted?

(a) Strategies that are best responses to each other are considered solutions regardless of the payoffs.
(b) When each agent is optimizing their own payoff, they always produce the best combined strategy as a solution.
(c) Every possible distinct set of actions, given information at that juncture, is assigned to a strategy.
(d) The use of strategy tables can only be used when both players have non-trivial strategies.
(e) The decision tables are changed so that agents' actions are considered over their set of strategies rather than individual actions.

9.3 If you solve a game and discover that there are multiple Bayesian Nash equilibria, how do you find the correct solution?

(a) One must then find the probability mixture over those strategies that provides the single best payoffs.
(b) The strategy combination with the greatest total payoff is the one selected.
(c) There is no single correct solution; all one can say within the model is that for each pair, those strategies are best responses to each other.
(d) There is no correct solution, if the process reveals multiple solutions then the game is unsolvable.

9.4 Consider the following multi-actor influence diagram that represents the situation facing two bidders for mining rights to a tract of land in Australia.

The bidders are mining companies who plan to mine the land for uranium. Each bidder must decide whether to *Bid low* or *high*, the *Uranium Deposit* is *rich* or *poor*, and the *Extraction Costs* are either *low* or *high*.

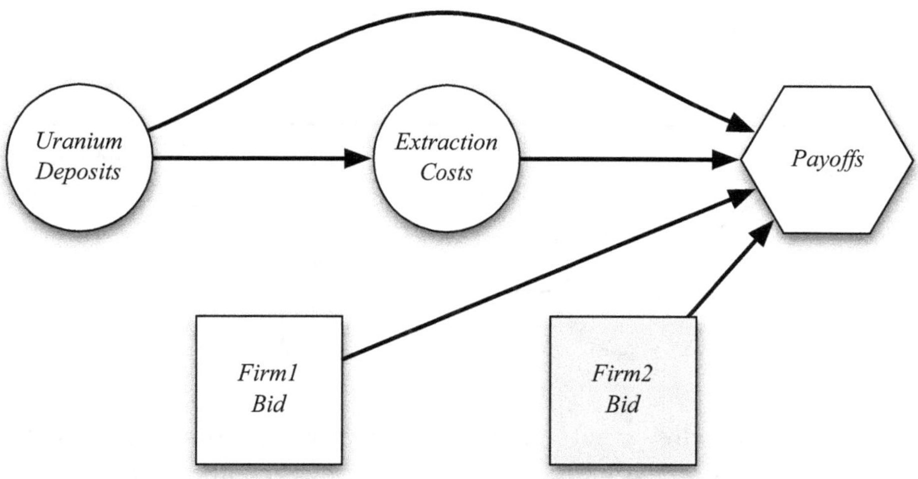

(a) This situation is a too simplistic representation of the decision problems faced by the firms. First, *Firm1* (the larger of the two) can decide to perform an *Assessment* of the *Uranium Deposit* which (if done) will result in an *Assessment Report* that indicates the quality of the *Uranium Deposits*. At the time of its bid, *Firm1* will have access to this *Assessment Report*. If *Firm1* decides to bid on the land, it must then decide whether or not to offer a *Press Release* indicating this decision. *Firm2* only observes the *Press Release*. Redraw the causal model diagram to represent the more sophisticated situation just described.

(b) Because this is not a game of perfect information, in order to solve the augmented model it must be first transformed into a multi-agent pure intervention model. Draw the diagram representing the transformed model.

9.5 Consider the following multi-agent influence diagram that represents the situation facing two producers of retail goods regarding whether to enter a market or not (i.e., decide *no* or *yes*). The *Entrant* is new to the market sector, and performs a simple *Market Test* (which can report *low* or *high*) to help determine whether its version of the product will succeed. The *Incumbent* seller runs a more sophisticated *Market Test* that can report *low, medium,* or *high,* and it uses this to inform its decision to enter the market again or not. The supply of goods in the market, combined with the actual *Market Demand,* determines the *Payoffs* to each company. This situation is represented in the diagram below.

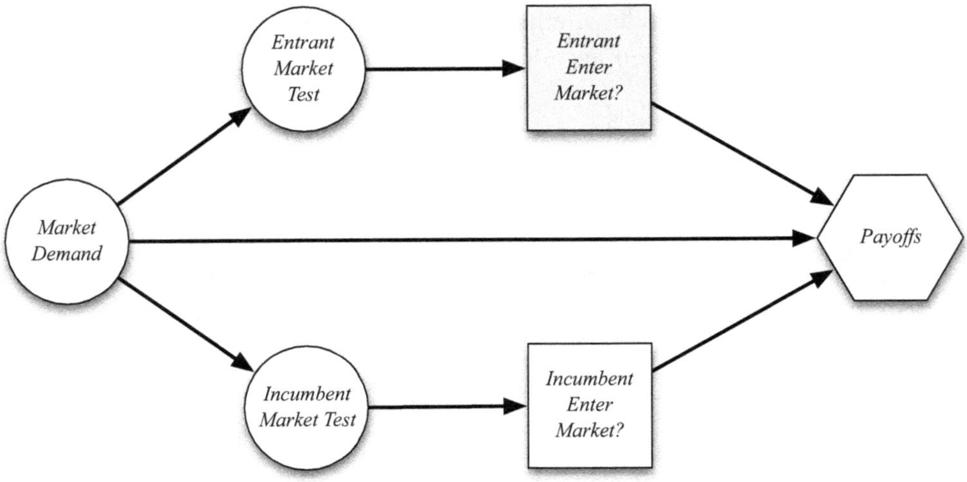

(a) Because this is not a game of perfect information, the model must be transformed into a multi-agent pure intervention model. Draw the diagram representing the transformed model.

(b) Write down the individual strategy table for each player.

(c) When determining the payoffs to each player for each strategy combination, it is convenient to use a joint decision table. Write down the structure of the joint decision table for this game.

9.6 An Insurance Company must decide whether or not to offer a policy to a potential Buyer. There is a 60% probability that the potential *Buyer's Risk* is *low*, but the Insurance Company does not know the true risk level of the buyer before making its offer. If the Insurance Company decides to make an offer, the potential Buyer is then able to decide whether or not to go ahead and make the purchase. The potential Buyer can observe the true state of his/her own *Risk* (*low* or *high*) when deciding whether or not to *Buy the Insurance Policy* offered. After this decision has been made, the Insurance Company and buyer receive their respective *Payoffs*.
If the Insurance Company decides not to *Offer a Policy*, then the game is over and both players receive $0. If the Insurance Company decides to go ahead and *Offer the Policy*, but the Buyer decides not to buy it, then they both receive $0. If the Buyer decides *yes* and is *low Risk*, then both the company and the Buyer receive $10. If the Insurance Company *Offer the Policy*, and the Buyer decides *yes* when *high Risk* is observed, then the company loses $5 and the buyer gains $15.

(a) Draw the causal model for the situation just described.

(b) This is a game of perfect information, and therefore could be solved using the step-wise backward optimization approach. Instead, use the more general technique for solving multi-agent decision problems and transform the model into a multi-agent pure intervention model. Draw the diagram representing the transformed model.

(c) Write down the individual strategy table for each player.
(d) Now construct the conditional probability table for the *Buy Insurance* node that is implied by the Buyer's strategy.
(e) Next write down the joint decision table for the Insurance Company and Buyer and use it to solve the game. Identify any and all equilibria of this game.

Chapter 10

Data-Driven Causal Modeling

As we have repeatedly emphasized, understanding key causal relationships is important in real world decision making. In particular, as we touched upon in Chapter 1 and explored in detail in Chapter 7, knowing the statistical correlations between the decision variables in a management *intervention* problem is not sufficient. A manager requires causal knowledge to accurately predict the consequences of his or her available decision options. A quantitative causal model generalizes purely probabilistic assessments and makes them more powerful.

In Chapters 7 and 9 the manager's decision variables were specified exogenously – that is, taken as given in the description of the problem. In many cases, however, part of the managerial problem is actually deciding *which* variables to influence among a large number of candidate variables. In any organization, there are typically innumerable variables over which a manager might attempt to exert control. This remains true even when one limits attention to the variables that are within the scope of a manager's formal decision authority.[1]

An *intervention selection problem* arises when the decision maker's intervention resources are sufficiently scarce (e.g., because her own time and energy are limited) that it is impossible to force the value of every variable under her control. In such situations, causal understanding once again comes to the fore. Given its importance, how does one attain such understanding? In some cases, causality can be observed directly. For example, kicking the copy machine may feel good, but one will quickly learn that such an intervention rarely causes it to work better. Causal relationships can also be learned from data. That is, if the causal structure of a

1 There are, of course, many gray areas between full decision rights and no decision rights. For example, a marketing manager may have full authority to determine the daily work assignments for those in her department and no authority to assign parking spaces in the corporate parking lot. However, hiring a new recruit from a business school into Marketing may be a "gray area" in the sense that she exerts a strong influence on this decision – but does not have total control (e.g., Human Resources may play an important role). For our discussion, we will consider cases in which the boundaries are clear.

system is unobserved but the variables themselves are observed, then it may be possible to infer the causal relationships from the data. This is the task to which we turn our attention in this chapter.

10.1 CAUSALITY VERSUS PROBABILITY

Let us illustrate the fundamental idea using a relatively simple example. The motivation for this comes from Moody (1996). Moody was a tech journalist who, in the mid-1990s decided to write a book on the management wisdom of Microsoft. To do this, he spent an entire year shadowing a new product development team (the project, code-named "Sendak", was later launched as *Encarta*). His book provides a fascinating chronicle of that experience.

The key departments in this story are *Marketing* (M), *Design* (D) and *Engineering* (E). Each department is composed of people who have their own motivations (which may or may not be aligned with those of the company) and beliefs (which may or may not be accurate). The project is to design, develop and commercialize an electronic encyclopedia. To keep things simple, assume that each department has two actions. The *Design* department can create an interface that is either *plain* or *ornate* ($D = p$ or $D = o$). The *Engineering* department can construct software that is either *lethargic* or *responsive* ($E = l$ or $E = r$). *Marketing* can launch the final product with an emphasis toward *university* or *business* users ($M = u$ or $M = b$).

As we know from organizational research, firms have formal and informal systems that result in different people and groups having varying degrees of organizational power. For example, Bob the Engineer may report directly to Ajay, the Engineering Manager. Yet, his software writing may be strongly influenced by the designs sent over from his counterpart in *Design* (even though the formal organization chart shows no link between them). We want to build a causal model that represents the true influence relationships in the organization, whether formally imposed by management or naturally arising out of work requirements, friendships, team assignments, and so on.

In this example, we focus upon the influence relationships between departments *Marketing, Engineering,* and *Design*. By *influence relationship,* we mean that the activities and outputs of one group partially drive the activities and outputs of another. In our stylized model, $M \rightarrow E$ means that *Marketing* has the power to decide whether the next software title will be *university*- or *business*-oriented in advance of E's activities. *Engineering* observes these decisions which, in turn, influences whether it decides to make the product *lethargic* or *responsive* (thus, in this case, the "influence" can be through providing potentially decision-affecting information to the next decision node).

Because these influences depend both upon formal and informal features of the organization, it may not be apparent to a senior manager how they actually flow. Moreover, keep in mind that what the department does is driven by the motivations and beliefs of its members. Thus, senior managers cannot rely upon the assumption that everyone is going to work in the best interests of the corporation. Even if everyone in the department was motivated to maximize corporate profits (which, in the real world, they almost certainly are not), they may not have the information at their local level to form accurate beliefs about what they should be doing to best achieve that goal. All of this will create variation in behavior and performance over time.

10.1.1 Probability View

Now suppose that you are the senior executive in charge of new product initiatives. You have data on departmental performance over many previous projects, both in terms of operating costs and the frequency with which different options were chosen. In the past, assume that the company has followed a decentralized policy – that is, the departments self-organized and proceeded without any intervention from senior management. The historic data is presented in Figure 10.1. The company recently appointed you to improve the profitability of new product launches. Unfortunately, given your busy schedule, the best you can do is pick an action for one department and monitor it – thereby guaranteeing that its action is the one you want. That is, you can set the value of one variable, M, D, or E, but do not have the resources to set more than one. Assuming the historic data is representative of the current situation, what should you do?

Seeing that profits tend to be higher when *Design* = *ornate*, *Engineering* = *responsive*, and *Marketing* = *business*, you might be tempted to look at the conditional expected profit under each and then set the one yielding the highest value. It should not be hard to see that, based upon the probabilities in Figure 10.1, the conditional probabilities for each of $D=o$, $E=r$ and $M=b$ are as shown in Figure 10.2. Thus, the expected profit (π) given $D=o$ is:

$$E(\pi|D=o)=0.32(\$50)+0.08(\$40)+0.24(\$60)+0.36(\$100)=\$69.6.$$

Outcomes				
D	*E*	*M*	Likelihood	Profit
p	*l*	*u*	12.0%	20.0
p	*l*	*b*	3.0%	10.0
p	*r*	*u*	6.0%	40.0
p	*r*	*b*	9.0%	30.0
o	*l*	*u*	22.4%	50.0
o	*l*	*b*	5.6%	40.0
o	*r*	*u*	16.8%	60.0
o	*r*	*b*	25.2%	100.0
		TTL/AVG	100.0%	56.5

FIGURE 10.1 Historic cost and performance data on past projects.

Likelihood	Profit	$P(E, M \mid D=o)$	$P(D, M \mid E=r)$	$P(D, E \mid M=b)$
12.0%	20.0	0.0%	0.0%	0.0%
3.0%	10.0	0.0%	0.0%	7.0%
6.0%	40.0	0.0%	10.5%	0.0%
9.0%	30.0	0.0%	15.8%	21.0%
22.4%	50.0	32.0%	0.0%	0.0%
5.6%	40.0	8.0%	0.0%	13.1%
16.8%	60.0	24.0%	29.5%	0.0%
25.2%	100.0	36.0%	44.2%	58.9%
100.0%		100.0%	100.0%	100.0%
	P(denom)=	70.0%	57.0%	42.8%
	E(Profit | Do ...)	$69.6	$70.8	$71.1

FIGURE 10.2 Conditional probabilities for $D=o$, $E=r$, $M=b$ based on likelihood.

Similarly, $E(\pi \mid E=r) = \$70.8$ and $E(\pi \mid M=b) = \$71.1$. Therefore, based upon this analysis, your best bet is to tell Marketing they *must* market this new product to the business segment and then monitor them carefully to make sure they do it.

10.1.2 Causality View

The problem with relying solely upon probabilities is that they do not encode causal information – and causality is a central concern in management intervention problems. To see this, suppose you hypothesize that the underlying causality is: *Design* drives the process by deciding interface type; this decision influences what the software *Engineers* decide to implement in terms of performance speed; finally, *Marketing* considers the performance speed (only) when choosing which segment to target. The model is shown in Figure 10.3.

Now suppose that, relying upon your calculations in the previous section, you decide to make the intervention required to guarantee *Marketing* = *business*. Let us denote this intervention as "**Do**($M=b$)."[2] Since *influence only flows downstream*,[3] **Do**($M=b$) has no effect on M's upstream causes. As a result, the actual change in expected profit is to go from \$56.5 with no intervention to \$59.2 under **Do**($M=b$) (see Figure 10.4). This is far below the outcome of \$71.1 predicted by the conditional probabilities.

What is going on here? Suppose the model in Figure 10.3 is true. When there has been no intervention, giving you information about the state of *Marketing* also gives you information about the likelihoods of the variables that influence *Marketing*, both directly and indirectly. However, when you intervene and set

2 To set the problem up as a single-agent intervention problem from the point of view of the managers, we would add a strategic option node to represent the intervention decision and add an objective node. The decision node would point to all the departments. Its states would enumerate the options for setting the values of the variables. For the purpose of seeing how the probabilities change due to causality, we keep things simple and use the "**Do**()" labeling.

3 It is only *inference* that flows in both directions.

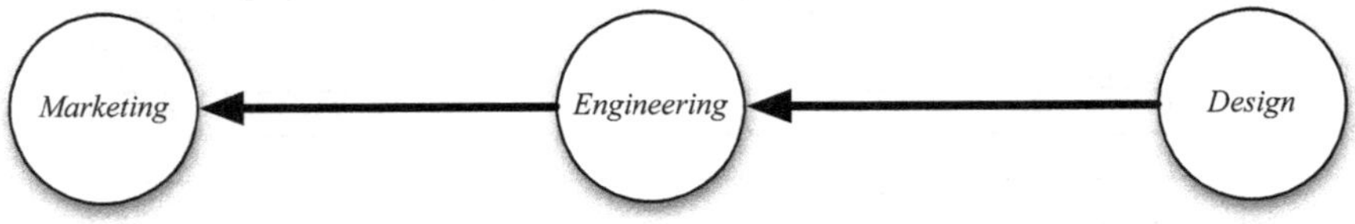

	Marketing	
Engineering	*university*	*business*
lethargic	0.8	0.2
responsive	0.4	0.6

	Engineering	
Design	*lethargic*	*responsive*
ornate	0.4	0.6
plain	0.5	0.5

Design	
ornate	*plain*
0.7	0.3

Outcomes			**Conditional Probabilities**				
D	***E***	***M***	**P(*D*)**	**P(*E* | *D*)**	**P(*M* | *E*)**	**P(Outcome)**	**Profit**
p	*l*	*u*	0.3	0.5	0.8	12.0%	20.0
p	*l*	*b*	0.3	0.5	0.2	3.0%	10.0
p	*r*	*u*	0.3	0.5	0.4	6.0%	40.0
p	*r*	*b*	0.3	0.5	0.6	9.0%	30.0
o	*l*	*u*	0.7	0.4	0.8	22.4%	50.0
o	*l*	*b*	0.7	0.4	0.2	5.6%	40.0
o	*r*	*u*	0.7	0.6	0.4	16.8%	60.0
o	*r*	*b*	0.7	0.6	0.6	25.2%	100.0
						100.0%	56.5

FIGURE 10.3 Parameterization of the model $D \rightarrow E \rightarrow M$.

	Marketing	
Engineering	*university*	*business*
lethargic	0	1
responsive	0	1

	Engineering	
Design	*lethargic*	*responsive*
ornate	0.4	0.6
plain	0.5	0.5

Design	
ornate	*plain*
0.7	0.3

Outcomes			**Conditional Probabilities**				
D	***E***	***M***	**P(*D*)**	**P(*E* | *D*)**	**P(*M* | *E*)**	**P(Outcome)**	**Profit**
p	*l*	*u*	0.3	0.5	0	0.0%	20.0
p	*l*	*b*	0.3	0.5	1	15.0%	10.0
p	*r*	*u*	0.3	0.5	0	0.0%	40.0
p	*r*	*b*	0.3	0.5	1	15.0%	30.0
o	*l*	*u*	0.7	0.4	0	0.0%	50.0
o	*l*	*b*	0.7	0.4	1	28.0%	40.0
o	*r*	*u*	0.7	0.6	0	0.0%	60.0
o	*r*	*b*	0.7	0.6	1	42.0%	100.0
						100.0%	59.2

FIGURE 10.4 Effect of **Do**$(M = b)$.

Marketing from **outside** the system, the links to upstream variables are broken (as shown in Figure 10.4). Thus, following **Do**($M = b$), *Design* and *Engineering* continue to behave as if there were no intervention (because they are not causally affected by *Marketing*).

Intuitively, fixing the value of a variable at the end of a causal chain frees it from other influences. Conversely, setting a variable at the top of a chain opens up the possibility of affecting not only the controlled variable but, indirectly, those further down the chain. Thus, if bad influences are coming in to a variable from many directions, it may be most economical to set the value of that variable directly. However, if a variable has the ability to spread good effects throughout the system, then this might be the variable to set.

In our example, suppose your intervention is **Do**($D = o$). This results are shown in Figure 10.5. In this case, the effect of the intervention – increasing expected profits from \$56.5 to \$69.6 – is the same as that predicted by conditional probabilities. This is because *Design* is a root node, meaning that all of its connections flow downstream. No links are broken by the intervention.

Finally, for the sake of completeness, consider **Do**($E = r$). The effect of this intervention is illustrated in Figure 10.6. This intervention results in increasing expected profits from \$56.5 to \$69.0. Therefore, assuming $D \rightarrow E \rightarrow M$ is the true causal model, the best intervention is therefore **Do**($D = o$), the one ranked worst by naive reliance upon conditional probabilities!

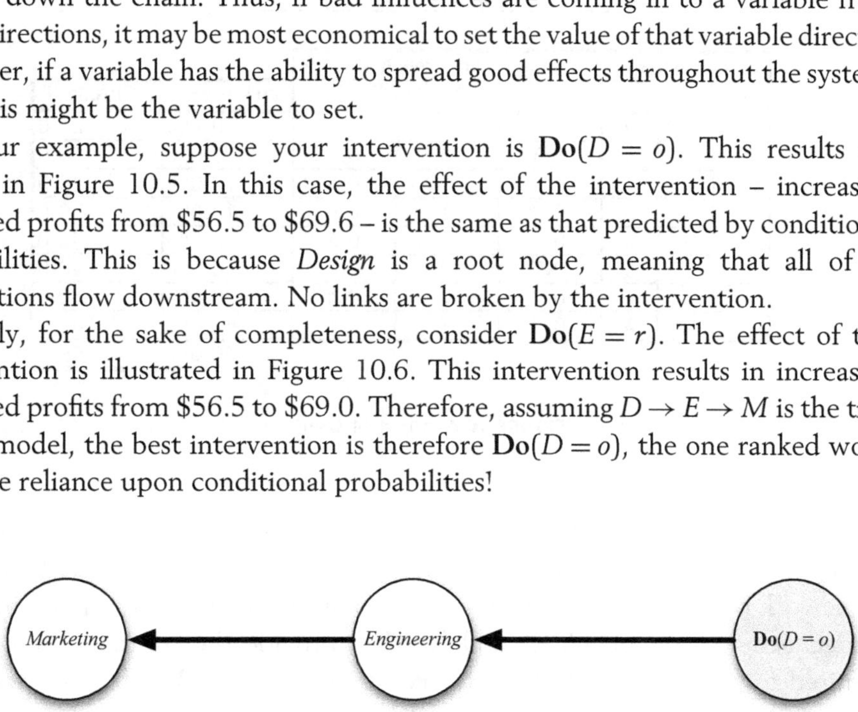

	Marketing	
Engineering	*university*	*business*
lethargic	0.8	0.2
responsive	0.4	0.6

	Engineering	
Design	*lethargic*	*responsive*
ornate	0.4	0.6
plain	0.5	0.5

Design	
ornate	*plain*
1	0

Outcomes			Conditional Probabilities				
D	*E*	*M*	P(*D*)	P(*E* | *D*)	P(*M* | *E*)	P(Outcome)	Profit
p	*l*	*u*	0	0.5	0.8	0.0%	20.0
p	*l*	*b*	0	0.5	0.2	0.0%	10.0
p	*r*	*u*	0	0.5	0.4	0.0%	40.0
p	*r*	*b*	0	0.5	0.6	0.0%	30.0
o	*l*	*u*	1	0.4	0.8	32.0%	50.0
o	*l*	*b*	1	0.4	0.2	8.0%	40.0
o	*r*	*u*	1	0.6	0.4	24.0%	60.0
o	*r*	*b*	1	0.6	0.6	36.0%	100.0
						100.0%	69.6

FIGURE 10.5 Effect of **Do**($D = o$).

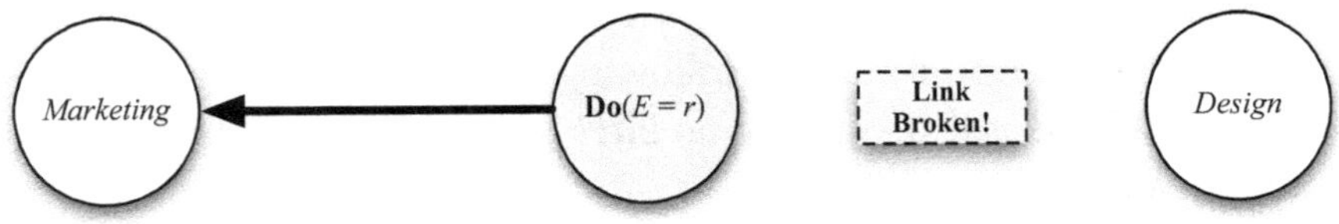

	Marketing	
Engineering	*university*	*business*
lethargic	0.8	0.2
responsive	0.4	0.6

	Engineering	
Design	*lethargic*	*responsive*
ornate	0	1
plain	0	1

Design	
ornate	*plain*
0.7	0.3

Outcomes			**Conditional Probabilities**				
D	***E***	***M***	**P(*D*)**	**P(*E* | *D*)**	**P(*M* | *E*)**	**P(Outcome)**	**Profit**
p	*l*	*u*	0.3	0	0.8	0.0%	20.0
p	*l*	*b*	0.3	0	0.2	0.0%	10.0
p	*r*	*u*	0.3	1	0.4	12.0%	40.0
p	*r*	*b*	0.3	1	0.6	18.0%	30.0
o	*l*	*u*	0.7	0	0.8	0.0%	50.0
o	*l*	*b*	0.7	0	0.2	0.0%	40.0
o	*r*	*u*	0.7	1	0.4	28.0%	60.0
o	*r*	*b*	0.7	1	0.6	42.0%	100.0
						100.0%	69.0

FIGURE 10.6 Effect of **Do**($E = r$).

The point that should be clearly understood here is that knowing probabilities is not the same as having a causal model. The latter provide a formal, semantic framework by which to encode both probabilistic and causal information. Causal models generalize probability models. They provide the additional information required to predict the effects of interventions in a causal system – information that is lacking in simple probabilistic information.

10.2 OBSERVATIONAL INDISTINGUISHABILITY

In the preceding sections, we saw that, given a causal model and the probability distribution on outcomes that it generates, two key issues in an intervention selection problem are deciding: (1) which variable is to receive the intervention; and (2) to which state that variable should be set. Suppose, however, that the causal model itself is unknown. That is, suppose that, after observing and collecting data on the unaltered system for a sufficient period of time, you know the probability distribution of Figure 10.1, but you do not know the underlying causal system that generated it. From the preceding example, you see that different causal models imply different interventions, even though they can generate the same probability distribution.

(a)

Marketing	
university	*business*
0.572	0.428

	Engineering	
Marketing	*lethargic*	*responsive*
university	0.6	0.4
business	0.2	0.8

	Design	
Engineering	*ornate*	*plain*
lethargic	0.349	0.651
responsive	0.263	0.737

(b)

	Marketing	
Engineering	*university*	*business*
lethargic	0.8	0.2
responsive	0.4	0.6

Engineering	
lethargic	*responsive*
0.43	0.57

	Design	
Engineering	*ornate*	*plain*
lethargic	0.349	0.651
responsive	0.263	0.737

FIGURE 10.7 Alternative models that are consistent with the data of Figure 10.1.

10.2.1 The Observational Indistinguishability Theorem

To see the problem clearly, look at the two models presented in Figure 10.7. These models have different causal structures and different node parameters. However, they both induce the same distribution over outcomes; specifically, the one shown in Figure 10.1. That is, even given that you know the actual distribution, there are at least three causal models that are consistent with it (the one analyzed in the previous sections and these two new models). The effects of the interventions we have been considering are detailed in Table 10.1.

It is important to keep in mind that we do not get to just choose the model we like. If that were true, we would pick the model illustrated in Figure 10.7(a) and Do($M = b$), giving us the biggest payoff in Table 10.1. Rather, the intervention

Table 10.1
Intervention predictions from the three models.

Intervention	Original	Figure 10.7(a)	Figure 10.7(b)
Do($D = o$)	**69.6**	68.5	68.5
Do($E = r$)	69.0	66.2	**70.8**
Do($M = b$)	59.2	**71.1**	59.2

problem here is to **figure out what the true model is** and then apply the correct intervention with respect to **that** model. What we have just shown is that the data in Figure 10.1 can be generated by an appropriate choice of parameters in each of the three models thus far considered.

At this point, you might be wondering just how many causal models there are with exactly three nodes. For those who are curious about such things, the number of different causal structures (directed acyclic graphs) that can be constructed with n nodes is given by $m(n)$ where:

$$m(n) = \sum_{k=1}^{n} (-1)^{k+1} \binom{n}{k} 2^{k(n-k)} m(n-k)$$

For $n = 3$, we find $m(3) = 25$.[4] All the possibilities are shown in Figure 10.8. Is there any way to infer the causality from the data? The answer is: to a certain extent.

For the following proposition, we need to define two graphical objects. First, given any directed graph, its **skeleton** is all of its links without regard to their directions. Thus, the models in Figures 10.3 and 10.7 all have the same skeletons. Second, within a directed graph, a **v-structure** is a set of three nodes and exactly two links such that the two connections both point to a center node, thus forming a "V" shape. This is the same structure we called a convergent triplet in section 2.3. The model in the third row, middle column of Figure 10.8 has one v-structure (i.e., $E \rightarrow D \leftarrow M$). The following proposition is due to Verma and Pearl (1990) (see also Pearl, 1988).

Proposition 1 *Suppose the states of variables $x_1, x_2, \ldots x_n$ are determined by a true causal model as defined in Chapter 4. Let P denote the joint probability distribution implied by the true model on the variables $x_1, x_2, \ldots x_n$. Then, set of causal models for which parameters exist such that the distribution P is also implied are those with the same: (1) nodes, (2) skeletons, and (3) v-structures as the true underlying model.*

Although it may not be obvious at first glance, this is a rather remarkable result.[5] It says that, if you can find *one* causal model that is consistent with a particular data set (and there is software available that does this), you can identify the entire set of causal models consistent with that data by mere visual inspection. We call these models *observationally indistinguishable* to one another because, even if you

4 Note: $m(0) = 1$ by definition.

5 The proposition, as we have written it, omits a requisite technical condition (see Verma and Pearl, 1990) on the relationship between the model and P. Essentially, the proposition holds for any finite model in which there are no trivial arrows – i.e., variation in the parent node should result in variation in the conditional probabilities governing the child node.

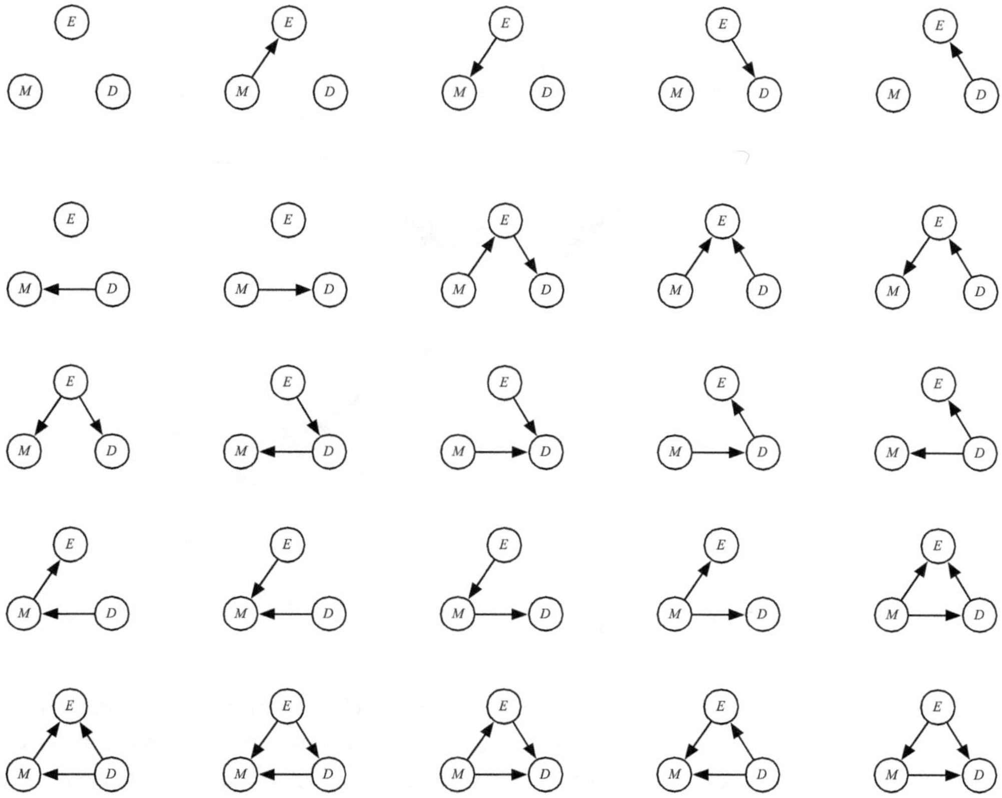

FIGURE 10.8 All 25 possible causal models with three nodes.

collected an infinite amount of data, you would never be able to narrow the set down to the true model.[6]

To see Proposition 1 in action, reconsider the model in Figure 10.3. It has the skeleton $M - E - D$ and no v-structures. What other causal explanations are stochastically consistent with this one? Exactly the two shown in Figure 10.7. If you add or subtract a link, then you no longer have the same skeleton. Therefore, the only other possibility is the model with $M \rightarrow E \leftarrow D$. However, we know this model cannot be parameterized in a way that generates the distribution in Figure 10.1 because this model creates a v-structure (whereas none exist in the model of Figure 10.3).

Consider another example, which we refer to as the "Gatekeeper" model. In it, *Design* and *Marketing* present projects for approval to *Finance* (the gatekeeper). If projects are approved, they are forwarded to *Production* and *Sales* for further action. The model is the one shown in Figure 10.9. Without telling you anything about the

6 At least, not by simply standing by and observing. You might, however, construct a few selective intervention experiments to reveal the true causal relations.

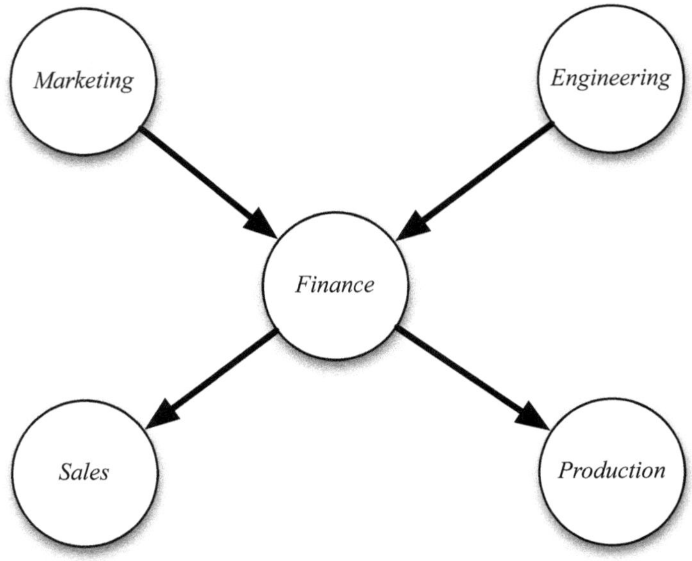

FIGURE 10.9 The Gatekeeper model.

actual probability distribution over actions generated by this model (or even telling you what those actions are), we know that there is no other model consistent with it; i.e., no other model capable of generating identical distributions. You should apply Proposition 1 yourself and make sure you can come to the same conclusion.

Sadly, with respect to our software development example, this means that knowing the distribution in Figure 10.1 only narrows the list of possible causal explanations down to the three observationally indistinguishable models detailed in Figures 10.3 and 10.7. (The good news is that at least you know it is only those three!) How should you proceed?

One way is to quantify your subjective beliefs about the true model and then undertake the intervention that delivers the maximum (subjective) expected value. Another is to experiment with different interventions until you determine which model is correct. For example, if you **Do**$(D=o)$ and find that the average payoff is 68.5, you know the correct model is one of the two in Figure 10.7. Therefore, you might try **Do**$(E=r)$. If the average payoff is 66.2, you know the correct model is (a), if not, you know it is (b). Finally, you might try a risk-minimizing intervention; i.e., implement **Do**$(D=o)$ because it results in fairly high payoffs under all models.

Definition 10.1 *Skeleton: the undirected connection structure of a causal model; the same set of edges, but without the directions.*

Definition 10.2 *V-structure: a converging triplet within a directed acyclic graph; two nodes both pointing to a third node.*

> **Definition 10.3** *Observational Indistinguishability: Two causal models are observationally indistinguishable from each other if any joint probability distribution implied by one can also be implied by the other given an appropriate selection of values for its probability tables.*

10.2.2 Causal Identification

Suppose that, due to cost, timing and other issues, you cannot establish accurate causal influences via experimental intervention. An important alternative to trial-and-error identification of the causal structure is the use of what econometricians call **instrumental variables**.[7] To see how this works, let us turn to a very simple example.

One day coming in to school, you bump into Professor Lethargic, a mid-50s university professor, all decked out in new athletic gear and stretching on some benches outside the school. You are somewhat surprised since this particular individual has never struck you as the exercising type. Curious, you stop and inquire what he is up to. He mentions a recent study showing that men who have intense exercise routines at age 50 live longer than those who do not. He explains that, even though he has never exercised a day in his life, this study has convinced him to get with the program and start jogging 4 days a week!

Suppose the actual data is as shown in Figure 10.2. Here, *Exercise Intensity* (E) can be either *low* (h) or *high* (h) and *Age at Death* (D) is either *young* (y) or *old* (o). Clearly, higher exercise intensity appears to be related to greater longevity. In an attempt to be helpful, and having nearly completed a course on causal modeling,

Table 10.2
Data on dividends and CEO bonuses.

	Outcomes	
Exercise	*Age of Death*	P(Outcome)
no	*low*	34%
no	*high*	16%
yes	*low*	16%
yes	*high*	34%
		100%

7 In scholarly empirical work, instrumental variables are used to help quantify the causal effect of one variable upon another via regression. Here, we are only interested in determining the direction of causality between two correlated variables.

Exercise Intensity at 50	
low	*high*
0.5	0.5

	Age at Death	
Exercise	*young*	*old*
low	0.68	0.32
high	0.32	0.68

FIGURE 10.10 Exercise causes longevity.

you show the professor the causal model in Figure 10.10 and explain that the node probabilities shown are the ones implied by the joint probability distribution from the study (Figure 10.2).

Professor Lethargic is delighted by this model because it implies that the intervention **Do**($E = h$) will increase the likelihood of living to a ripe old age from 32% to 68%! Knowing the model consistency result (Proposition 1), you feel obligated to dampen his excitement by pointing out that the model $E \rightarrow D$ is observationally indistinguishable from $E \leftarrow D$. The details of this observationally indistinguishable alternative model are presented in Figure 10.11. Both the model in Figure 10.10 and the one in Figure 10.11 result in the joint distribution on outcomes shown in Figure 10.2. You carefully point out that if the true model is the one in Figure 10.11, then **Do**($E = h$) will have no effect on longevity – his odds of dying young will be 50/50.

Presented with the alternative model, Professor Lethargic laughs it off. He explains – somewhat patronizingly – that the study only looked at the mortality of men 51 or older. Since the mortality variable is only measured *after* the exercise

	Exercise Intensity	
Death	*low*	*high*
young	0.68	0.32
old	0.32	0.68

Age at Death	
young	*old*
0.5	0.5

FIGURE 10.11 Longevity causes exercise.

intensity variable, the true model must be $E \rightarrow D$. He says, "It simply defies logic that death somehow causes exercise intensity." You have to agree with this and, hence, rule out the model of Figure 10.11 on the basis of common sense. With a smug smile, Lethargic "thanks" you and sends you on your way.

Something about being treated with such condescension spurs you to further action. You think about it some more. The study clearly found a correlation between *Exercise Intensity* and *Age at Death*, but might there be other explanatory variables not included in the study. Suppose some people have a genetic predisposition for good health. Such people would certainly be more likely to live longer and, you reason, might also be more likely to be vigorously exercising at age 50.

Since the researchers who conducted the original study did not measure genetic predisposition, you can only create a hypothetical model to illustrate to Professor Lethargic why he might want to rethink the efficacy of his exercise decision. You return to his office for a visit and show him the model in Figure 10.12. It is not hard to demonstrate that this model generates the same distribution on E and D (the observed variables) as shown in the original study (Figure 10.2).

This is troubling news to Professor Lethargic indeed. The problem is that the only available experiment is for him to $\mathbf{Do}(E = h)$ and wait to see when he dies. Not a very appealing option. He wants to know if there is any way to determine which model is correct. Perhaps there is.

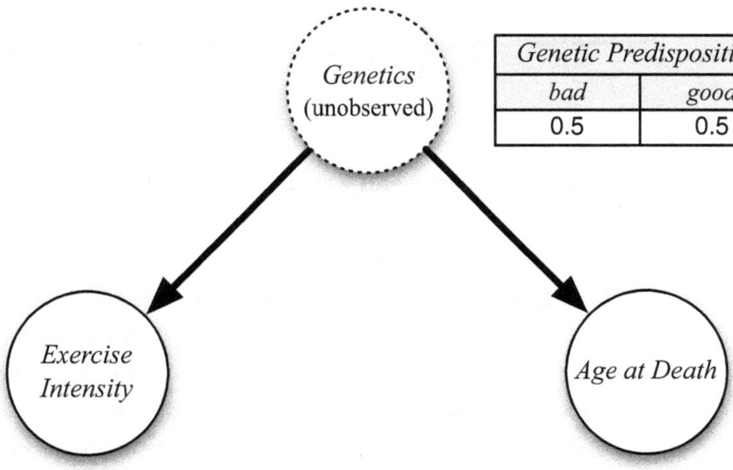

Genetic Predisposition	
bad	*good*
0.5	0.5

	Exercise Intensity	
Death	*low*	*high*
young	0.8	0.2
old	0.2	0.8

	Age at Death	
Exercise	*young*	*old*
low	0.8	0.2
high	0.2	0.8

FIGURE 10.12 Exercise intensity and longevity caused by unobserved factors.

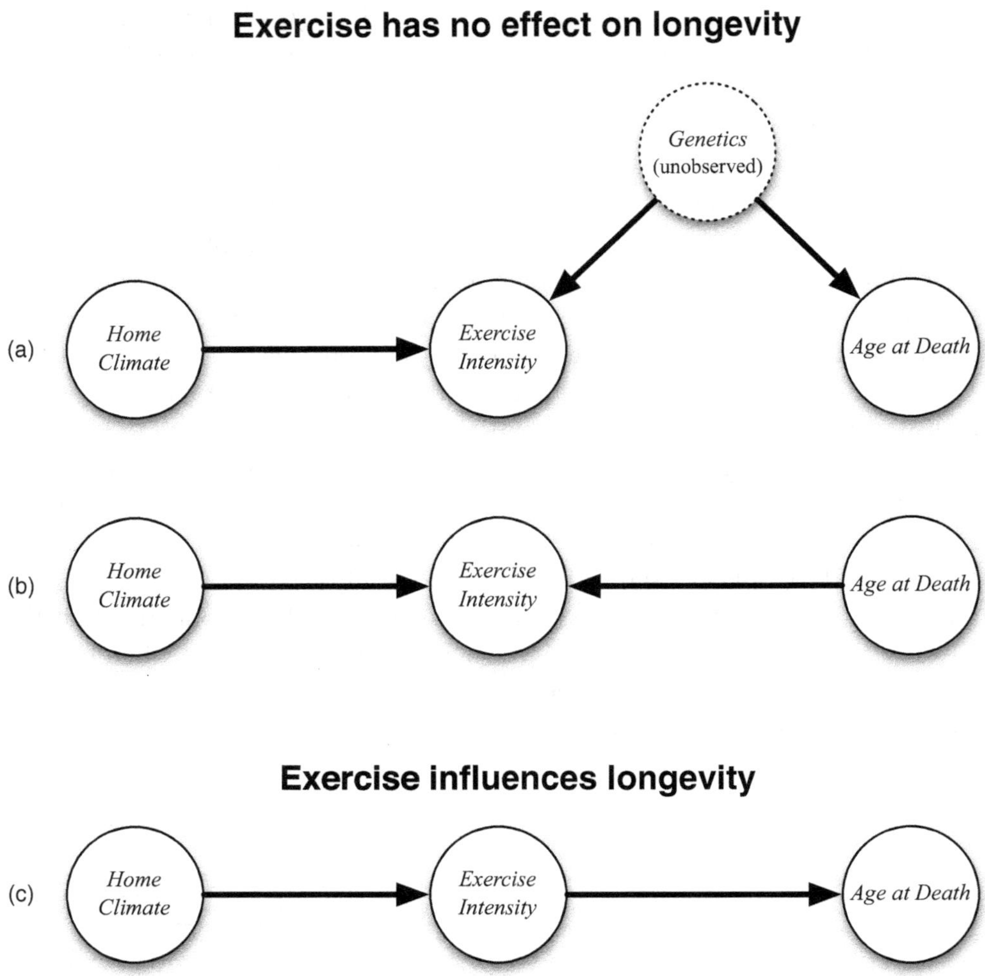

FIGURE 10.13 Two hypotheses with an instrumental variable.

A very popular tool for identifying causality in social systems is the **instrumental variable.**[8] To see how this works, suppose that you are able to collect data on the city locations of all the subjects in the longevity study. Assume that the weather has no effect on longevity (this is a big assumption, but go with it for the moment). Clearly, however, it does have an effect on exercise intensity – that is, the nicer the weather, the more likely people are to exercise. Therefore, you classify the *Climates* (C) of the subjects' cities as *mild* (m) or *nasty* (n). The addition of this variable (called the *instrument*), results in the three hypothetical models shown in Figure 10.13.

8 Recall, in most social sciences the ability to conduct truly controlled experiments is almost non-existent. As a result, it is very difficult to identify causality in these systems. The method of instrumental variables was developed as an alternative.

How can we use the instrumental variable as shown in these models? If either of the top models (a) or (b) are correct, then *Climate* should be correlated with *Exercise Intensity* **but not** *Age at Death*. More precisely, referring back to Chapters 4 and 5, in the joint probability distribution generated by these models, *Climate* and *Age at Death* should be marginally independent, but become dependent conditional upon *Exercise Intensity*.

If (c) is correct, then variation in *Climate* is correlated with *Exercise* and, via *Exercise*, it is also correlated with *Age at Death*. Using Chapter 4 to put it more precisely, in the joint probability distribution generated by this model, *Climate* and *Age at Death* should be marginally dependent, but become independent conditional upon *Exercise Intensity*.

Therefore, without conducting any controlled experiments – just by collecting data on an appropriate variable – an analyst can say something about causality! If *Climate* is not correlated with *Age at Death*, then *Exercise* does not influence *Age at Death*. Note: this test does not distinguish between the hypotheses (a) that *Exercise* and *Age at Death* have a common cause and (b) that *Age at Death* influences *Exercise*. But it does tell us that if *Climate* is correlated with *Age at Death*, then *Exercise* must influence *Age at Death* as an intermediate step. Only in the last case (c) would Professor Lethargic's exercise routine have the intended effect.

A serious word of caution: some key premises here were that, (1) *Climate* was a root node (i.e., was not influenced by other variables in the model); and (2) *Climate* only influenced *Exercise*. This may not have been a very good assumption. For example, in areas where it snows in the winter, car accidents are much more prevalent. Also, sunny climates may lead to sunny dispositions, which may lead to longer lives. Unfortunately, if this is true, climate will be positively correlated in all cases, and hence there will be no way to distinguish the effect of exercise. This situation is shown in Figure 10.14.

Finding convincing instrumental variables is something of a creative art form. Scholars expend great effort in finding convincing instruments. We say "convincing" because whether you believe an instrument-based claim of causal influence will depend upon whether you believe that the instrument meets the requirements for identification. Just as with the other variables being assessed in an analysis of this kind, the instrument's causal links cannot actually be observed. Therefore, the story that supports the variable's use as an instrument must be compelling.

Finally, it is worth pointing out that the instruments described here are very simple in terms of how they interact with the system of interest – i.e., a *root node* with a *single influence relation* going *from it to one of the correlated variables*. A variable with more complex linkages still works as an identifying device so long as adding it to the system allows one to conduct marginal and conditional probability tests capable of distinguishing the existence of an influence relationship in a particular direction. The problem with more complicated situations is the one raised in the preceding paragraph: it is hard to tell a convincing story of simple relationships,

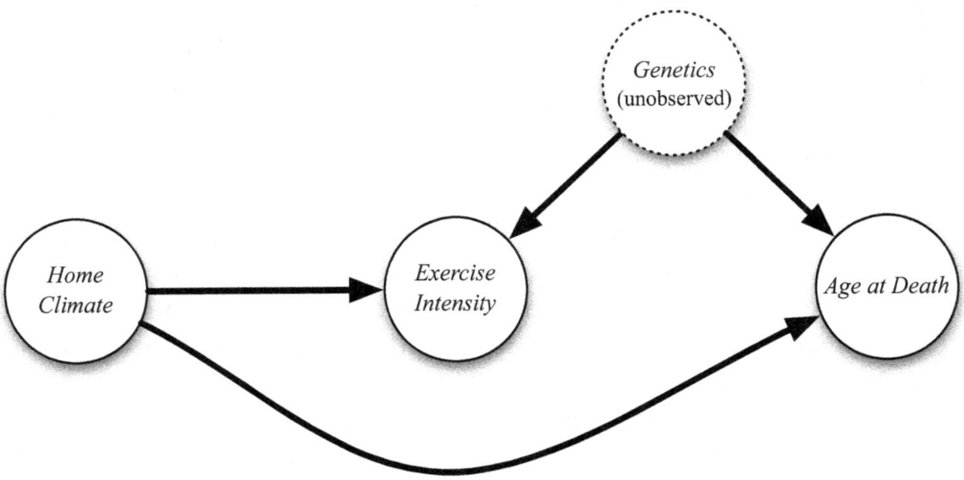

FIGURE 10.14 If climate directly influences longevity, it is useless as an instrument.

and much more so if the relationships are complicated. This brings us back to the most important message of this book: correlation does not imply causation, but causation does imply correlation. Instrumental variables are helpful in teasing these apart, but it is often nuanced and difficult to do so in practice.

Epilogue

The week after your discussion about the relationship between exercise and longevity, you run into Professor Lethargic (who is no longer exercising intensively) at a local cafe. He greets you enthusiastically with interesting news. He has just read a study showing a negative correlation between the *Survival Rates* of entrepreneurial ventures and the *Experience* (in years) of the CEO. Inspired by your recent discussions, he shows you his qualitative causal model (Figure 10.15).

He tells you, "This counterintuitive finding clearly demonstrates something I have believed all along: CEOs with lots of experience in big companies do not know how to be entrepreneurial." He goes on to claim "Clearly, venture capitalists should avoid hiring older CEOs who are incapable of thinking outside the box! I am going to contact all my ex-students who are now working in venture capital and inform them they should put younger, less experienced managers in charge

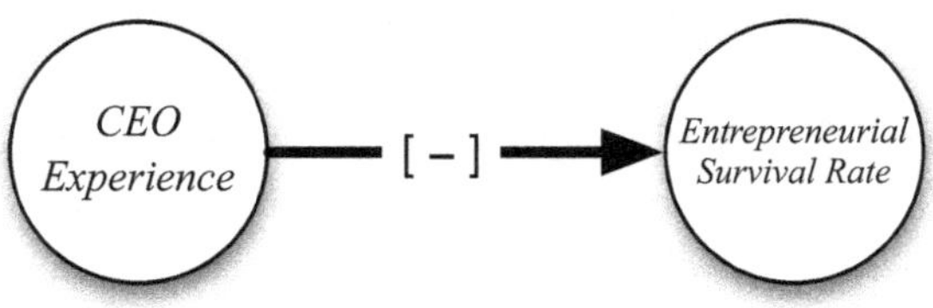

FIGURE 10.15 *CEO Experience* increases the rate of entrepreneurial failure.

of their ventures!" Hearing this oversimplification makes you cringe, "*Oh, brother. Here we go again ...*"

> **Definition 10.4** *Simple instrumental variable: a root node whose only influence relation is from it to one of two correlated variables. The purpose is to determine whether or not the variable it influences in turn influences the other variable with which it is correlated.*

10.3 BUILDING PREDICTIVE REGRESSION MODELS

Many readers will have already studied regression analysis in other courses. Typically, such courses cover basic methods designed to calibrate a regression model and then test its predictive ability. Chief among these is using a portion of a data set to estimate your regression (the "learning" sample) and then checking the accuracy of the estimated model on the remaining data (the "holdout" or "testing" sample). If the fit with the holdout sample is good, you will have greater confidence in your out-of-sample predictions.

10.3.1 From Structural Equation to Causal Models

As hinted at in earlier discussions, causal models can be used to represent functional models as well as quantitative causal models (i.e., directed, acyclic graphs with appropriate probability parameters associated with each node). A functional model is one that consists of a set of equations

$$y_i = f_i(\phi_i, \varepsilon_i), \quad i = 1, \ldots, n$$

where y_i is a dependent variable that is a function of a set of independent variables, $\phi_i = \{x_{i1}, x_{i2}, \ldots, x_{ik}\}$ are called the "parents" of y_i, and the ε_i variables, referred to as "noise", account for unobserved, random elements. When the functions f_i are linear (i.e., sums of the products of constants and variables to the first power), then we have a *structural equation model* (SEM) of the form:

$$x_k = a_k + b_{kl}x_l + \cdots + b_{kw}x_w + \varepsilon_k, \quad k = 1, \ldots, n$$

A specific example would be the following SEM with six variables:

$$x_3 = a_3 + b_{13}x_1 + \varepsilon_3 \tag{10.1}$$

$$x_4 = a_4 + b_{14}x_1 + b_{24}x_2 + \varepsilon_4 \tag{10.2}$$

$$x_5 = a_5 + b_{45}x_4 + b_{25}x_2 + \varepsilon_5 \tag{10.3}$$

$$x_6 = a_6 + b_{36}x_3 + b_{46}x_4 + b_{56}x_5 + \varepsilon_6 \tag{10.4}$$

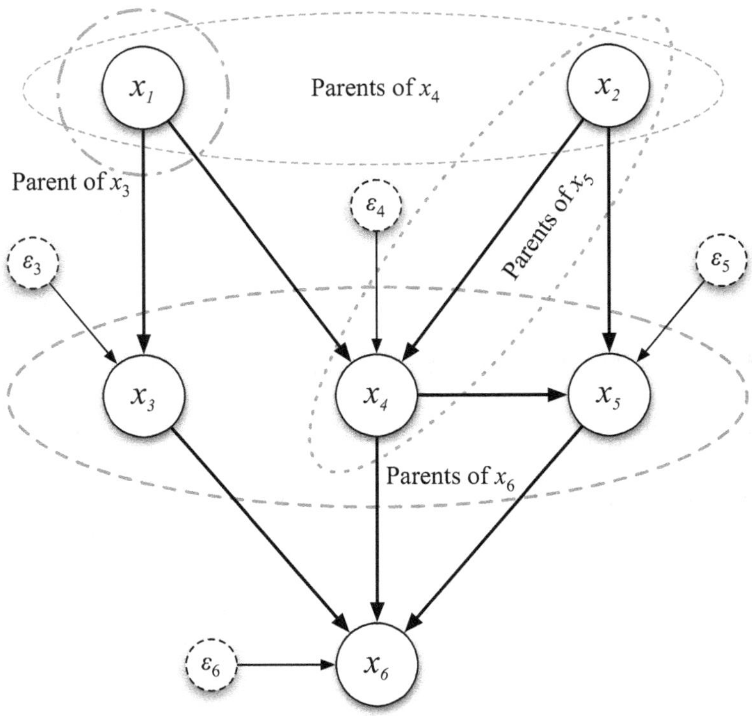

FIGURE 10.16 Parent relationships for x_3, x_4, x_5, and x_6 according to 10.1–10.4.

It is easy enough to build a *qualitative* causal model from a SEM: simply build a graph in which the dependent variables each have a direct link pointing from their parent independent variables. The causal model corresponding to equations 10.1–10.4 is shown in Figure 10.16. Technically, we can always build a graph that uses arrows to indicate functional relations – that is, we can follow the mechanical translation process whether or not the SEM is intended to show causal relationships. In most of the work we do, however, the SEM is used as a semantic device to indicate real-world causal relations.

When the structure of an SEM is intended to indicate causality, we can construct a corresponding quantitative causal model with the coefficients shown directly on the corresponding links (as in Figure 10.17). For example, the interpretation of the network model in Figure 10.17 is that a one-unit change in x_2 results in: a b_{24}-unit change in x_4; a b_{25}-unit change in x_5; and a $(b_{24}b_{46} + b_{25}b_{56} + b_{24}b_{45}b_{56})$-unit change in x_6. Alternatively, a one-unit change in x_5 results in a b_{56}-unit change in x_6 ... and *no other changes*.

10.3.2 Linear Regressions and Causal Models

Clearly, regression estimation is closely related to SEMs and, hence, to causal models. Without any causal analysis, one would typically estimate the following

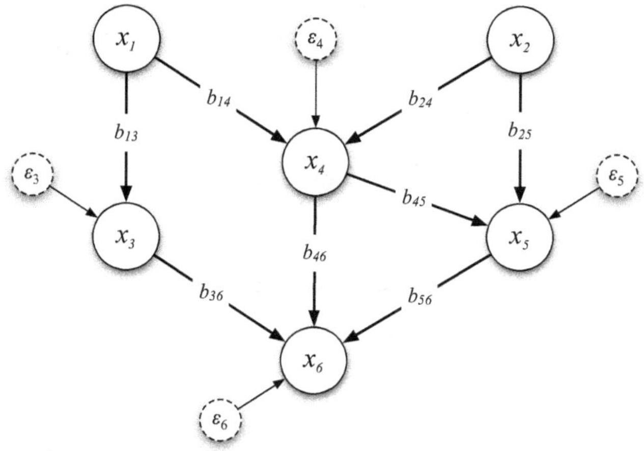

FIGURE 10.17 Linear coefficients shown on network links.

regression model (RM) for the situation in Figure 10.17:

$$x_6 = a'_6 + b'_{16}x_1 + b'_{26}x_2 + b'_{36}x_3 + b'_{46}x_4 + b'_{56}x_5 + \varepsilon'_6 \tag{10.5}$$

The primes indicate that these are the coefficients (and an error parameter) pertaining to this particular model. Although the effects of x_1 and x_2 are indirect (according to the unobserved causal model), the coefficient estimates here will be unbiased and consistent; i.e., will indicate the true effects on the expected value of x_6 given changes in the dependent variables.

If the true process is as shown by the causal model in Figure 10.17, then the problem with the regression model 10.5 is that the dependent variables are highly correlated; e.g., because x_2 influences *both* x_4 and x_5, all of x_2, x_4 and x_5 will be correlated with one another. In statistical terminology, regression model 10.5 has a **multicollinearity** problem. While the predicted values of x_6 estimated from this model are fine, the modeler is **typically unable to use standard significance tests to reliably determine which independent variables exert the most influence**.

One solution to this problem is stepwise regression – that is, checking through various combinations of independent variables to find the smallest subset that yields the highest accuracy. The usual procedure is to start with the independent variable exhibiting the highest correlation with the dependent variable. Then, from the remaining independent variables, add the one exhibiting the highest partial correlation with the dependent variable (controlling for the first independent variable), and so on. The modeler continues to add variables until there are either none left or none remaining that significantly increase R^2.

For example, our SEM suggests the following estimation would result in an unbiased, consistent model with no multicollinearity problems:

$$x_6 = a_6'' + b_{16}''x_1 + b_{26}''x_2 + \varepsilon_6'' \tag{10.6}$$

Indeed, by substituting into the SEM and simplifying, it is relatively easy to show that in the new regression model 10.6 we have:

$$a_6'' = a_6 + a_3b_{36} + a_4b_{46} + a_5b_{56} + a_4b_{45}b_{56}$$
$$b_{16}'' = b_{13}b_{36} + b_{14}b_{46} + b_{14}b_{45}b_{56}$$
$$b_{26}'' = b_{24}b_{46} + b_{25}b_{56} + b_{24}b_{45}b_{56}$$

where the right-hand sides of the equations show the true parameters. As previously indicated, regression model 10.6 is not expected to result in better predictions than regression model 10.5. However, the coefficients are meaningful in the sense that they really measure the true effects of changes in the two independent variables.

If one believes the SEM defined by equations 10.1–10.4 and wants to estimate the corresponding coefficients, it can be done via a **two-stage regression**. In Stage I, the modeler estimates the following regressions:

$$x_3 = a_3 + b_{13}x_1 + \varepsilon_3$$
$$x_4 = a_4 + b_{14}x_1 + b_{24}x_2 + \varepsilon_4$$
$$x_5 = a_5 + b_{45}x_4 + b_{25}x_2 + \varepsilon_5$$

In Stage II, the modeler uses the estimates from Stage I to estimate:

$$x_6 = a_6 + b_{36}\hat{x}_3 + b_{46}\hat{x}_4 + b_{56}\hat{x}_5 + \varepsilon_6$$

where the "$\hat{x}_i$" indicates the estimated value of the independent variable obtained from the Stage I regression. For the purposes of prediction, this more involved procedure should not improve accuracy beyond that obtained in regression model 10.6. However, it will provide estimates of all the coefficients if, for some reason (e.g., management interventions), one feels it is useful to have them.

When we add causal knowledge, it turns out there is more that we can say about the predictive accuracy of various regression estimations. For example, if the causal model in Figure 10.17 is true, then providing information about x_3, x_4, and x_5 tells you everything you need to know in order to estimate x_6 (by c-independence). That is, once you know the values of x_3, x_4, and x_5, providing additional information about x_1 and x_2 are of *zero* assistance in predicting x_6.

Why is this significant? Because it implies, for example, that one cannot improve upon the accuracy of the following estimation:

$$x_6 = a_6 + b_{36}x_3 + b_{46}x_4 + b_{56}x_5 + \varepsilon_6 \tag{10.7}$$

Indeed, because the actual values of x_3, x_4, and x_5 pick up the associated errors, regression model 10.7 should be more accurate than regression models 10.5 or 10.6.[9] It is worth pointing out that this increased accuracy implies that, in this case, using the stepwise regression procedure to figure out the best set of independent variables is most likely to arrive at regression model 10.7 as the best.

Definition 10.5 *Multicollinearity: the problem of having correlated independent variables in a multiple regression model. Predictions are not affected, but assessments of coefficient significance are made less clear.*

Definition 10.6 *Two-stage regression: if some independent variables are linear functions of the others, use the latter to estimate the former with a regression analysis; then, regress the dependent variable on the estimated dependent variable values obtained in the first step.*

10.3.3 Good Causal Models Imply Good Predictions

Thus far we have considered a specific case of a structural equation model for which regression estimates are good (i.e., unbiased and consistent). The previous section shows that in such cases, there are many ways to get good predictions (as long as the test set shares the relationships found in the learning set). In particular, we saw that the use of multi-stage regression in the running example would most likely lead to the selection of a model in complete accordance with the underlying causal mechanism.

Effect, not Cause

Now, consider the model in Figure 10.18 (we have removed the error terms and coefficient labels to keep things focused on the causality). Assume the goal is to estimate a regression model to predict Y and, at the same time, that the modeler does not know this true causal structure. The difference between the situation described by Figure 10.18 and the one described by Figure 10.17 is that we now have a variable (x_3) that is actually an *effect* of the variable we wish to predict.[10]

Now, suppose (without knowing the true underlying structure), we estimate

$$Y = a + b_1x_1 + b_2x_2 + b_3x_3 + b_4x_4 + b_5x_5 + \varepsilon_Y \qquad (10.8)$$

9 Notice that RM 10.7 uses the actual values of x_3, x_4, and x_5, whereas RM 10.6 uses estimated values from the Stage I regression.

10 The examples in this section can be found in Spirtes et al. (2001, Ch. 8).

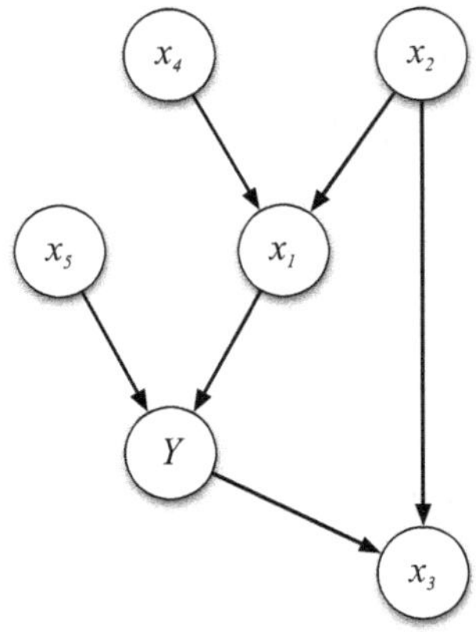

FIGURE 10.18 Here, x_3 is an *effect* of Y.

In large samples, linear regression will give *all* variables in regression model 10.8 significance, even though x_3 does not affect Y! The estimated regression coefficient b_3 in this case will be incorrect – changes in x_3 are not (actually cannot be) predictive of changes in Y.

Moreover, this problem is not generally solved by any standard regression procedure. For example, Spirtes et al. (2001, p. 194) used a data-mining technique called "stepwise regression" to generate simulation data using 20 versions of the the causal model in Figure 10.18 (each version used a different set of values for the true coefficients).[11] A total of 5,000 observations were generated for each of the 20 versions. Once the data were generated, the STEPWISE procedure in MINITAB was applied to identify the set of regressors with the most significant coefficients. In every case, $\{x_1, x_2, x_3, x_5\}$ were selected as the most significant variables. Notice that adding more variables does not fix this problem and, indeed, may actually compound it (i.e., if they, too, are caused by Y rather than causes of Y). As Spirtes et al. (2001, p. 192) point out, "It is easy to construct cases in which a variable with no influence on the outcome variable has a standardized regression coefficient larger than any other single regressor."

One way to approach this problem is first to estimate the maximum likelihood causal structure consistent with one's actual data. There are a large number of software solutions available, both free and commercial, that can be used to conduct this analysis.[12] For example, using a Bayesian network estimator on the simulation data generated by the 20 models mentioned above, x_1 and x_5 were correctly

11 Stepwise regression is not related to multi-stage regression demonstrated above. It is a semi-automated approach to building models by iteratively adding and removing variables until the fit of the model to the data is (locally) optimized. We will not cover this technique, but it is useful for exploring the space of possible supported models when the underlying structure is unknown.

12 Performing this analysis, like the stepwise regression technique, is beyond the scope of this text, but an internet search for "causal model estimation" and/or "Bayesian network software" will assuredly provide a wealth of useful reference material.

identified as the only variables with a direct influence on Y. This identification of variables with direct influence is very important for predictive accuracy.

Consider another example; this one from Timberlake and Williams (1984). They studied the effect of foreign investment in third-world countries. They examined the effects of *Foreign Investment* penetration in 1973 (F), *Energy Development* in 1975 (E) and *Civil Liberties* (C) on the degree of *Political Exclusion* (P). They estimate the following model:

$$P = a + b_1 F + b_2 E + b_3 C$$

The causal version of this model and their coefficient estimates are illustrated in Figure 10.19. This result was interpreted in a way that made it shocking: apparently, foreign investment was permitting dictatorships to remain in power, and to continue to oppress their people. However, inferring a Bayesian network from the data (even using an algorithm that allowed for the possibility of hidden causes) resulted in the causal model shown in Figure 10.20. As explained by Spirtes et al. (2001, p. 201), this model says

> ... *that foreign investment and energy consumption have a common cause, as do foreign investment and civil liberties, that energy development has no influence on political exclusion, but political exclusion may have a negative effect on energy development, and that foreign investment has no influence, direct or indirect, on political exclusion. [emphasis added]*

These results highlight once again the importance of thinking about causality in relation to one's predictions. In the political example, getting the causality wrong creates two problems. First, as we saw in the last chapter, it may lead to bad

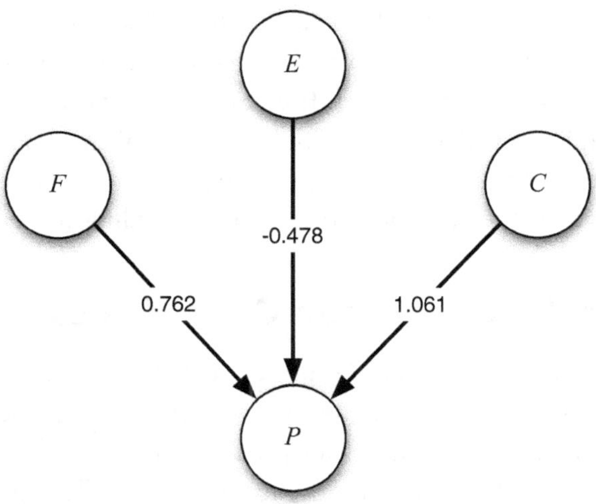

FIGURE 10.19 Does foreign investment cause political oppression?

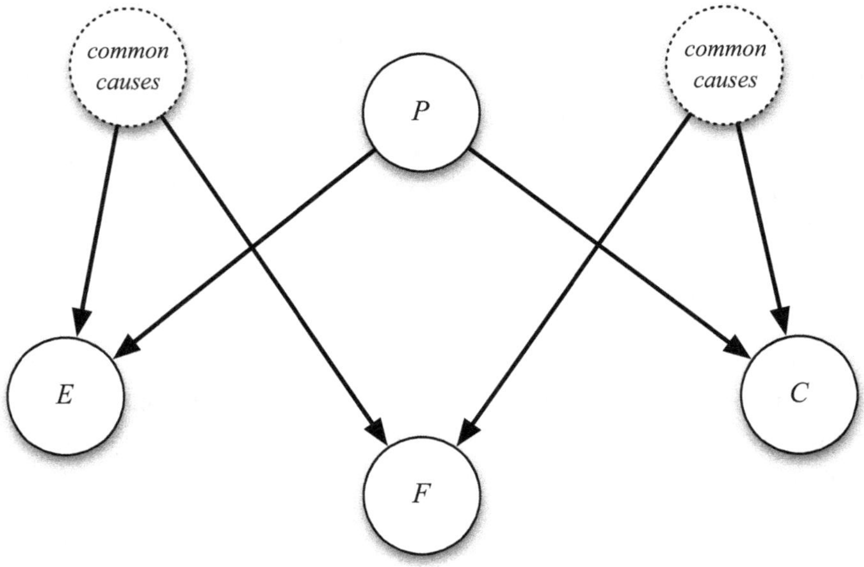

FIGURE 10.20 The causal structure inferred from the data.

interventions (e.g., trying to reduce foreign investment in a misguided attempt to change the degree of political exclusion). Second, it may lead to bad predictions of the unperturbed system (e.g., observing an increase in a country's foreign investment and predicting an increase in political exclusion).

Ideally, you should consider estimating the maximum likelihood causal model from your data as a first step; i.e., prior to running regressions. If this is not possible, the problem does not go away! A complementary solution is to ask experts to help you understand the causal structure underlying the situation. If you have time, identifying instrumental variables or conducting experiments can also help. Barring any of that, if you have strong beliefs and/or special knowledge about the causal relations driving the data, then you must consider incorporating them into the specification of your regression model. If you do, and if you are correct, then your predictions should be improved by the incorporation of this knowledge. However, don't cling to your beliefs too strongly, regressing with creative variations in the model may reveal that the relationships are not according to your best guess.

10.3.4 A Brief Note on Box Office Gross

Finally, we will briefly discuss an academic paper that estimates certain attributes of movies believed to influence ticket sales. Our purpose is to present some variables and the hypothesized relationships between them that will spur you along in thinking causally about large data prediction problems. The paper, Waguespack and Sorenson (2010), is concerned with the determinants of MPAA content ratings (e.g., G, PG, R) and the effect this has on ticket sales. Briefly, the paper finds that

MPAA ratings affect ticket sales and, moreover, the power and status of individuals involved with a particular movie (such as the director and producer) appear to influence ratings.

In the US, movies receive an MPAA Rating prior to being put on the screen (or sent straight to video). This rating is assigned by the Classification and Rating Administration (CARA). From least to most restrictive, the categories are: G (general audiences; all ages admitted), PG (parental guidance suggested; some material not suitable for children), PG-13 (parents strongly cautioned; some material may be inappropriate for children under 13), R (restricted; children under 17 require accompanying parent of adult guardian), and NC-17 (no-one 17 and under admitted). These categories have been in effect since 1990, though similar categories have existed since the first self-regulatory board for movies was formed in 1922.

Filmmakers have expressed frustration with the process by which films are rated because the outcomes of this process are hard to predict. An example of the (seeming) inconsistency and (definite) frustration comes from Adam Green, the writer and director of the film *Hatchet*: "Joleigh still gets the belt sander put in her face. That was okay, but you can't impale her on a shovel handle. Explain that one to me!" (Waguespack and Sorenson, 2010, p. 14).

Even controlling for content, MPAA ratings appear to have a substantial effect on ticket sales. Within the sample of movies used in the study, "PG and PG-13 titles delivered, on average, 76% more revenue than R titles. Past studies on more extensive samples and controlling for a host of other factors have found similar results." This is not to say that more mature content is, *per se*, the problem. Rather, "More mature content increases sales, but a more restrictive rating hurts performance" (Waguespack and Sorenson, 2010, p. 25–6).

This raises the question of bias in the assignment of MPAA ratings. Again in the words of Adam Green, "if you're a studio film and you can pay them off, you can do whatever you want. But if you're me, and you're walking in there for your trial by yourself, and you don't have any money, they'll come down on you" (Waguespack and Sorenson, 2010, p. 14). The suggestion is that central players may, on average, receive softer treatment by the MPAA ratings organization. As a result, the US ratings system may erect barriers that prevent independent and lower status producers from entering the most profitable market segments.

Our interpretation of the model hypothesized by Waguespack and Sorenson (2010) is shown in Figure 10.21. Again, the focus of the paper is not on ticket sales *per se* (hence, that part of the model is not well developed). Rather, it is on the factors influencing whether a movie gets an MPAA rating of R rather than the less restrictive G, PG, or PG-13. This latter part of the model is quite sophisticated. Let us elaborate upon that part now.

First, it is recognized that the actual violence, sexuality, and profanity in a movie is unobserved. The authors use data from the Kids-in-Mind website to

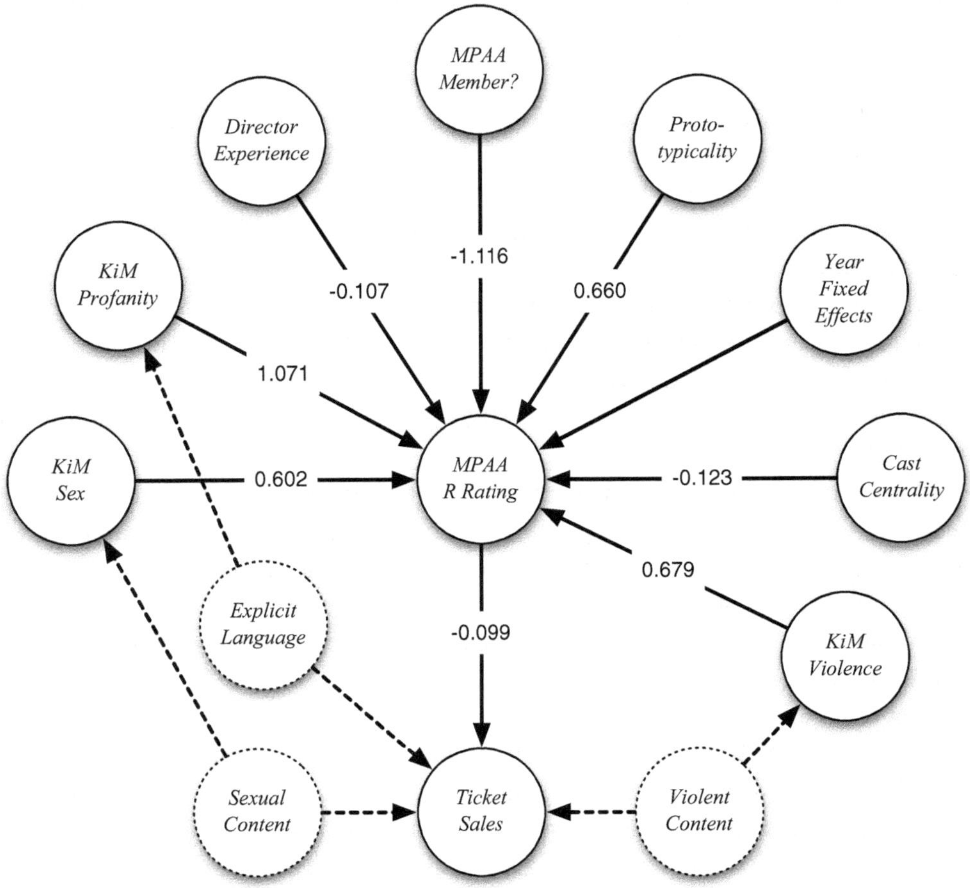

FIGURE 10.21 Explaining the likelihood of getting an MPAA rating of R.

provide a more nuanced measure of content containing violence, sex, and explicit language. Presumably, the true degree of mature content in a movie must exert some influence on the likelihood that a movie gets an R rating.

If actual content were the only thing influencing such outcomes, Mr Green's complaints would be unfounded. The authors hypothesize that there are factors leading to the ratings other than content. Three factors are considered in order to assess bias in ratings assignments: *MPAA Membership*, *Cast Centrality*, and *Prototypicality*. MPAA members include the largest, oldest, and most prestigious firms in the movie business. It is possible that these firms may directly or indirectly influence the decisions of CARA. *Cast Centrality* is a measure of the density of affiliations in past film projects between the directors and producers on a film and other directors and producers over the prior three years. Directors and producers with high centrality are powerful, high-status individuals who may provide a "halo" effect to a movie during the rating process. *Prototypicality* is the proportion of prior films made by the movie's director that also got an R rating. The idea is that directors may develop a reputation for making movies that get certain kinds of ratings (for example, Tarantino has a score of 1 on this measure – all his movies are rated R).

As we can see from the estimated coefficients included in Figure 10.21, mature content has the expected directed effect on MPAA ratings (more mature content makes it more likely the rating will be R). Independent of the content, the greater a director's reputation as an R-movie maker, the greater the likelihood his or her future movies will also be rated R. Most interestingly, a film that is backed by an MPAA member is less likely – content held constant – to get an R rating. Similarly, if the movie is associated with high-profile cast members, it is also less likely to get an R rating.

Constructing the causal model allows us to see these attributes and their influence as independent, and estimate their effects in isolation from each other. The result of the analysis confirms Adam Greene's suspicions and complaints on a point-by-point basis. Other statistical techniques can also provide this view, and make equally good predictions, but cannot correctly account for any changes in MPAA ratings that would result from an intervention on one of these variables. Of course, that assumes that the causal model in Figure 10.21 is the correct causal model.

As already mentioned, the automated estimation techniques are highly sophisticated and include the instrumental variable and observational indistinguishability considerations. The tools are polished and powerful, and there to be used. The critical point becomes that *you* understand what the tools are doing for you; e.g., by understanding and internalizing the contents of this book. In order to correctly interpret your analysis results, you must understand the formal notions of causation, the tracks of influence, and the flow of information. More sophisticated techniques require more training and more effort, but they reward you with more accurate and more useful results. And furthermore, the causal modeling approach is specifically useful when the future is different from the past (due to structural changes and/or interventions) because it captures not just how much variables change in response to each other, but also the particular influence patterns.

KEY CONCEPTS: CHAPTER 10

- Regression models with different structures can make the same predictions from the same data, but react differently to changes in values, interventions, and the introduction of new factors.
- Due to limited resources, managers (and others) must choose which variable(s) under their control to set, and understanding the causal structure is essential to properly predict what would happen under each scenario.
- Interventions typically have greater effect when operated on root nodes, and/or nodes with many downstream influenced variables.

- Causal relationships can be learned/inferred from data in many cases, although multiple causal models are often compatible with a given probability distribution over the variables.
- The observationally indistinguishability theorem says that you can find parameters for a model to match the probability distribution generated by the true model if it has the same nodes, skeleton, and v-structures.
- Instrumental variables can be used to test relationships among correlated variables and tease out the true causal structure as long as the instrumental variables is correlated with one, but not the other variable of interest.
- A structural equation model can be translated into a causal model by simply making the conditional probability tables work like functions relating variables, however this assumes that the relationships among variables capture causality.
- A multi-stage regression utilizes the parent–child relationships in the model to estimate the parameters of the model one child at a time, and usually produces more accurate predictions.
- A regression model will report a significant influence from a variable even when it is an effect of the dependent variable under analysis. To avoid these mistakes, it is best to first try to estimate the maximum likelihood causal model.
- A causal model can produce better prediction by capturing how much each variable effects the likelihoods across others; and as a result, it also provides superior explanations of the observed phenomena.

CHAPTER 10 REVIEW QUESTIONS

10.1 Which of the following statements are true?

(a) Multiple causal models can generate the same probability distribution over outcomes.

(b) Two causal models are observationally indistinguishable if they have the same nodes, number of root nodes, and skeletons.

(c) A hypothetical common cause can be used as an instrumental variable to determine the causal relationship between two known correlated test variables.

(d) If a proposed instrumental variable directly causes both test variables, then it is useless as an instrument.

10.2 Which of the following are true of the relationship between regression analyses and causation?

(a) Causal models with different structures will only match different collections of observed correlations found in a data set.

(b) If A is a cause of B, then there must be a positive correlation in the values of A and B.

(c) A causal model is especially helpful to understand the effects of an intervention in the data-generating process.
(d) Multi-stage regressions can partially incorporate the causal structure and this often improves predictions.

10.3 Using a set of training data, an analyst produces five distinct causal structures that are capable of generating the same distribution of values over the variables. What must be true regarding these causal models.

(a) They will produce different predictions when run on a new set of data.
(b) Interventions on the same variable in the different models will generate different results.
(c) Given the result of an intervention in one model, there is some intervention that will produce the same result as that in the other four models.
(d) The five models are observationally indistinguishable.
(e) All five models have the same number of nodes.

10.4 Examine the set of equations below and draw a causal model diagram that captures the same relationships.

$$x_2 = a_2 + b_{12}$$

$$x_3 = a_3 + b_{13}x_1$$

$$x_4 = a_4 + b_{14}x_1 + b_{24}x_2 + b_{34}x_3$$

$$x_5 = a_5 + b_{25}x_2 + b_{35}x_3$$

$$x_6 = a_6 + b_{36}x_3 + b_{46}x_4 + b_{56}x_5$$

$$x_7 = a_7 + b_{17}x_1 + b_{27}x_2$$

10.5 Examine the causal model presented in the diagram below and draw a causal model that is observationally indistinguishable from the one presented.

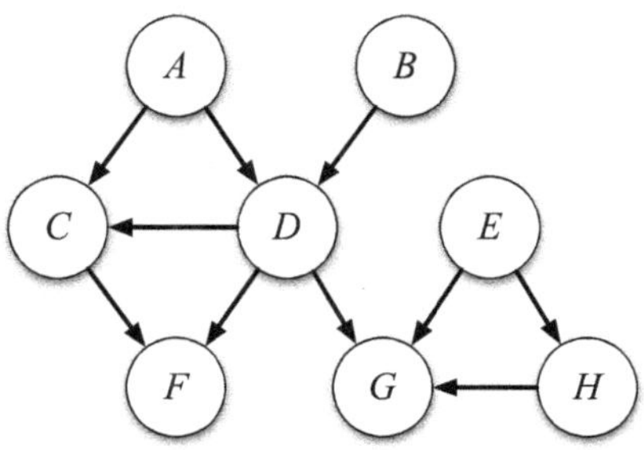

10.6 You believe that the variables A and B are correlated because they share a common cause C; i.e., not because either causes the other. Could you use instrumental variables to test your belief? If so, draw the causal model for this test and if not, explain/show why it is not possible.

10.7 If you are trying to test the hypothesis that A is a cause of B, in which of the following diagrams does the gray shaded node represent a useful instrumental variable?

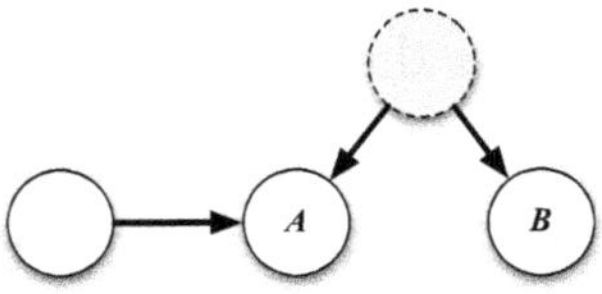

(I)

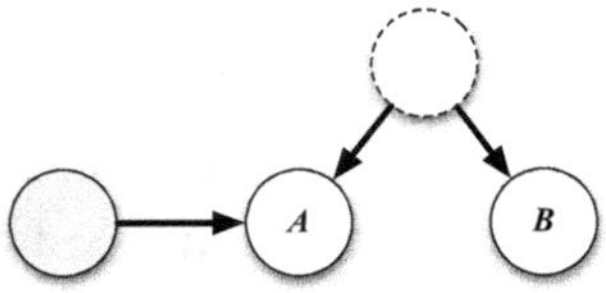

(II)

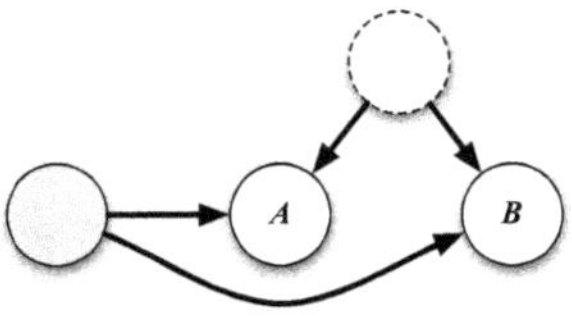

(III)

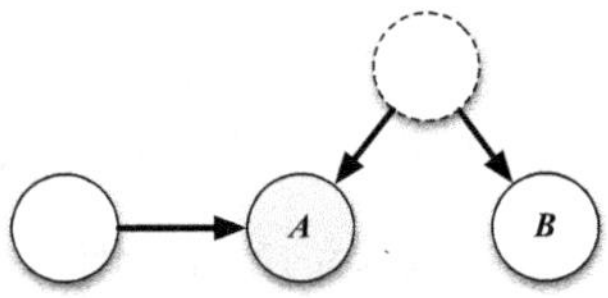

(IV)

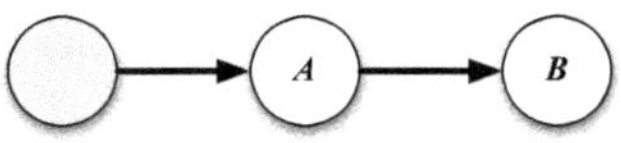

(V)

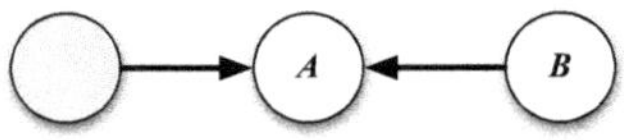

(VI)

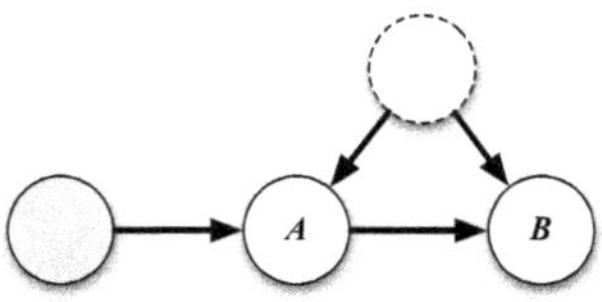

(VII)

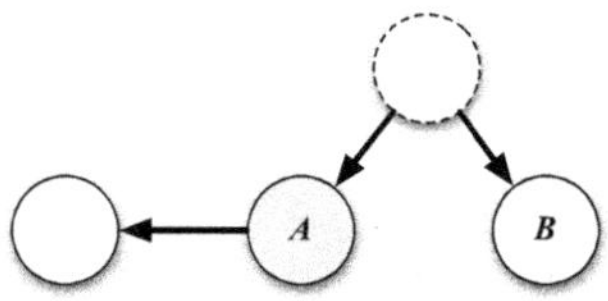

(VIII)

Bibliography

Cheng, V. (2012). *Case interview secrets: A former McKinsey interviewer reveals how to get multiple job offers in consulting.* Seattle, WA: Innovation Press.

Conrady, S. and L. Louffe (2011). Paradoxes and fallacies: Resolving some well-known puzzles with bayesian networks. *Technical report,* Conrady Applied Science, LLC. Franklin, TN.

Cosentino, M. (2007). *Case in point: Complete case interview preparation* (5th ed.). Needham, MA: Burgee Press.

Dobby, C. (2012, October 31). Taxi-hailing apps draw fire. *Financial Post.*

Grady, D., S. Rubin, D. Petitti, C. Fox, D. Black, B. Ettinger, V. Ernster, and S. Cummings (1992). Hormone therapy to prevent disease and prolong life in postmenopausal women. *Annals of Internal Medicine 117*(12), 1016.

Koller, D. and B. Milch (2003). Multi-agent influence diagrams for representing and solving games. *Games and Economic Behavior,* pp. 181–221.

Moody, F. (1996). *I sing the body electronic: A year with Microsoft on the multimedia frontier.* New York, NY: Penguin.

Pearl, J. (1988). *Probabilistic reasoning in intelligent systems: networks of plausible inference* (2nd ed.). San Francisco, CA: Morgan Kaufmann Publishers, Inc.

Penalva, J. and M. Ryall (2008). Empirical implications of information structure in Finite Extensive Form Games. *The BE Journal of Theoretical Economics 8*(1).

Porter, M. (2002). What is strategy? *Strategy for Business: A Reader.*

Ryall, M. (2009). Causal ambiguity, complexity, and capability-based advantage. *Management Science 55*(3), 389–403.

Shachter, R. (1986). Evaluating influence diagrams. *Operations Research 34*(6), 871–882.

Spirtes, P., C. Glymour, and R. Scheines (2001). *Causation, prediction, and search.* Cambridge, MA: The MIT Press.

Tellis, G. (1988). Advertising exposure, loyalty, and brand purchase: a two-stage model of choice. *Journal of Marketing Research 25*(2), 134–144.

Timberlake, M. and K. Williams (1984). Dependence, political exclusion, and government repression: Some cross-national evidence. *American Sociological Review,* 141–146.

Verkaik, R. (2006, March). The mystery of Maxwell's death.

Verma, T. and J. Pearl (1990). Equivalence and synthesis of causal models. In *Proceedings of the Sixth Annual Conference on Uncertainty in Artificial Intelligence,* pp. 270. Amsterdam, NL: Elsevier Science Inc.

Vickrey, D. and D. Koller (2002). Multi-agent algorithms for solving graphical games. *Proceeding of the Eighteenth National Conference on Artificial Intelligence.* American Association for Artificial Intelligence. Menlo Park, CA. pp. 345–351.

Waguespack, D. M. and O. Sorenson (2010). The ratings game: Asymmetry in classification. *Working Paper*.
Wikipedia (2012). Uber (company). Wikipedia, the free encyclopedia. [Online; accessed November 28, 2012].
Women's Health Initiative (2002). Risks and benefits of estrogen plus progestin in healthy postmenopausal women: principal results from the Women's Health Initiative randomized controlled trial. *JAMA 288*(3), 321.

Index

Note: Page numbers in *italics* indicate figures, page numbers in **bold** indicate tables. 'n' following a page number indicates a footnote, the number following 'n' indicates the footnote number where there is more than one footnote on the page.

advertising 6, 9–10, *28*, 29, *30*
aging population 73
allocations, of resource decisions 177–181
annual financial forecast 133
Apple Inc, decision to launch a new tablet in Canada 152–164
applications: of qualitative causal modeling 44–77; of situational analysis 132–150; taxi-hailing application for smartphone users 176–188
artificial intelligence 14

barren nodes 157
basic causal reasoning 25–27
Bayes' Rule 86, 110, 117
BayesiaLab 132, 179n
Bayesian Nash equilibrium 215
Bayesian networks 14, 17, 150, 247; definition of 17; probabilistic nodes 18
Bayesian networks estimator 246
beliefs: posterior 114; prior 114
best reply strategy 194
best response 215
box office gross 248–251
brand loyalty 9
BuenoBrains Incorporated 204, 205; strategy table 207, **208**

c-independence 31–39, 103, 114, 159
cardiovascular disease (CVD), hormone replacement therapy (HRT) and 5
case interviews: asking questions 52; business problem 45; case material 47; *Firm Profit* 46; framework 44; structure of 46
causal analysis, introduction to 1–13
causal dependence 32, 37
causal dilemma 30–31
causal form games 195–197
causal identification 235–241
causal independence 31–39, 103, 114, 159; converging triplets 34–36; definition of 32, 37; diverging triplets 33–34; serial triplets 32–33
causal inference 17
causal influence relationships 21
causal modeling 14; data-driven 224–254; software 216
causal modeling software 216; screenshots 132
causal models 17, 97, 100–103, 135–140, 151; building 135–140; continuous variables 18n; data-driven 224–254; decision nodes 18, 50, 153, 195; directed graphs 17; evidence in 114–121; extended 154–155; inference in 114–121; linear regressions and 242–245; for market volume *26*; mirror-image 217; nodes *see* nodes; and predictions 245–248; quantitative 78–108, 109, 224; skeleton 232, 234; software-based 177; solving games with 197–203; summary information for *141*; without strategic agents 17
causal relations 16
causality 8, 15, 224; versus probability 225–230
cause of death 1–2, 3
Classification and Rating Administration (CARA) 249

competition 54, 68, 70, 109
competitor legal responses 185–188
conditional certainty tables 23, *24*
conditional probability 83, 92; marginal probability and 111–114; for parents of objective nodes 158–159
conditional probability tables 83, 84, 139, **148**, 155, 187, 209
contingency planning 134
continuous variables 18n
convergent connections 113–114, 123–126; setup 123–124; updating 125–126
converging triplets 34–36
correlation 8
correlation coefficients 6
coupons 10, 29
customer satisfaction 52, 53

data-driven causal modeling 224–254
decision nodes 18, 50, 153, 195
decision problems 17; elements of 190–191; extensive forms 190; pure 153
decision rights 224n
decision strategies 162
decision tables 155
decisions 152–153; general solution procedure 156–164; multiple 153; resource-allocation 4; single 152–153
directed graphs 17, 232; v-structure 232, 234
divergent connections 112, 121–123; setup 121–122; updating 122–123
diverging triplets 33–34, 121

economic conditions 52
effect of variables 245
Eizenman, Erez 1
entry games 190–191, 198–203
environmental factors 16
environmental variables 26, 197
environmentally friendly ink 146–150
events 79–80; definition of 81; intersection of 80, 81; multiple 80; union of 80, 81
evidence 3; in causal models 114–121; passing down the chain/downstream 118; passing up the chain/upstream 118, 119–120
evidence-based management 4
evidence-based reasoning 17, 114, 115
evidential reasoning *see* evidence-based reasoning
exercising, and longevity 235–239

factorization 84, 96, 97
Feather Touch (FT) 5–11; advertising 6; advertising model 94–95; *Brand Choice* 90; brand loyalty 9; consumer choice *93*; correlation coefficients 6; *Customer Loyalty* 90, 91; probabilities from count data 91–92; regression coefficients for unit sales 7; true causal relationships 9
feedback loops 14
film ratings 249
Financial Analysts 133
financial forecasting 132
financial models 133; spreadsheets 134–135
focus groups, in marketing 109
form games: extensive 190; normal 190
frequency-based probability 81, 82
functional models 241
functions of nodes 137

game theory 190
games 17; causal form 195–197; composition of 192; definition of 192; elements of 190–191; entry games 190–191, 198–203; extensive form 190; model setup for non-trivial strategies 205–209; moves 190; Nash solutions 192–194; Nature's moves 190; with non-trivial strategies 204–216; normal form 190; payoffs 190; players 190; solving with causal models 197–203
Gans, Joshua 177
Gatekeeper model 233, *234*
GrandeGadgets Corporation 204, 205; strategy table 207, **208**
graphs, directed 17, 232
guns, trauma caused by 2

Hack-a-Hack Corporation 177–181; average cost per car, per year 178; competitor legal responses 185–188; drivers 178, *179*, *180*, *182*; market research 181–185; price uncertainty 181–185; profits 180–181; tech-savvy markets 183
hormone replacement therapy (HRT), and cardiovascular disease (CVD) 5
Hubris Health, interview case study 47–52; Account Management 54, 55, 71–73; Administrative Staff 58–59, 61–62; age of employees using services of 64, 65, 73; average income of population enrolled with 64; and BioSoftMed 69–70, 73; *Brand Image* 48, 50, 54, 55, 68; business model 56–57; clients using 67–68; Competition 48, 54, 68, 75; *Costs* 49, 50, 57, 65; Customer Satisfaction 48, 50, 52, 53, 54, 55, 68; doctors 57, 58, 62–65, 73; economic conditions 48, 52; employee satisfaction 54; *Enrolled Employees* 49, 57, 58, 62; *Fixed Fee* 49, 53; *New Accounts* 48, 54, 55, 57, 62, 68; *Number of Client Companies* 62; *Number of Staff Members* 50; *Profit* 49, 64; recommendations 71–75; referrals 58, 63, 64, 65; renewals 48, 52, 54, 55, 57, 62, 65, 68; *Revenue* 49, 57; service staff 61, 62; *Service Staff Members* 58; situational assessment of 48, *49*; specific questions to ask 59–65; treatments 58, 65; and young people 75

inference 17, 114–121, 227n.3
influence 227
influence diagrams 14, 17, 18, 27; definition of 17; extended 28; nodes 18
influence links 21–25
information: passing down the chain/downstream 118; passing up the chain/upstream 118, 119–120; synthesization of 122
information nodes 158
informational links 20, 21
instrumental variables 235, 238, 239, 240; simple 241
interactive decision problem 189
interactive decision problems; *see also* games
intersection of events 80, 81
intervention problems, pure 153
intervention selection problem 224
interventions: multi-agent 189–223; predictions **231**; single-agent 151–175
interview case analysis 44, 45; combining all information 65–71; questions to ask 59–65; recommendations 71–76; significant drivers 52–55; sources of common problems 56–59
issue trees 1

joint probability 84–85, 96–97
joint probability distribution (JPD) 102, 103, 110
joint probability tables 93, 95

lawsuits 185–188
likelihood ratios 117
limousine business 176
linear regressions 10; and causal models 242–245
links between nodes 19–25; [−] 158; [+] 21; directed links 19; influence links 19; informational 20, 21, 158; non-informational 20

management, evidence-based 4
managerial intervention 4–11, 25, 151
manufacturer's suggested retail price (MSRP) 28
marginal probability 84–85, 99–100; from conditional probabilities 111–114
marginal probability tables 90
marginalization 97–100, 114
market research 181–185
market volume, causal model for *26*
marketing 143–146
marketing effectiveness (ME): marginal probabilities of *145*; updated priors on *145*
marketing variables, correlation coefficients of **6**
markets, tech-savvy 183
mathematical notation 22

Maxwell, Robert, cause of death 1–2, 3
McKinsey Case Competition 1
MECE (mutually exclusive and collectively exhaustive) 1, 3, 18, 154
Microsoft 225
models: causal 14, 135–140, 151; data-driven causal 224–254; functional 241; Gatekeeper model 233, *234*; with market uncertainty and research options *184*; multi-stage regression 11; Office Equipment model 78, 79; predictive regression 241–251; qualitative causal 14–43, 78–108; quantitative causal 78–108, 109, 224; regression 243; software-based causal 177; structural equation models (SEM) 241, 242, 244
monotonic relationships 23, 25; conditional certainty tables *24*
moves 190, 192
movies, attributes which influence ticket sales 248–251
MPAA content ratings 248–251
MSRP (manufacturer's suggested retail price) 28
multi-agent interventions 189–223
multi-stage regression 10, 11
multicollinearity, definition of 245
mutually exclusive and collectively exhaustive (MECE) 1, 3, 18, 154

Nash equilibrium 192, 194, 202, 218
Nash solutions 192–194; best reply strategy 194; strategies 192
net present value (NPV) 152
network links, linear coefficients on *243*
nodes 17–19; ancestors 20; barren 157; child 19, 20, 23, 24; decision 18, 50, 153, 195; descendent 20; directed/causal paths 20; function of 137; leaf node 20; links between 19–25; objective 18, 46, 49, 152, 158–159, 195; parents 19, 20, 23, 24; path 19, 20; probabilistic 18; removal 161–162; replacement 161–162; root 1, 20, 97; states of 21–22; strategic option 18, 158, 207
non-monotonic relationships 25

objective nodes 18, 46, 49, 152, 195; conditional probabilities for parents of 158–159; expected value of 159
observational indistinguishability 230–241
observational indistinguishability theorem 231–234

payoffs 152, 195, 201
"peak-load" pricing 181
posterior beliefs 114
posterior probability 86
predictive regression models 241–251

price: "peak-load" pricing 181; and profit 9; and unit sales 9
price setting 181
price uncertainty 181–185
prior beliefs 114
prior probability 86
probabilistic independence 103
probabilistic nodes 18
probability 78–86, 226–227; causality versus 225–230; conditional 83, 92, 111–114, 158–159; from count data 91–92; events 79–80; factorization 84, 96; frequency-based 81, 82; fundamental rule in 84, 96; joint 84–85, 96–97; marginal 84–85, 99–100, 111–114; posterior 86; prior 86; refined approximations 89; subjective 81, 82; of a system state 103; variable states 78–79
probability tables 90; conditional 83, 84, 139, **148**, 155, 187, 209; joint 93, 95; marginal 90; system-level joint probability table 85
profit 46; price and 9
Prosecutor's Fallacy 146

QID (quantitative influence diagram) *see* quantitative influence diagram (QID)
qualitative causal models 14–43; application of 44–77; basic causal reasoning 25–27; building 17–31; causal relations 16; construction of 15; criteria for adding links 19; decision makers 15; environmental factors 16; examples 25–31; Feather Touch (FT) 90; influence diagrams 27, 28; objectives 15–16; quantifying 86–89; strategic agents 15, 16; strategic options 16
quality 22
quality control (QC) manager 109
quantitative causal models 78–108, 224; approximations 89; uses of 109; working with 90
quantitative influence diagram (QID) 151–152; extended model 154–155; general solution procedure 156–164; information nodes 158; multiple decisions with information 153–164; one decision with no information 152–153; strategic option nodes 158

regression: linear 10; models 243; multi-stage 10, 11; single-stage 11; stepwise 243, 246n.11; two-stage 244, 245
regression coefficients **7**
regression estimation **11**, 242
regression parameters 6
relationships, monotonic 23
resources decisions 4, 177–181
retail marketing *144*
retail units: influence of marketing effectiveness on *143*; marginal probabilities of *145*
root nodes 1, 20, 97

Salomao, Paulo 1
semiconductor manufacturing *98*, 99, 102–103; *Batch* as a common cause *121*; convergent connections *124*; divergent connections *122*; marginal probabilities for 99–100, *112*, *114*; multiple, converging causes *124*; serial connection example *115*; system-level joint probabilities **99**; updating beliefs when *Batch* = *bad 126*; updating beliefs when *Batch* = *bad* and *Error* = *no 126*;
sensitivity analysis 134
serial connections 111–112, 115–121; setup 115–117; updating 117–119
serial triplets 32–33
signals 203
simple instrumental variables 241
Simpson's Paradox 148–149
single-agent interventions 151–175, 227n.2
single-stage regression 11
situational analysis 109–131
situational assessment 1–4, 25, 45, 48
skeleton 232, 234
Slimtree Publishing Inc. 133; calculating CPT for *Ink Type* node **148**; causal diagram for the determinants of *Profit 136*; CPT for *Run Cost 148*; CPT generated for *RM 140*; effect of *Ink Type* on *Run Cost* when *Title* is known *149*; effect of *Ink Type* on *Run Cost* when *Title* is unknown *149*; effectiveness of Retail Marketing *144*; environmentally friendly ink 146–150; equal probabilities of the two *RU* states *138*; forecast profit and loss statement *134*; inferences from operating expenses *142*; influence of marketing effectiveness on retail units *143*; marginal probabilities of *ME* and *RU 145*; marketing 143–146; *ME* given *RU* = *high 145*; parent states associated with *RM* states via a function *140*; production runs *147*; profit and loss spreadsheet *134*; *RM* states corresponding to variable ranges *139*; run costs **147**; sales volume 143; summary information for initial causal model *141*; variable ranges for *RU* (retail units sold) *137*
smartphone technologies 176
software-based causal models 177
software, causal modeling 216
software solutions 216–218
spreadsheets 134–135
states: system 85; variable 18, 78–79

statistical correlation, inferring causation from 5
stepwise regression 243, 246n.11
stochastical independence 83
strategic agents 15, 16, 17, 21
strategic option nodes 158
strategic options 16, 21
strategic tables 207
strategies 192; definition of 194; games with non-trivial strategies 204–216
strategy nodes 207
strategy tables 207
structural equation models (SEM) 241, 242, 244
subjective probability 81, 82
system-level joint distribution and factorization 96–97
system-level joint probability distribution (JPD) *98*, **99**, 100, 110
system-level joint probability table 85
system states 85, 103

Taxi Business 176–188
taxis, taxi-hailing application for smart phones 176
tech-savvy markets 183
triplets: converging 34–36, 113; diverging 33–34, 112, 121; serial 32–33
TruSmartz Consulting 5–6
two-stage regression 244, 245

UberCab 176–177
union of events 80, 81
unit sales, price and 9

v-structure 232, 234
variable nodes 18
variables 18; continuous 18n; discrete 79; effect of 245; environmental 197; filter 197; functional relationship among **36**; instrumental 235, 238, 239, 240, 241; states 18, 78–79

For Product Safety Concerns and Information please contact our EU representative GPSR@taylorandfrancis.com
Taylor & Francis Verlag GmbH, Kaufingerstraße 24, 80331 München, Germany

www.ingramcontent.com/pod-product-compliance
Lightning Source LLC
LaVergne TN
LVHW080808011026
849160LV00015B/1572

9780415657600